Public Reason and Bioethics

Hon-Lam Li • Michael Campbell
Editors

Public Reason and Bioethics

Three Perspectives

palgrave
macmillan

Editors
Hon-Lam Li
Department of Philosophy and CUHK Centre for Bioethics
The Chinese University of Hong Kong
Hong Kong

Michael Campbell
Department of Philosophy
Kyoto University
Kyoto, Japan

ISBN 978-3-030-61169-9 ISBN 978-3-030-61170-5 (eBook)
https://doi.org/10.1007/978-3-030-61170-5

This Palgrave Macmillan imprint is published by the registered company Springer Nature Switzerland AG.
The registered company address is: Gewerbestrasse 11, 6330 Cham, Switzerland

Acknowledgments

The editors would like to thank all contributors of this volume, and are especially grateful to Ruiping Fan, Dominic Farrell, and Joseph Tham, without whose commitment and support from the start this volume would not have existed. Special thanks are also owed to Alastair Campbell and Terence Hua Tai for their commitment and unfailing support from the start.

Special thanks are due to Dr. Edgar Cheng, Dr. Derrick Au, and the CUHK Centre for Bioethics for their unwavering support. Thanks are also due to Lauree Wang for having organized two workshops for this project. The editors would like to thank the Department of Philosophy of the Chinese University of Hong Kong for its continuous support.

The editors are grateful to Brendan George of Palgrave Macmillan for his patient work overseeing this project from the beginning to the end. Thanks to Rebecca Hinsley for her coordination. For excellent professional editing, we want to thank Hemalatha Arumugam and her team.

The research project on Public Reason and Bioethics was partially supported by a grant from the Research Grants Council of the Hong Kong Special Administrative Region, China (Project No. CUHK No. 14660716). Hon-Lam Li was the principal investigator of this project, and Alastair Campbell, Michael Campbell, Ruiping Fan, Dominic Farrell, Terence Hua Tai, and Joseph Tham were the co-investigators.

Michael Campbell's work was supported by the following grant: Operational Programme Research, Development and Education (OPVVV/OP RDE), "Centre for Ethics as Study in Human Value," registration No. CZ.02.1.01/0.0/0.0/15_003/0000425, co-financed by the European Regional Development Fund and the state budget of the Czech Republic.

Contents

Notes on Contributors

Alastair V. Campbell is Visiting Professor of Medical Ethics and the Emeritus Director of the Centre for Biomedical Ethics in the Yong Loo Lin School of Medicine of the National University of Singapore. Previously he was Professor of Biomedical Ethics, University of Otago, and then Professor of Ethics in Medicine, University of Bristol. He is a former president of the International Association of Bioethics. He is a recipient of the HK Beecher Award, a fellow of the Hastings Centre, New York, and of the Ethox Centre, University of Oxford, and an elected Fellow of the Royal Society of Edinburgh. His recent books include *Health as Liberation* (1996), *Medical Ethics* (with D.G. Jones and G. Gillet, 3rd Edition 2005), *The Body in Bioethics* (2009), and *Bioethics: The Basics* (second edition 2017). For the past few years he has been an advisor to the CUHK Centre for Bioethics in the Chinese University of Hong Kong.

Michael Campbell is Visiting Assistant Professor in the Department of Philosophy at Kyoto University. He is the co-editor of *Wittgenstein and Perception* (with M O'Sullivan, 2016), *Ethics, Society and Politics: Themes from the Philosophy of Peter Winch* (with L Reid, Springer 2020), and *Spinoza on Ethics and Understanding* (with S. Tropper, Anthem 2020). His articles have appeared in journals including *Asian Bioethics Review*, *The Journal of Value Inquiry*, and *Philosophical Investigations*.

Ruiping Fan is Professor of Philosophy in the Department of Public Policy at the City University of Hong Kong. He serves as associate editor of the *Journal of Medicine & Philosophy*, associate editor of the *Chinese Medical Ethics*, and co-editor of the *International Journal of Chinese & Comparative Philosophy of Medicine*. He has published extensively on Confucian bioethics and comparative social and political philosophy.

Dominic Farrell LC is Professor of Philosophy at the Pontifical Athenaeum *Regina Apostolorum*, Rome. His research focuses on ethics, political philosophy, and Thomas Aquinas. His articles have appeared in *The Review of Metaphysics* and *The Journal for Markets and Morality*, and the book *The Ends of the Moral Virtues and the First Principles of Practical Reason in Thomas Aquinas* was published in 2012.

Nancy S. Jecker PhD, is Professor of Bioethics and Humanities at the University of Washington School of Medicine. Jecker's research focuses on individual and societal aging, justice, human dignity, medical futility, and global perspectives in philosophy and bioethics. More than 200 of his articles have been published and also 4 books. Her most recent book is *Ending Midlife. Bias: New Values for Old Age* (2020).

Hon-Lam Li, BA (Princeton), MA (*Oxon.*), and PhD (Cornell), was supervised by Thomas Nagel and T. M. Scanlon at Princeton, tutored by John Mackie, John McDowell, Jonathan Glover and John Finnis at Oxford, and supervised by Allen Wood at Cornell. His areas of research include practical ethics, ethics, political philosophy, and philosophy of law. He had practiced as a barrister-at-law in Hong Kong before teaching at the Colorado College, and is now Professor in the Department of Philosophy, and Deputy Director at the Centre for Bioethics, the Chinese University of Hong Kong. His articles have appeared in international journals, such as *The American Journal of Bioethics*, *Public Affairs Quarterly*, *Criminal Justice Ethics*, and the *Journal of Medicine and Philosophy*. He has edited (with A. Yeung) *New Essays in Applied Ethics: Animal Rights, Personhood, and the Ethics of Killing* (UK: Palgrave Macmillan, 2007). He is working on a book on contractualism and practical ethics.

Terence Hua Tai PhD in philosophy, Cornell University, 1989, specializes in Kant, philosophy of mind, moral and political philosophy, and bioethics. He is the chair professor at the Department of Philosophy, National Chung Cheng University, Taiwan. Representative works include "Equality and Community in Public Deliberation: Genetic Democracy in Taiwan", co-authored with Wen-Tsong Chiou, in V. Launis and J. Räikkä, eds., *Genetic Democracy*, Springer, 2007; "Kant's Transcendental Strategy in the First *Critique*," in Margit Ruffing, et al., eds., *Kant und die Philosophie in Weltbürgerlicher Absicht: Akten des XI Kant-Kongresses 2010*, de Gruyter, 2013; and "Kant's Critical Objection to the Rationalists in the B-Deduction," *Kant-Studien*, 2020.

Yingying Tang is professor in the Institute of Humanities and Social Sciences at Shandong University, Qingdao. She holds a PhD in philosophy from the University of Southern California. Her areas of research interests are political philosophy, moral philosophy, and comparative philosophy. She is working on a research project on distributive justice supported by a General Research Grant from the National Social Science Fund of China.

Joseph Tham LC graduated from Medical School and has worked as a family physician, and he entered the seminary of the Legionaries of Christ and was ordained a priest in 2004. He has obtained his degrees in philosophy and theology at Rome's *Regina Apostolorum* Pontifical University. He is the author and editor of numerous articles and books, including *The Missing Cornerstone* (2004), *The Secularization of Bioethics* (2007) and *Bioetica al Futuro* [*Bioethics of the Future*] (2010), *Religious Perspectives on Human Vulnerability in Bioethics* (2014), *Religious Perspectives on Bioethics and Human Rights* (2017), *Sexuality, Gender and Education* (2018), *Religious Perspectives on Social Responsibility in Health* (2018), and *Interreligious Perspectives on Mind, Genes and the Self* (2018).

Lei Zhong is an associate professor in the Department of Philosophy at the Chinese University of Hong Kong. He has obtained a PhD degree in philosophy from the University of Michigan, Ann Arbor. His areas of specialization are philosophy of mind and moral philosophy. His work has appeared in leading philosophy journals, including *Journal of Philosophy*, *Philosophy and Phenomenological Research*, *Analysis*, *Australasian Journal of Philosophy*, and *Philosophical Studies*.

Introduction

Contemporary society is beset with conflicting world views and pluralistic values. In any cosmopolitan society there can be found Catholics, Muslims, Buddhists, Protestants, Confucians, Hindus, Jews, Taoists, Sikhs, agnostics, atheists, and people of other religious faiths. Throughout history and up until the present day, religious conflict remains a serious threat to peace and stability. During the European religious wars in the seventeenth century, millions of people were killed. Even though different religious groups today may be more tolerant of each other, religious faith often shapes an individual's moral outlook in ways that place him or her at odds with contrasting viewpoints and ways of life.

The kind of conflict between a religion and its non-believers in contemporary society is usually not a direct clash of self- or group-interest, but is rather a clash of conceptions of the good. Even though a religion aims to influence the beliefs of its believers or perhaps even regulate their conduct, it tends to extend beyond its own domain not only when it seeks to spread its own particular gospel or doctrine but, more importantly, when it strives to have the prohibitions within its moral outlook enacted as positive laws. For instance, a religious group might resist movements seeking to lift the ban on medical assistance in dying,[1] stem-cell research, reproductive technology, genetic enhancement, and even homosexual marriage. This clashes with the liberal ideal of toleration, namely, that the law should not criminalize a citizen unless they harm a third party. On the other hand, liberals would do the exact opposite, trying to lobby legislators to pass laws to allow these activities.

Conflicts also occur among different people's moral and philosophical beliefs. In the moral domain, clashes over controversial issues occur between utilitarians, deontologists, virtue ethicists, Confucians, egoists, and skeptics. Over issues in political ethics, the conservative right and the liberal left collide. Abortion clinic bombings and sabotage of animal labs are not unheard of. Disagreement over medical assistance in dying among peers, friends, and even family members is commonplace.

How, then, should controversial ethical issues be discussed, let alone resolved, in a way that is civil and reasonable? In *Political Liberalism*, John Rawls takes up this question. First, he affirms "the fact of reasonable pluralism." By this, he means that rational and reasonable people can reasonably disagree over religious, moral, and philosophical issues, which are controversial and complex.[2] He holds that different groups in society embrace different "comprehensive doctrines," which are sets of beliefs affirmed by citizens regarding a range of values, including religious, philosophical, and moral commitments.[3] A comprehensive doctrine is *reasonable* if and only if it is theoretically coherent, organizes values in a consistent way, and is tolerant of other reasonable comprehensive doctrines.

---

[1] In this book, different authors prefer different terms, such as medically assisted death, medically assisted suicide (MAS), physician-assisted suicide (PAS), physician-assisted death (PAD), and medical assistance in dying (MAID). All of these terms refer to the same thing.

[2] Rawls attributes the cause of "the fact of reasonable pluralism" to "the burden of judgements" (John Rawls, *Political Liberalism*, (New York: Columbia University Press, 1993), 54-58). Reasonable and rational people disagree over religious, moral, and philosophical matters because these matters involve evidence that is hard to assess, concepts that are vague, conflicting considerations, and difficulty in setting our priorities.

[3] Comprehensive doctrines also include beliefs about personal virtues, political beliefs about how society ought to be arranged, and also "what is of value in life, the ideals of personal character, as well as ideals of friendship and familial and associational relationships, and much else that is to inform our conduct, and in the limit to our life as a whole" (ibid., 13).

Rawls holds that in a well-ordered society,[4] free and equal citizens[5] embracing different reasonable comprehensive doctrines can agree to a political conception of justice which protects their basic liberties and rights, and which guarantees fair equality of opportunity and an all-purpose means to realize their freedoms. This is so even if their reasons for reaching this "overlapping consensus" stem from their different doctrines, because they recognize the value of toleration. Such a consensus is based on normative grounds and is not merely a convenient *modus vivendi*. The political conception of justice is *self-standing* in the sense that it does not rely on any comprehensive doctrine.

Rawls contends that in a public forum where "constitutional essentials and matters of basic justice"[6] are debated, legislators, government officials, judges, and candidates for high office should discourse by appealing to only the political conception of justice and political values of public reason. Religious, moral, or philosophical doctrines should be avoided, unless the "proviso" that a justification in terms of public reasons can subsequently be provided is satisfied.

Importantly, Rawls later clarified that not only "constitutional essentials and matters of basic justice," but also issues that "border on" constitutional essentials and "are politically divisive" should be discoursed in terms of public reasons.[7] Rawls thinks that abortion is a good example,

---

[4] The idea of a well-ordered society is a "highly idealized concept" (*Political Liberalism*, 35). A society is well-ordered if (a) "it is a society in which everyone accepts, and knows that everyone else accepts, the very same principles of justice," (b) "its basic structure—that is, its main political and social institutions...—is publicly known ... to satisfy these principles," and (c) "its citizens have a normally effective sense of justice and so they regularly comply with society's basic institutions, which they regard as just" (ibid.) In a well-ordered society "the publicly recognized conception of justice establishes a shared point of view from which citizens' claim on society can be adjudicated" (ibid.).

[5] Rawls holds that when persons have realized their capacity for a sense of justice and for a conception of the good, and have the powers of reason, they are free. Their having these powers to the requisite minimum degree to be fully cooperating members of society makes them equal (*Political Liberalism* 19). Free and equal citizens in a liberal constitutional democracy have an enduring desire to honor fair terms of cooperation.

[6] Constitutional essentials have to do with citizens' basic liberties and rights, as well as constitutional power and procedure. Basic justice refers to matters of equal opportunity, the social minimum, and an all-purpose means by which to realize citizens' freedoms.

[7] John Rawls, *Justice as Fairness: A Restatement*, (Cambridge, MA: Harvard University Press, 2001), 117. The concept of "constitutional essentials" is rather narrow and include "(a) fundamental principles that specify the general structure of government and the political process" and "(b) equal basic rights and liberties of citizenship that legislative majorities are to respect: such as the right to vote and to participate in politics, liberty of conscience, freedom of thought and of association, as well as the protections of the rule of law" (PL 227).

but other issues that "can be the cause of deep conflict"[8] would include medical assistance in dying and same-sex marriage.

Discourse in terms of public reasons, as Rawls points out, by no means guarantees consensus. So long as controversial issues are debated in the public forum in terms of public reasons, Rawls maintains that enactment of a law according to the majority view in the legislature will be *legitimate*. Rawls would also think that decisions of the appellant and supreme courts, typically couched in terms of public reasons, are legitimate.

However, Rawls has never clearly indicated how discussion in public reason is to be carried out. He says that with respect to a particular issue, say, abortion, political values relevant to this issue should be ordered and be reasonably balanced. He claims, without providing any detail, that any reasonable ordering and balancing of the relevant political values must allow abortion during the first trimester.[9]

In Chap. 1, Part I, of this volume, **Hon-Lam Li** articulates Rawls' theory of political liberalism and raises pertinent issues. Though sympathetic to Rawls' view, Li argues that Rawls' idea of reasonable ordering of political values is too vague. He contends that self-standing *arguments* should be included as public reasons in public discourse. In this regard, he argues that T. M. Scanlon's contractualism—especially Scanlonian principles and reasoning—could be incorporated for three reasons. First, Scanlon and Rawls have mutually influenced each other's view over a fairly long period of time and their terms of reference are similar or congenial. Second, Scanlon's principles are self-standing and do not rely on any comprehensive doctrine. Finally, his theory is worked out in detail theoretically and has been applied to practical issues.[10]

---

[8] Ibid.

[9] According to Rawls, the three important political values relevant to abortion are the due respect of human life, the ordered reproduction of political society over time, and the equality of women as equal citizens (*Political Liberalism*, 243n; *The Law of Peoples*, (Cambridge, MA: Harvard University Press, 1999), 169).

[10] See Corey Brettschneider, "The Rights of the Guilty: Punishment and Political Legitimacy." *Political Theory* 35, no. 2 (2007): 175–99; Hon-Lam Li, "Contractualism and the Death Penalty." *Criminal Justice Ethics* 36, no. 2 (2017): 152-182; Hon-Lam Li, "What We Owe to Terminally Ill Patients: The Option of Physician-Assisted Suicide." *Asian Bioethics Review* 8, no. 3 (2016): 224-43; Hon-Lam Li, Nancy S. Jecker, Roger Yat-Nork Chung, "Reopening Economies during the COVID-19 Pandemic: Reasoning about Value Tradeoffs." *The American Journal of Bioethics* 20, no.

In his argument on decriminalization of medical assistance in dying, Li brings contractualism to bear on some of the difficult issues. In particular, Li argues that contractualism offers a more plausible response than utilitarianism to the challenge that if medical assistance in dying is decriminalized, some elderly patients may feel obligated to die.

In Chap. 2, **Fr. Dominic Farrell** and **Fr. Joseph Tham** expound upon the natural law theory (NLT) and its conception of public reason. After explaining how the NLT conceives of the process of public reason, they go on to consider how the NLT bears on the question of medically assisted suicide.[11] According to NLT, there exists an objective normative order of human goods, founded upon the *telos* (i.e., ultimate end) of human beings. Three features of such an approach are especially important in this context. Firstly, an act is morally permissible, or a positive law is just and legitimate, only if it is consistent with this objective order of human goods. Secondly, each good for a human being is achievable only through a community and is therefore a communal good. Finally, whereas the natural law is concerned with the full range of virtuous actions, a polity's (positive) law should be limited to sanctioning actions that are necessary for the conservation of political society and the protection of the common good.

According to the NLT, the ultimate source of political authority is the conformity of institutions with the right order of human nature. Hence, the values possessed by autonomy and the results of autonomous decision-making are not absolute but are instead relative to their being appropriately related to the objective order of human goods. Farrell and Tham argue that we should not accept the liberal presumption that we either can or should bracket comprehensive doctrines, including substantive and potentially controversial beliefs concerning human nature. Contra Rawls, public reason is not a "self-standing" framework independent of moral and political conviction but is instead simply reasoning *in public* on issues of the common good. As reasoning, it is predicated on the idea of being a truth-directed inquiry. Thus, in the political sphere, we ought

---

7 (2020), 136-8; Hon-Lam Li, "Contractualism and Punishment." *Criminal Justice Ethics* 34, no. 2 (2015): 177-209.

[11] Farrell and Tham, as well as Fan, prefer the term "medically assisted suicide." It refers to the same thing as "medical assistance in dying" by Li.

to see each other as engaged in a collective enterprise to grasp the truth concerning how best to live.

Drawing on the work of Alasdair MacIntyre, Farrell and Tham argue that we can derive, from reflection on the preconditions of carrying out a collective truth-oriented enterprise, the beginnings of a framework for a natural order of goods. They demonstrate that what may be taken to be requirements of rational inquiry correspond to certain divine commandments, which are, on their account, precepts of natural law. Turning then to bioethical questions, they consider in particular arguments for and against the legalization of medically assisted suicide. They argue that medically assisted suicide should be proscribed, since as it involves directly and intentionally ending the life of an innocent person, it constitutes murder. As well as providing considerations which speak against assisted suicide, Farrell and Tham argue that the nature of the discussion also demonstrates the impossibility of conducting debate over such matters while prescinding from controversial philosophical subjects, such as the nature of personhood.

The third perspective that features in this book is the Confucian tradition as interpreted by **Ruiping Fan**. In Chap. 3, Fan argues that Rawls' idea of public reason cannot apply to nations (or parts thereof) that are broadly Confucian, such as China, Japan, South Korea, Taiwan, Hong Kong, and Singapore. Fan holds that Rawls' theory does not apply to these countries because of significant cultural and historical differences between them and European countries. The Reformation and the Thirty Years' Religious War had a major impact on the subsequent development of religious toleration in Europe. Yet, the various different philosophies or religions in China—namely Confucianism, Taoism, and Buddhism—were quite harmonious with, and tolerant of, each other. Thus, although Rawls' theory of public reason might be of importance to Western societies which were beset by religious conflict, it is not suitable for Confucian societies today.

This cultural divergence is at its sharpest in the individualism which Fan argues is latent in Rawls' idea of public reason. An individualistic conception of the nature of value is unsuited, Fan claims, for societies which are predicated on the Confucian virtues such as *ren* (humanity), *yi* (appropriateness), *li* (propriety), *xiao* (filial piety), and *he* (harmony).

These are communitarian values, and according to Fan, Rawls would not regard them as suitable candidates for public reasons. Against this, Fan argues that whether some consideration is a public or a private reason should be understood as a matter of degree rather than of kind; public reasons are just those which are capable of engaging the whole of a populace. Thus, culturally specific beliefs concerning the nature and importance of the family may provide public reasons when they express convictions that are sufficiently deeply embedded with a particular society.

Fan puts forward a theory that comprises (1) Confucian virtues, (2) a list of basic liberties and rights that are developed from Confucian virtues or values, and (3) principles of political justice that integrate the elements at these lower levels into a coherent whole. Fan acknowledges that the rights to life, to liberty, and to formal equality are important, but argues that these basic liberties and rights are not intrinsically liberal. Finally, Fan applies this view to medically assisted suicide. He argues that to decide on this matter one must balance the following considerations: (1) the importance of a patient's autonomy, (2) the requirement to cherish life, (3) a physician's role as a healer, and (4) the broader social consequences of legally permitting assisted suicide. He concludes that a reasonable balancing of these considerations should point to the conclusion that assisted suicide should not be legalized in any of the Confucian countries in East Asia.

In Chaps. 4, 5, and 6, Li, Farrell and Tham, and Fan critique each other's view. Chapters 7, 8, and 9 are further responses by these authors to Chaps. 4, 5, and 6. Apart from responding to rival conceptions of public reason, each group reflects on possible deficiency of their own theories and whether they are drawn to each other a little bit more during the course of deliberation and debate. These chapters were written sequentially, and reflect the results of robust debate between the participants.

This book comprises two parts. Whereas Part I consists of nine chapters from three different traditions (where the question of which is the best conception of public reason is debated), Part II contains five essays that supplement the debate, by situating the debate within its conceptual and historical background and by reflecting on the ways in which the concept of public reason might be developed in the future.

In Chap. 10, **Terence Hua Tai** interrogates the historical origins of the concept of public reason by examining in detail Kant's conception of public reason. Tai shows that Kant's conception differs from Rawls' formulation in several key respects. In particular, whereas Rawls sees the concept of public reason as a formal demand which governs debate within a pluralistic society, for Kant public reason is fundamentally connected to the preconditions for being an autonomous agent within a society of coercive rules. The public use of reason allows us to convert certain laws of the state—which are "laws of heteronomy" in the sense of being imposed on us by an external authority—into "laws of autonomy," that is, into freely willed constraints, obedience to which constitutes an act of genuine self-expression.

Tai contrasts his reading with Onora O'Neill's account of Kantian public reason, showing that in order to grasp Kant's position we must stress not only the law-like form and universal scope of laws of public reason but also their connection with our common humanity and their falling within the "domain of right." The public use of reason is its uses by members of the state when they speak as scholars rather than as engaged in some or other public office. Public reasoning therefore faces in two directions, being designed both to justify our institutions and to point citizens the way toward intellectual maturity. Once they have gained this maturity, they will then be able to stand back from the laws that they live under and see not only that they must be followed but why or to what extent. In this way they can convert laws of heteronomy into laws of autonomy.

At the end, the principles of right upon which our state is based must be justified by reference to the humanity of the citizens which the laws serve. But for Kant, the justifiability of our laws must be distinguished from the act of justifying them. Justification can be undertaken only by those who have educated themselves in the ways of correct reasoning, and this education goes on through studying the examples of those who debate over the justifiability or otherwise of some particular law. By showing the educative function of public reason, as well as the connection between the structures of law and the common humanity of the citizens who are governed by it, Tai's reconstruction shows how the concept of public reason can play an important role in bioethical deliberation when it is embedded in richer conceptions of education, autonomy, and the public good.

In Chap. 11, **Yingying Tang** and **Lei Zhong** consider the nature of autonomy and its status as a moral and political value. They argue that autonomy plays a crucial role in making the pursuit of a way of life valuable. On their conception, ways of life have intrinsic value, but this value can only be manifested if the way of life is chosen for the right reason, namely, because of a sense of it as valuable in itself. However, such a choice can only be made autonomously; those who act under coercion or manipulation may pursue a valuable way of life, but they cannot make manifest the values inherent in it.

For Tang and Lei, autonomy consists of two components: independence from domination or coercion, and coherence between one's "deep" and "superficial" selves. A truly autonomous person both reflectively endorses her desires and interests and has the capacity to form and pursue them free from domination by others. Drawing on this distinction, they then go on to consider whether the value of autonomy is consistent with either perfectionism or neutralism concerning the legitimate role of the government.

According to perfectionism, the state is sometimes justified in compromising individual autonomy in order to promote a particular conception of the good life. According to neutralism, the government is never justified in acting in ways which favor some particular way(s) of life over others. Tang and Lei argue that the relation between the value of ways of life and the value of autonomy means that perfectionism is self-defeating—by requiring conformity to a way of life, that way of life is deprived of its potential value. Accordingly, it is impermissible for the state to coerce or manipulate its citizens in ways which favor some way(s) of life over others. At the same time, they argue that, since not all of the means available to the government for promoting some way(s) of life are coercive, neutralism cannot be justified simply by reference to the concept of autonomy, since not all means for promoting particular patterns of thought or behavior involve coercion or manipulation. They conclude therefore that either the state is sometimes justified in promoting some way(s) of life over others using non-coercive means, or else there are other grounds, rooted in values such as "respect," which could justify a stricter conception of the requirement that the state be neutral.

In Chap. 12, **Nancy Jecker** expounds and argues for an account of public reason which grounds our obligations to others in facts concerning our shared humanity. Drawing on the so-called capabilities approach, she argues that the rules which govern our shared moral and political life must be grounded in facts concerning the central capabilities of human life. On such an approach, the demands of justice are determined through a two-step process. Firstly, a heuristic is drawn up for measuring individual well-being through establishing, for a given individual, what range of opportunities is open to her to realize those capacities which constitute the goods of human life. Secondly, through public deliberation, thresholds are then set for what opportunities or functionings are sufficient to meet the demands of justice.

Jecker contrasts this approach with the contractualist foundations for public reason offered by those working within the liberal tradition. If the demands of justice are grounded in our ability to reason and form agreements with others, then those whose capacities to participate in such reasoning is limited, are in danger of being relegated to the margins of the theory, or else excluded entirely from the claims of justice. Moreover, contractualist approaches seem to tie the requirements of justice to political communities, whether actual or imagined. In the absence of a workable conception of a global community, that would render claims of justice bound to particular states, with the corresponding dangers of instability and parochialism. By contrast, Jecker argues, by seeing our obligations to others as stemming from natural facts concerning our shared human nature, we can devise a heuristic for determining what we owe to others which is both more inclusive and may more easily be extended across national boundaries.

Jecker's approach thus adopts the Rawlsian conception of public reason but, *contra* Rawls, grounds it in a substantive theory of what kinds of things are good for human beings. Against the charge that such an approach is culturally imperialistic, she gives a twofold reply. First of all, she argues that the basic composition of such a list is minimal, describing generic categories and reflecting those features of our lives which are shared across different cultures. Secondly, the more specific lists and thresholds which constitute the particular form which the requirements of justice take, in particular milieus, are to be set by public deliberation by the relevant group, and so may, within limits, reflect the cultural and historical facts of the time.

In Chap. 13, **Michael Campbell** considers the development of public reason as an off-shoot of the liberal conception of the state. Taking John Rawls' conception of public reason in *Political Liberalism*, Campbell goes on to consider the merits of extending Rawls' theory to entitlements to healthcare, focusing in particular on the work of Normal Daniels. He ends by considering some challenges to such an approach and suggests that the concept of public reason must continue to evolve if it is to meet the ethical challenges raised by the diversity and complexity of the modern world.

Finally, in Chap. 14, **Alastair Campbell** provides an overview of the history of bioethics, describing its genesis in the work of theologians writing in the 1970s and its ensuing professionalization, internationalization, and secularization. As Campbell shows, bioethics must continue to evolve in order to meet the challenges of the twenty-first century. Some kind of concept of public reason, in the sense of a set of shared standards for clear and rational debate, must be operative in any discipline which seeks consensus on controversial questions. But the precise nature of this concept, as well as the eventual direction that bioethics will take, is not something that can be laid down in advance. In particular, bioethics must find a way to become politically involved if it is to identify and confront endemic abuses of economic power within the healthcare industries. At the same time, it must take care to preserve its academic integrity. Just how these twin imperatives can be reconciled remains a challenge with which the discipline is yet fully to come to grips.

Department of Philosophy
and CUHK Centre for Bioethics
The Chinese University of Hong Kong,
Hong Kong

Hon-Lam Li

Department of Philosophy
Kyoto University,
Kyoto, Japan

Michael Campbell

Part I

The Three Perspectives

1

Rawlsian Political Liberalism, Public Reason, and Bioethics

Hon-Lam Li

In their debate as vice-presidential hopefuls in 2012, Paul Ryan and Joe Biden responded differently to a question as to how their religions have influenced their personal views about abortion. Ryan said that he did not see how people can separate their public life from their private life and religious faith. He further said that life begins at birth, and that although he respected people who do not agree with his pro-life view, the policy of a Romney administration would be against abortion, except for rape, incest, and the life of the mother. Biden, however, responded that his

I am grateful to Tom Nagel for a discussion on some of the issues raised in Part A of this chapter on May 28, 2019. I am indebted to Tim Scanlon for showing me his unpublished paper, "Some Main Points in Rawls' Theory of Justice," and for discussing some issues via emails. I am grateful to Bonnie Steinbock and John G. Bennett for particularly helpful comments on an earlier version of Part (B). I thank Alexandre Erler, Peter Chau, Michael Campbell, and Win-chiat Lee for commenting on earlier drafts. I thank Carolyn P. Neuhaus for discussion, and the Hastings Center for hosting me in May 2019, as well as the CUHK Centre for Bioethics for making this trip to Hastings possible. A version of this chapter was presented as the Featured Speech at a philosophy conference hosted by the National Taiwan University on December 15, 2020.

H.-L. Li (✉)
Department of Philosophy and CUHK Centre for Bioethics, The Chinese University of Hong Kong, Shatin, Hong Kong

H.-L. Li, M. Campbell (eds.), *Public Reason and Bioethics*,
https://doi.org/10.1007/978-3-030-61170-5_1

Catholic faith defined who he was, and that he personally accepted that life begins at birth. However, he refused to impose his personal view on equally devout Christians, Muslims, Jews, as well as others. There is perhaps nothing unusual about Ryan's reply, for who can blame an official who is a Catholic pursuing a policy when the alternative would clash with his religious and moral beliefs? Yet, when Biden said that he should distance himself from his own religious and personal moral view when making public policy, he was adopting a position grounded in the idea of public reason, according to which a public official should engage in discourse on public policy matters of fundamental importance[1] in terms of political, rather than personal, values.

The most important proponent of this approach is John Rawls, although Thomas Nagel has also argued for a congenial but stronger view.[2] In this chapter (and also in Chap. 7), I shall articulate a qualified defense of John Rawls' political liberalism and his idea of public reason. This chapter is divided into two parts. In the first part, I explain the foundational differences between *A Theory of Justice* and *Political Liberalism*, despite the fact that Rawls maintains the Two Principles of Justice in both works. I explain why, in view of the fact that reasonable people would subscribe to different comprehensive religious, philosophical, and moral doctrines, Rawls needs to seek a new foundation for social stability in a constitutional liberal democracy. I explain the connection between Rawls' ideas of overlapping consensus, political conception, and public reason. I also explain Rawls' idea of "duty of civility" and the condition under which a controversial issue can be legitimately resolved by way of voting. Further, instead of pursuing Rawls' idea that bioethical issues could be resolved by way of "reasonable balancing" of "political values," I turn to T. M. Scanlon's contractualist approach, which is congenial to Rawlsian political liberalism and can resolve moral, political, and bioethical issues. In the second part of this chapter, I try to resolve the problem of medical

[1] John Rawls limited the scope of public reason to "constitutional essentials and matters of basic justice" in *Political Liberalism*, but he clarifies in *Justice as Fairness: A Restatement* ("Restatement") that issues that "border" on a constitutional essential and are political divisive ought to be covered. See *Restatement*, 41, 117.

[2] See Thomas Nagel, "Moral Conflict and Political Legitimacy," *Philosophy and Public Affairs*, vol.16, no. 3 (Summer, 1987): 215–240, and also in *Equality and Partiality*, 163, n49.

assistance in dying (henceforth "MAID") by using Thomson's/Scanlon's idea that permissibility is not affected by intention. I deal with the Slippery Slope Arguments, the argument from abuse, and the argument from pressure on elderly patients who might feel obligated to die. Finally, I argue that Scanlonian contractualism is preferable to utilitarianism (whether or not utilitarianism is self-standing or not). I argue that we should make use of Scanlonian contractualism to develop a richer conception of public reason, and to employ it to argue for the decriminalization of MAID.

(A) Political Liberalism

John Rawls

Contemporary democratic society is beset with conflicting pluralist values. Citizens have different religious faiths (including atheism), and disparate moral and political allegiances. Disagreement erupts from time to time, sometimes ending in bombings of abortion clinics or sabotage of animal research laboratories, as well as social protests. John Rawls has coined the term "the fact of reasonable pluralism" (*Political Liberalism*, henceforth "PL," 24n) to describe the phenomenon that we cannot reasonably expect even rational and reasonable people to agree on controversial issues, such as abortion and MAID. He believes "reasonable pluralism" to be a "permanent fact" of constitutional democracy (*Justice as Fairness: A Restatement*, henceforth "Restatement," 33–34). Rawls holds that the causes of such reasonable disagreement, "the burdens of judgment," have to do with the difficulty in assessing empirical evidence and in reaching normative conclusions, as well as our disparate life experiences (PL 56–57).

Rawls is the main contemporary proponent of the idea of public reason. To understand this notion further would require us to take into account his earlier and later theories, as well as the reasons why he considered it necessary to make the political turn. In his earlier work, *A Theory of Justice* (henceforth "*Theory*"), Rawls dealt with the question of how citizens can be free and equal,[3] in spite of the fact that they have unequal starting points in life, with different natural endowments and family backgrounds. Rawls solves the problem by putting forward a theory of justice that he aims to apply to the fundamental institutions, or the "basic structure of society." In *Theory*, Rawls defends a Kantian comprehensive doctrine—in particular the Kantian conception of autonomy—and derives his Two Principles of Justice from the original position.[4] These two principles (revised in *Political Liberalism*) read as follows:

1. Each person has an equal right to a fully adequate scheme of equal basic liberties which is compatible with a similar scheme of liberties for all.
2. Social and economic inequalities are to satisfy two conditions. First, they must be attached to offices and positions open to all under conditions of fair equality of opportunity; and second, they must be to the greatest benefit of the least advantaged members of society. (PL 291)

In his second book, *Political Liberalism* ("PL"), Rawls pursues the question of how the liberal form of society can be stable over time *for the right reasons*, and not only because of a *modus vivendi*, given that people hold different and irreconcilable religious, moral, and philosophical doctrines. Rawls' solution to this question, in *Theory*, was that every citizen, or at least the great majority of citizens, will be persuaded to subscribe to these two principles. Yet, in *PL*, Rawls realizes that even if every citizen subscribes to one comprehensive doctrine—religious or moral—the use of oppressive state power remains necessary to maintain the stability of society. As Rawls writes:

---

[3] Persons are free when they have realized their capacity for a sense of justice and for a conception of the good, and have the powers of reason. When they have these powers to the requisite minimum degree as fully cooperating members of society, they are equal (*Political Liberalism* 19). Free and equal citizens in a liberal constitutional democracy, Rawls says, have an enduring desire to honor fair terms of cooperation.

[4] The original position is a hypothetical construct in which representatives from different social classes are placed behind a "veil of ignorance" and therefore are ignorant of their own attributes, including their genders, their natural endowments, religious beliefs, whether they are rich or poor, as well as their conceptions of the good and particular information about their society. Rawls argues that because everyone is deprived of such information, they being rational and self-interested would choose the two principles of justice.

> [A] continuing shared understanding on one comprehensive religious, philosophical, or moral doctrine can be maintained only by the oppressive use of state power. If we think of political society as a community united in affirming one and the same comprehensive doctrine, then the oppressive use of state power is necessary for political community. (PL 37)

As Burton Dreben points out, this is a "shocking" claim, because it entails that even if the whole society agrees on the liberalism of Kant or Mill, oppressive use of state power is necessary. This explains why Rawls later concluded that his stability argument in *Theory* was flawed, and why he has devoted 20 years to rewriting the central argument on stability in his later works.[5] Because of this view, Rawls holds that the Inquisition was not an accident:

> In the society of the Middle Ages, more or less united in affirming the Catholic faith, the Inquisition was not an accident; its suppression of heresy was needed to preserve that shared religious belief. The same holds, I believe, for any reasonable comprehensive philosophical or moral doctrine, whether religious or nonreligious. A society united on a reasonable form of utilitarianism, or on the reasonable liberalisms of Kant or Mill, would likewise require the sanction of state power to remain so. Call this "the fact of oppression." (PL 37)

To understand Rawls' point further, we need to understand the distinction drawn by Rawls between comprehensive doctrines and political conceptions.

Comprehensive Doctrine

In *Theory*, Rawls derives the two principles of justice from the original position, which relies on a Kantian comprehensive doctrine. A comprehensive doctrine is a set of beliefs affirmed by citizens regarding a range of values, including religious, philosophical, and moral commitments, as

[5] Burton Dreben, "On Rawls and Political Liberalism," in *Cambridge Companion to Rawls*, ed. Samuel Freeman (Cambridge University Press, 2006), 317.

well as beliefs about personal virtues, and political beliefs about how society ought to be arranged, and in particular about "what is of value in life, the ideals of personal character, as well as ideals of friendship and of familial and associational relationships, and much else that is to inform our conduct, and in the limit to our life as a whole" (PL 13). Moreover, an individual's conception of the good, which is usually part of a comprehensive doctrine, involves the meaning of life, and life-plans for which it is worth living.

In *PL*, Rawls acknowledges that he cannot expect every citizen to accept the same comprehensive doctrine.[6] This is because it is a fact about comprehensive doctrines that it is difficult to achieve any consensus or agreement about them. It is a permanent feature of "constitutional democracy"—by which Rawls means constitutional liberal democracy[7]—that there exists a plurality of comprehensive doctrines. A comprehensive doctrine is reasonable only if (a) it is theoretically more or less coherent, (b) it organizes values in a consistent way, and (c) it normally belongs to a tradition (PL 59). Moreover, it must also not seek to endorse political power to prevent other citizens from affirming their own comprehensive doctrines as long as their doctrines are also reasonable.[8] Because *Theory* relies on a Kantian comprehensive doctrine, Rawls has to seek a different basis on which to ground the stability of liberal institutions.

Stability and "Reasonable Pluralism"

Whether or not the liberal basic structure in a society is stable over time is an empirical question.[9] Religious toleration emerged from the estab-

[6] As I said above, Rawls holds that consensus cannot be obtained among rational and reasonable people because of the "burdens of judgment." Moreover, he holds that even if everyone in society agreed to the same comprehensive doctrine, the use of political power is necessary to maintain stability.

[7] See Burton Dreben, "On Rawls and Political Liberalism," *The Cambridge Companion to Rawls*, 322.

[8] See John Rawls, *Restatement*, 191–192; T. M. Scanlon, "Some Main Points in Rawls' Theory of Justice": https://www.academia.edu/44023083/Some_Main_Points_in_Rawls_Theory_of_Justice_1.

[9] T. M. Scanlon makes this point in his paper, "Some Main Points in Rawls' Theory of Justice." He also points out that the problem of stability (for the right reasons) is, for Rawls, a normative issue.

lishment of a balance of power after the sixteenth- and seventeenth-century European wars of religion. Rawls is, however, not interested in the question whether stability can be obtained merely as a result of a *modus vivendi*. Instead, he wants to see whether stability can be secured "for the right reasons," that is, whether stability can be secured for normative reasons. One reason is, I believe, that a *modus vivendi* is by its very nature unstable (PL 147).

Attempting to explain how constitutional liberal democracy and conflicting comprehensive doctrines are compatible, Rawls narrows the kind of society he considers to be a society of "reasonable pluralism." On his view, it is a permanent feature of constitutional liberal democracy that a diversity of reasonable comprehensive doctrines exist such that citizens can reasonably subscribe to different views over inherently difficult or controversial issues (*Restatement*, 33–34).

The Reasonable

What does Rawls mean by "the reasonable"? First, the willingness to propose fair terms of cooperation and to abide by them provided that others do so (PL 54). Second, the willingness to recognize the "burdens of judgment," which consist of empirical and scientific evidential factors bearing on theoretical questions, as well as the force of normative considerations relevant to opposing views on a moral or political issue (PL 56–57). Given the burdens of judgment, citizens can reasonably disagree about religious faith and biblical revelation, conceptions of the good, the meaning of life, metaphysical issues, as well as moral and political doctrines.

Given the permanent fact of pluralism, how can people of different religious faiths, or moral and political allegiances, discourse meaningfully on political issues, let alone come to any agreement? To overcome this problem, Rawls proposes a "political conception of justice." "Political conception" and "political values," on the one hand, are to be contrasted with "comprehensive doctrines" and "doctrines," on the other. By the political conception of justice, Rawls means a theory (1) which is *self-standing* (or free-standing), that is, not dependent on any comprehensive doctrine, (2) which focuses on the "constitutional essentials and matters of basic justice" regarding the basic structure of society, and (3) which is articulated in the publicly shared ideas of a democratic society, such that citizens who subscribe to any reasonable comprehensive doctrine could still agree to political values.

On Rawls' view, a moral view is either a political conception (or a part thereof), or else a comprehensive doctrine (or its component).[10] On the one hand, a moral view can be political in Rawls' sense if it is self-standing. Thus, for Rawls, various moral values, namely, the due respect for human lives, the ordered reproduction of political society over time, and the equality of women as equal citizens, are political values. (PL 243). On the other hand, a moral view can be a moral doctrine, which is by definition dependent on, or is part of, a comprehensive doctrine. Rawls points out that utilitarianism is a moral doctrine. (More on this below.) So are Kant's and Mill's liberalisms.

Is justice-as-fairness, as Rawls' theory of justice is usually known, a moral doctrine or a political conception? In reading *Political Liberalism*, the reader may encounter some unnecessary confusion as a result of Rawls' failure to clarify this point. As Burton Dreben points out, justice-as-fairness was a (partially comprehensive) moral doctrine in *Theory*, but in *Political Liberalism* it has become a (liberal) political conception, as it is self-standing and does not rely on any comprehensive doctrine.[11] The conclusion of justice-as-fairness (consisting in the two principles of justice) remains the same, but the grounds on which it stands in the two books are totally different.

In *PL*, justice-as-fairness relies on two grounds: first, that it is "latent" in the public political culture; and second, that it satisfies the criterion of reciprocity (see John Rawls, "The Idea of Public Reason Revisited," in his *The Law of Peoples*, henceforth "LP," 141). However, if these two

[10] Rawls points out that a comprehensive doctrine is either fully or partially comprehensive: "A doctrine is fully comprehensive when it covers all recognized values and virtues within one rather precisely articulated scheme of thought; whereas a doctrine is only partially comprehensive when it comprises certain (but not all) nonpolitical values and virtues and is rather loosely articulated. Note that, by definition, for a conception to be even partially comprehensive, it must extend beyond the political and include nonpolitical values and virtues" (PL 175).

[11] See Burton Dreben, "On Rawls and Political Liberalism, 332, 345. Burton says that justice-as-fairness in *Theory* depends on a comprehensive doctrine of autonomy (345). Moreover, Paul Weithman points out that Rawls was dissatisfied with his "congruence argument" (presented in Part III of *Theory*), which relies on a Kantian comprehensive doctrine. See Paul Weithman, *Why Political Liberalism? On John Rawls's Political Turn* (Oxford: Oxford University Press, 2010), Chap. VIII.

considerations are the only grounds for justice-as-fairness in political liberalism, they do not uniquely support justice-as-fairness, because they do not uniquely support the Difference Principle.[12] That is why, I believe, in "The Idea of Public Reason Revisited," Rawls points out that justice-as-fairness is not basic to political liberalism and public reason (LP 142, fn 27).

Political Values

For Rawls, political values are also moral values. These politically moral values are political in three senses. First, they differ from moral doctrines (which are comprehensive doctrines) in that they are self-standing.

Second, moral doctrines may be relevant to all life situations and belong to the background culture in civil society, even in how churches and universities, learned and scientific societies, and clubs and teams should be run. But political values do not govern these associations (except indirectly, e.g., the Church cannot prevent people from leaving it). On the other hand, political values are relevant for the public forum, and indeed applicable only to discourse about the basic structure of society. (Rawls clarifies that controversial issues, such as abortion, that "concern or border on" constitutional essentials and are "politically divisive" or "the cause of deep conflict" are to be resolved in terms of public reasons; see *Restatement*, 41, 117.) Because citizens are born into the basic structure of society and exit only upon death, the structure could be coercive and how it is organized requires justification.

This justification has to be public in the sense that it has to be, or is in good faith believed to be, acceptable to those who subscribe to different comprehensive religious, philosophical, metaphysical, and moral doctrines by satisfying the "criterion of reciprocity." This criterion requires that "when those terms are proposed as the most reasonable terms of fair cooperation, those proposing them must also think it at least reasonable for others to accept them, as free and equal citizens,[13] and not as

---

[12] The Difference Principle says that "social and economic inequalities are to be arranged so that they are (a) to the greatest benefit of the least advantaged and (b) attached to offices and positions open to all under conditions of fair equality of opportunity" (*Theory*, 83).

[13] As to what Rawls means by "free" and "equal," see note 3 above.

dominated or manipulated, or under the pressure of an inferior political or social position" (LP 137).[14]

Third, the public values are the shared, "fundamental intuitive ideas viewed as the public political culture of a democratic society" (John Rawls, *Collected Papers*, 480). In a democratic society with a tradition of democratic thought, they are "at least familiar and intelligible to the educated common sense of citizens generally" (PL 14), and are "formulated … in terms of certain fundamental ideas viewed as latent in the public political culture of a democratic society" (PL 175). The main institutions in society and their accepted forms of interpretation provide a source of implicitly shared ideas and principles (PL 14). Rawls mentions with reference to the problem of abortion, "the due respect for human life," "the ordered reproduction of political society over time," and "the equality of women as equal citizens" as three political values bearing on the issue (PL 243).[15]

Overlapping Consensus and Public Reason

The problem Rawls wrestles with in *PL* is the "torturing question in the contemporary world, namely: Can democracy and comprehensive doctrines, religious or nonreligious, be compatible? And if so, how?" (LP 175).

[14] Page references falling within 131–180 of *The Law of Peoples* refer to "The Idea of Public Reason Revisited."

[15] Rawls has not offered us an exhaustive list of political values, but they include (related to the family) the equality of children as future citizens, the value of the family in securing the orderly production, and reproduction of society and of its culture from one generation to the next (PL 163–164); the values of equal political and civil liberty; fair equality of opportunity; the values of economic reciprocity; the social bases of mutual respect between citizen (PL 139); examples of political values those in the preamble of the United States Constitution, namely, a more perfect union, justice, domestic tranquility, a common defense, the general welfare, and the blessings of liberty for ourselves and our posterity (LP 144); political values, such as equal basic liberties, equality of opportunity, ideals concerning the distribution of income and taxation, as well as efficiency and effectively (LP 144); political values related to animals and nature are the good of preserving the natural order; to foster species of animals and plants for the sake of biological and medical knowledge; to protect the beauties of nature for purposes of public recreation and the pleasures of a deeper understanding of the world (PL 245). Finally, liberty of conscience (LP 151), freedom of association (LP 158), and the freedom of religion (LP 163), including the freedom to affirm no religion (LP 145) are political values. (This list is not a complete one.) The "transcendent values" of salvation and eternal life—the *Visio Dei*—are, however, not political values. See Chap. 7 for further discussion.

To approach this problem, Rawls points out that different comprehensive doctrines have *reasons* for achieving a common ground on which to discourse with each other on issues regarding constitutional essentials[16] and matters of basic (distributive) justice,[17] even though these "reasons" might be different for different comprehensive doctrines. Citizens are willing to treat each other fairly insofar as they are fairly treated in return, even though their reasons for doing so may stem from their different comprehensive doctrines and hence may be different. Despite their different comprehensive doctrines, they have reasons (stemming from their different comprehensive doctrines) for agreeing on a political conception of justice. Rawls calls this agreement on the political principle of justice the "overlapping consensus."

A political conception of justice has three features. First, it is a moral conception worked out for a specific kind of subject, namely, a society's main political, social, and economic institutions ("the basic structure") of society (PL 11). Second, a political conception of justice is a self-standing view; it is not a comprehensive doctrine, nor is it parasitic or dependent on any comprehensive doctrine (PL 12). Third, its content is expressed in terms of certain fundamental ideas seen as implicit in the public political culture of a democratic society (PL 13). Although each feature is exemplified by justice-as-fairness, any version of liberalism would qualify as a political conception of justice.

In order for citizens of different comprehensive doctrines to discourse on "constitutional essentials and matters of basic justice," the idea of public reason is needed. On Rawls' view, the idea of public reason is closely connected with the idea of the political conception of justice. (More on this below). The idea of public reason is that those who subscribe to

[16] Constitutional essentials include (a) "fundamental principles that specify the general structure of government and the political process: the powers of the legislature, executive and the judiciary; the scope of the majority rule" and (b) "equal basic rights and liberties of citizenship that legislative majority are to respect: such as the right to vote and to participate in politics, liberty of conscience, freedom of thought and association, as well as the protections of the rule of law" (PL 227), or "questions about what political rights and liberties, say, may reasonably be included in a written constitution, when assuming the constitution may be interpreted by a supreme court, or some similar body" (PL 442, n.7).

[17] Matters of basic justice include "principles regulating basic matters of distributive justice, such as freedom of movement [including free choice of occupation] and equality of opportunity, social and economic inequality, and the social bases of self-respect" (PL 228).

different comprehensive doctrines—whether religious or nonreligious—can carry out discourse in the public forum regarding "constitutional essentials and matters of basic justice"[18] in terms of the political. Being self-standing and not parasitic on any comprehensive doctrine, public reasons form a proper subset of all normative reasons available. They transcend comprehensive doctrines and provide a common vocabulary in which discourse can be carried out.

Public Reasons and Admissibility

Public reason in public discourse is analogous to admissible evidence in a criminal court trial. In such trials, not every conceivable piece of evidence is admissible. Hearsay evidence is generally inadmissible (PL 218). Analogously, reasons that are *essentially* tied to comprehensive doctrines, such as those that arise from churches, universities, and associations, cannot be admitted as public reasons in the public forum where constitutional essentials and matters of basic justice, including politically divisive issues bordering on constitutional essentials, are debated, unless they can be arrived at and articulated in terms of public reasons.

Two features of public reason are particularly relevant to our concern with controversial bioethical issues. First, any public reason must be publicly accessible such that every citizen will be able to understand it. Second, public reason must satisfy the criterion of reciprocity, that is, those making a proposal must in good faith believe that others (regardless of their different comprehensive doctrines) can reasonably accept it. To take one example, can Christians in good faith think that non-Christians not only can understand the Christian concepts of sin and human dignity (understood as the worth bestowed on us by God),[19] but will also accept them? I am afraid that non-Christians, even if they can understand them, will not be able to accept them as their own.[20]

---

[18] See Note 1.

[19] McCarthy, Michael; Mary Homan; Michael Rozier, "There's No Harm in Talking: Re-establishing the Relationship between Theological and Secular Bioethics," *The American Journal of Bioethics*, 20 (2020): 12, 5–13, DOI: 10.1080/15265161.2020.1832611.

[20] See Hon-Lam Li, "Public Reason as The Way for Dialogue," *The American Journal of Bioethics*, 20 (2020): 12, 29–31, DOI: 10.1080/15265161.2020.1832618.

This does not mean that all religious doctrines must be excluded at the public forum for discussion. Rawls points out that "reasonable comprehensive doctrines—religious or nonreligious—may be introduced in the public political forum at any time, provided that in due course proper political reasons—and not reasons given solely by comprehensive doctrines—are presented that are sufficient to support whatever the comprehensive doctrines introduced are said to support" (LP 152). One obvious example is the biblical story of the Good Samaritan, who helps an injured person in dire need. This example can be stated in nonreligious terms. T. M. Scanlon takes it as an obvious principle that we should prevent something very bad from happening, or help someone in dire need, if the cost to us is slight (or even moderate). He calls this the Rescue Principle (T. M. Scanlon, *What We Owe to Each Other*, henceforth "WWO," 224–228).

Yet public reason should not be confused with secular reason. The latter is too broad and covers comprehensive moral doctrines. Thus, Rawls points out, if an argument against homosexuality is based on the view that homosexuality precludes what is a worthy human life, this argument is secular but is also a comprehensive moral doctrine. If, on the other hand, homosexuality is defended on the grounds that legislations that punish homosexuality infringe "the civil right of free and equal citizens," then these grounds would be (or at least could be) part of a political conception of justice (LP 147–148).

In *A Theory of Justice*, where the consenting parties through the veil of ignorance (in the original position) do not know of their social positions, genders, religions, natural endowments, and conceptions of the good, Rawls holds that the "primary social goods" are equally useful for, and hence neutral among, different conceptions of the good and life-plans. In *Political Liberalism*, there is no need for the veil of ignorance, because public values are the shared, intuitive ideas of democracy's public, political culture. In *PL*, then, public reason is neutral among different reasonable comprehensive doctrines. The only comprehensive doctrines toward which public reason is not neutral are those that do not tolerate other reasonable comprehensive doctrines.

Although the idea of public reason is to screen off doctrines from public discourse regarding constitutional essentials and matters of basic justice, it does not follow that public reasons must be good reasons, just as not every piece of admissible evidence in a trial is a weighty or credible piece of evidence. In a trial, pieces of admissible evidence might strengthen each other, just as they might undermine each other. By the same token, in the public forum, public reasons might reinforce or undercut one another.

Public Reason and Political Conceptions

According to Rawls, public reason has five different aspects.

(a) Scope of application

Public reason applies to basic institutions of society (hence "constitutional essentials and matters of basic justice"). Although it is unclear whether abortion falls under a consitutional essential, it certainly "borders on" one and is "the cause of deep conflict" (*Restatement*, 41, 117). Rawls holds that public reason should be brought to bear on abortion. For similar reasons, public reason is relevant to a discourse on the constitutional right to assisted death, or MAID.[21] Distribution of healthcare resources should fall within "matters of basic justice" and hence should fall within the scope of public reason.

(b) Persons to whom public reason applies

Public reason applies to legislators, judges (especially those serving in appellant and supreme courts), chief executives, and candidates for public office. Controversial bioethical issues, such as abortion and MAID, must be debated in the legislature, the court, by chief executives, as well as in the public. Because the coercive force of the law will be used, it would be fair to the citizens only if debates be conducted in term of public reasons, viz., terms and premises that they can understand and which they can reasonably be expected to accept.

[21] Dworkin et al. "Assisted Suicide: Philosophers' Brief," *New York Review of Books*, March 27, 1997.

(c) Content of public reason

The content of public reason is given by a family of reasonable political conceptions of justice, that is, a family of liberalisms,[22] and not by a single liberalism. Of these, justice-as-fairness is but one. Rawls also includes as liberalisms Habermas' discourse conception of legitimacy, as well as Catholic views of the common good and solidarity when they are expressed in terms of political values by, for example, John Finnis (LP 141–142). I shall argue that T. M. Scanlon's contractualism also qualifies as a liberal conception, and that his view and Rawls' are, in various respects, very congenial. (More on Scanlon later in the chapter.)

How do we determine whether a political conception of justice is liberal or not? Rawls says that the "limiting feature of" forms of political conception of justice is the "criterion of reciprocity" (*LP* 141). This criterion says that citizens offering reasonable terms of cooperation to others must also think in good faith that it is the most reasonable terms to others, as free and equal citizens, to accept. This condition specifies the minimally necessary condition on which various forms of political conception can be liberal. (I shall return to this later.)

As Rawls says, there are many liberalisms, and therefore many forms of public reason specified by a family of political conceptions of justice (LP 142). These conceptions can be "worked out from fundamental ideas seen as implicit in the public political culture of a constitutional regime, such as the conceptions of citizens as free and equal persons, and of society as a fair system of cooperation" (LP 142, 143). However, libertarianism (which allows extreme inequality) is excluded because it fails the criterion of reciprocity (LP 49)[23]

[22] Rawls has relaxed this from justice-as-fairness to a family of liberalisms. Why? Although he still talks about the two principles of justice arrived at in the original position, reasoning behind the veil of ignorance has been replaced in PL by the principle of reciprocity and the public political culture (from which to arrive at a political conception of justice which people espoused to reasonable comprehensive doctrines would endorse). But the content of political conception of justice has to be less specific than the two principles of justice, since this depends on the public political culture. It is surely easier to argue for a family of liberalisms.

[23] The idea is that the poor will not accept extreme inequality that is allowed by libertarianism, and hence that the rich cannot believe in good faith that the poor can reasonably accept their terms of cooperation.

and also because a society grounded in libertarianism will not be stable for the right reason (LP 50). Similarly, free and equal citizens would reject utilitarianism, which would prescribe that the less advantaged accept over the whole of their life fewer economic and social advantages for the sake of greater advantages for the more advantaged, and the less advantaged can reasonably reject this demand (*Restatement* 127).[24]

According to Rawls, conceptions of justice are characterized by three main features:

> First, a list of certain basic rights, liberties, and opportunities (such as those familiar from constitutional regimes);
> Second, an assignment of special priority to those rights, liberties, and opportunities, especially with respect to the claims of the general good and perfectionist values; and
> Third, measures ensuring for all citizens adequate all-purpose means to make effective use of their freedoms. (LP 141)

Rawls holds that the content of public reason is given by "principles and values of the family of liberal political conceptions of justice meeting these conditions" (LP 143). To engage in public reason is to appeal to one of these political conceptions—to "their ideals and principles, standards and values" (LP 144)—when debating fundamental political questions.[25]

(d) Legitimate law

These reasonable political conceptions—which are part of public reasons—are to be applied during discussions of coercive norms to be enacted in the form of legitimate laws for a democratic society (LP 133).

(e) Role of citizens

Finally, citizens should check that principles derived from their conceptions of justice satisfy the criterion of reciprocity (LP 133).

[24] Rawls believes that utilitarianism has other problems, and that a society grounded in utilitarianism would be unstable (without oppression). See *Restatement*, 126–130, 110, 115.

[25] I shall discuss more of this below.

The Duty of Civility

Rawls holds that the requirement that only public reasons should be used in the public forum regarding constitutional essentials and matters of basic justice applies to public officials.[26] They should discourse only in terms of public reasons and explain to other citizens their positions on matters of political justice in terms of principles that they regard as the most reasonable. In doing this, they fulfill what Rawls calls the moral "duty of civility."

Ideally, Rawls maintains, citizens who are not government officials should put themselves in the place of legislators and ask themselves what sort of public reason they can offer for or against a political issue. They can then support or repudiate the positions taken by government officials and candidates for public office on political issues, depending on whether those positions are taken according to public reason or not (and whether the reasons adduced are sound or not). When they hold government officials on account of public reason, and when they vote on political matters according to what they regard as the most reasonable conception of political justice (instead of their own personal view or individual advantage), they have fulfilled their duty of civility. Rawls plausibly holds that where resolution cannot be achieved after a discourse in public reasons, "the legal enactment expressing the opinion of the majority is legitimate law" (LP 137).

Disagreement Over Religious, Philosophical, and Moral Matters

Religious disagreement is the most obvious case in which disagreement cannot be resolved by reference only to comprehensive doctrines, for example, the holy scripts of different religions. (This is partly because revelation and religious faith are immune from falsification.) So religious disagreement is an example *par excellence* to show where the burdens of judgment can leave us.

[26] These include judges (particularly judges in appellant and supreme courts), and also legislators, chief executives, and other government officials, as well as candidates seeking public office.

The conception of the good is (often) tied to religious faith insofar as many believers hold that a life that promotes or coheres with the idea of "eternal salvation" is the best form of life. Moreover, if one has chosen a life of pleasure over a life of pure contemplation or a political life, for instance, one is unlikely to be persuaded that another form of life is more meaningful or better.[27] If citizens disagree over metaphysical, philosophical, and moral doctrines, as well as religious faiths, the idea of public reason says that we should try to resolve the issue in terms of political principles and values, instead of trying to resolve such disagreements in their original form.

To indicate how an argument on abortion might be developed in terms of public reasons without resolving the metaphysical problem of when a life begins, Rawls holds that we should try to resolve the disagreement "in terms of a reasonable balance of public political values" (PL 243). Thus, Rawls writes:

> Suppose … that we consider this question in terms of three important political questions: the due respect for human life, the ordered reproduction of political society over time, including the family in some form, and finally the equality of women as equal citizens. (There are, of course, other important political values besides these.) Now I believe any reasonable balance of these three values will give a woman a duly qualified right to decide whether or not to end her pregnancy during the first trimester. The reason for this is that at this early stage of pregnancy the political value of the equality of women is overriding, and this right is required to give it substance and force. (PL 243n)

Rawls subsequently clarifies that this is not an argument, but only an "illustration" of how a reasonable balancing of public political values might look like (LP 169 n80). Yet, Rawls' strategy is clear: with regard to the political problem involving social ethics in general, and abortion in

[27] Aristotle considers these forms of life in *Nichomachean Ethics*, and concludes that a life of pure contemplation is the best form of life, whereas the life spent in seeking pleasure is the worst. Aristotle's argument is unlikely to persuade one who is not already inclined toward leading a life of pure contemplation.

particular, since we cannot resolve—and certainly cannot expect all parties to agree upon—all the difficult metaphysical, philosophical, or moral issues involved, let us sidestep these issues and try to resolve the problem with a "reasonable balancing" of the various relevant political values at play.

Rawls has not elaborated his idea of "reasonable balancing" of political values. His view is probably as follows: we should first identify all relevant political values—that is, those relevant to the particular issue we are facing. However, it is by no means clear that everyone will agree on which political values are relevant. The second step would, however, be to balance these values. But it is an even more daunting task to "balance" the relevant political values involved. Someone who holds that a fetus is a person might think that "respect for life" is the paramount political value, and that other political values must be subordinated to it. Those who believe that a fetus is only a *potential* person would, however, strongly disagree. Hence it seems that "balancing" may not be the right way to consider the problem of abortion.

In "The Idea of Public Reason Revisited," however, Rawls says that to engage in public reason is to appeal to "ideals and principles, standards and values" inherent in one of the liberal political conceptions of justice.[28] It seems that Rawls allows (at least) principles, in addition to values.

In his later work, *Justice as Fairness: A Restatement*, the idea of public reason can be understood more broadly. Public reason consists of two kinds of political values. The first kind comprises "the values of equal political and civil liberty; fair equality of opportunity; social equality and reciprocity (expressed by the difference principle) and so on" (*Restatement*, 91). These are the shared, "fundamental intuitive ideas viewed as the public political culture of a democratic society" (*Collected Papers*, 480). The second kind includes "not only the appropriate use of the fundamental concepts of judgment, inference, and evidence, but also the virtues of reasonableness and fair-mindedness as shown in the adherence to the

[28] Dreben explains that when Rawls uses the words "political conception of justice," Rawls means liberal political conception of justice.

criteria and procedures of commonsense knowledge and to the methods and conclusions of science when not controversial" (*Restatement*, 91–92). (I shall discuss Rawls' endorsement of Judith Thomson's article, "Abortion,"[29] in Chap. 7.)

Thomas Nagel's Idea of Public Justification

One way in which personal views are suppressed in the face of a controversial issue is by means of the idea of public justification. Thomas Nagel's argument is illuminating, complex, and controversial (Thomas Nagel, "Moral Conflict and Political Legitimacy," *Philosophy & Public Affairs*, 16:3, Summer, 1987, henceforth "Moral Conflict," 230). It involves the interplay of the subjective and the objective, which permeates nearly all of Nagel's works.[30] In a nutshell, he argues that when we have a religious faith, we believe it to be true. Yet, liberalism requires that there be "a highest-order framework of moral reasoning (not the whole of morality) which takes us outside ourselves to a standpoint that is independent of who we are" ("Moral Conflict," 229). The kind of public justification that true liberalism requires, according to Nagel, "first, preparedness to submit one's reasons to the criticism of others, and to find that the exercise of a common critical rationality and consideration of evidence that can be shared will reveal that one is mistaken" ("Moral Conflict," 232). That means, it is possible "to present to others the basis of your own beliefs, so that once you have done so, *they have what you have*, and can arrive at a judgment on the same basis. That is not possible if part of the source of your conviction is personal faith or revelation" (ibid.). Public justification also requires "second, an expectation that if others who do not share your belief are wrong, there is probably an explanation of their error which is not circular" (ibid.).

[29] Judith Thomson, "Abortion," *Boston Review*, Summer 1995.

[30] The most important works where Nagel discusses the clash of the subjective and the objective are "What is it like to be a bat?" and "Subjective and Objective," both in his *Mortal Questions* (New York: Cambridge University Press, 1979) as well as in *The View from Nowhere* (Oxford: Oxford University Press, 1986) and *Equality and Partiality* (Oxford: Oxford University Press, 1995).

As Nagel argues:

> The idea is that when we look at certain of our convictions from outside, however justified they may be from within, the appeal to their truth must be seen merely as an appeal to our beliefs, and should be treated as such unless those beliefs can be shown to be justifiable from a more impersonal standpoint. If not, they have to remain, for the purpose of a certain kind of moral argument, features of a personal perspective—to be respected as such but no more than that.
>
> This does not mean we have to stop believing them—that is, believing them to be *true*. Considered as individual beliefs they may be adequately grounded, or at least not unreasonable: the standards of individual rationality are different from the standards of epistemological ethics. It means only that from the perspective of political argument we may have to regard certain of our beliefs, whether moral or religious or even historical or scientific, simply as someone's beliefs, rather than as truths—unless they can be given the kind of impersonal justification appropriate to that perspective, in which case they may be appealed to as truths without qualification. ("Moral Conflict," 230)

Nagel argues that there exists "a kind of epistemological division between the private and the public domains: in certain contexts I am constrained to consider my beliefs merely as beliefs rather than as truths, however convinced I may be that they are true, and that I know it" (ibid.).

Nagel's primary targets are religious beliefs based on faith and revelation, but he also argues that the same reasoning applies in controversial issues of private morality, for example, on abortion, sexual conduct, and killing animals for food ("Moral Conflict," 233). This is because Nagel holds that disagreements over these moral issues come down "to a pure confrontation between personal moral convictions," which is different from "a disagreement in judgment over the preponderant weight of reasons bearing on an issue" (ibid).

Although I am sympathetic to Nagel's idea of public justification,[31] in particular his idea that liberalism requires a higher-order impartiality, the

[31] I think that one of Rawls' central points can also be couched in terms of the clash between the subjective and the objective. Specifically, I have in mind Rawls' ideas of the comprehensive doctrines (which are in the Nagelian sense subjective) and the political conception of justice reached

following questions might be raised: why should deeply religious people accede to the liberal idea of higher-order impartiality, if they believe that they *know* God and have been in constant communication with Him? Why isn't religious faith or revelation like our knowledge of qualia, which is essentially subjective (in Nagel's sense) and is not open to confirmation?[32] In any case, it would be difficult to make the deeply religiously devout realize that what they believe are only "merely beliefs, rather than … truths" ("Moral Conflict," 230). In other words, how do we move deeply religious people to accept that their faiths are subject to the requirement of higher-order impartiality? Wouldn't doing so violate Rawls' criterion of reciprocity?

It will be similarly difficult to convince those championing animal liberation to agree that they are acting out of their "personal moral convictions" ("Moral Conflict," 233). Why can't the disagreement over animal liberation be "a disagreement in judgment over the preponderant weight of reasons bearing on [the] issue" ("Moral Conflict," 233).[33] A similar point may be made about abortion and sexual conduct. (I shall return to the clash of religions and the political at the end of this chapter.)

In summary, Nagel's point is that when it is your personal belief versus mine over a controversial issue—when it becomes almost a matter of personal taste—the only reasonable conclusion is that no law should be passed on it, whether it is abortion, killing animals for food, or sexual ethics. Nagel has since conceded that his epistemic argument in "Moral Conflict and Political Legitimacy" for public justification has been refuted

via an overlapping consensus (which is objective or "impersonal"). Rawls holds that with respect to fundamental questions about constitutional essentials and matters of basic justice, "the political values expressed by its principles and ideals normally have sufficient weight to override all other values that may come in conflict with them" (PL 138).

[32] Nagel's view is that qualia (e.g., our *taste* of chocolate, *sight* of a red rose, or *hearing* of Beethoven's Fifth Symphony) is subjective but real, and this example serves as a counterexample to the idea that all knowledge is objective. See his "What Is It Like to Be a Bat" in his *Mortal Questions*, and his *The View from Nowhere*, chap. 2.

[33] I would like to think that my arguments on animal research and quasi-vegetarianism are grounded on reasons or reasonable arguments. See "Animal research, Non-vegetarianism, and the Moral Status of Animals—Understanding the Impasse of the Animal Rights Problem," *Journal of Medicine and Philosophy*, vol. 27, No. 5 (2002), 589–615; "Toward Quasi-vegetarianism," in Hon-Lam Li and Anthony Yeung, eds., *New Essays in Applied Ethics* (UK: Palgrave Macmillan, 2007), 64–90.

by Joseph Raz.[34] Yet he still maintains that his conclusion that no law should be passed regarding an issue which is private and controversial is correct. On the other hand, he holds that while nuclear disarmament and capital punishment are controversial, they are public issues that must be settled.

Nagel is careful to distinguish his liberal position from other "pro-choice" positions. The liberal argument, as he calls it, is to not pass any law on controversial issues such as abortion, because any such legislation would be grounded in a controversial view. The "pro choice" view, on the other hand, holds that abortion should be legal because there are good reasons for justifying abortion. I have three concerns here.

My first concern: one way to understand Nagel's idea of not passing any law on abortion is that he meant we should freeze the existing legal position on abortion. Thus understood, such a moratorium of the current law (which could be pro-choice or pro-life) can be *ad hoc* and unfair. This is because in a Catholic country where abortion is illegal, a legal moratorium which froze the law on abortion would continue to make the practice a criminal offense. On the other hand, in another country where abortion is legal, a moratorium would have the opposite effect. Thus, a moratorium would not necessarily achieve the liberal ambition of realizing women's right to abortion. Alternatively, Nagel might have meant that all laws on abortion should be repealed. But this also might not achieve the liberal intention, because the constitution of a country may actually protect the right to abortion, and repealing the law would undermine, rather than reinforce, the liberal ambition. My point is that applying Nagel's point in any existing society by repealing the law would have consequences that are a matter of chance.

My second concern: in *Equality and Partiality*, Nagel says that although he has given up his (epistemic) argument advanced in "Moral Conflict and Political Legitimacy," he still maintains that no law should be passed on any issues that are controversial. We might query whether controversial issues of public policy are adequately addressed if we simply bracket the discussion of controversial issues. For one thing, the issue of slavery

[34] Joseph Raz, "Facing Diversity: The Case of Epistemic Abstinence," *Philosophy & Public Affairs* 19, no. 1 (Winter, 1990): 3–46.

was controversial in mid-nineteen-century America. But should the issue have been bracketed from discussion? Moreover, capital punishment is also a controversial issue. Should Nagel's liberal argument apply to both issues? Nagel's reply would be that *public* issues should not be bracketed, and consequently that slavery[35] and capital punishment should not be bracketed in the way that abortion, killing animals for food, and sexual ethics should be bracketed. This response raises two questions.

First, how are we to determine when an issue is a public one, and why? Second, why should the normative relation between (1) an issue being controversial and (2) it being bracketed have anything to do with its being public or self-regarding? In other words, if a public issue can be resolved via deliberation and voting—that is, deliberative democracy—so can the issues of abortion, killing animals for food, and sexual ethics. Nagel's approach should be contrasted with Rawls': Rawls allows that issues that border on constitutional essentials and matters of basic justice[36] and are politically divisive to be debated in terms of public reasons in the public forum. Even if it is impossible to reach agreement, once public officials and other citizens have fulfilled their duty of civility, the legal enactment expressing the view of the majority view is legitimate law.[37]

Moreover, Nagel's view notwithstanding, it does not seem obvious that issues that appear to be self-regarding, such as incest between consenting adult siblings (who have been sterilized), should automatically be bracketed and immune from legislative debate and enactment.

My third concern: a legislative moratorium is problematic in another way. Suppose that in a Catholic country abortion is illegal. As I said, a moratorium would preserve this country's anti-abortion laws, whereas in another country (such as China) women's legal right to have an abortion is much greater. If the liberal call for a legislative moratorium is allowed, stakeholders would see in advance what this would mean for them.

[35] Nagel actually refers to nuclear disarmament and capital punishment, not slavery. In personal conversation, Nagel said that his argument does not cover the case of slavery, which (unlike abortion, killing animals for food, and sexual ethics) is a public issue.

[36] Such issues include abortion, killing animals for food, sexual ethics, MAID, capital punishment, issues in reproductive technology, and just distribution of healthcare.

[37] For reasons to be explained, I think that Rawls' approach is more defensible.

Liberal women in the Catholic country would likely object, as well as Catholics in China. The same is true with other social issues, such as sexual ethics and killing animals for food.

In this connection, consider the case of the *Queen vs. Dudley and Stephens* (1884), concerning four sailors who drifted for several weeks in a boat in the Atlantic after a shipwreck. After many days without food and water, the captain and another sailor murdered the cabin boy and devoured him.[38] Prior to the killing, various lot-drawing options were discussed, but no agreement was reached. One ground on which the cabin boy was killed, according to the defense counsel in the murder trial that ensued, was that he had no dependents. How good is this ground? The problem with this is that good reasons could also have been invoked for choosing one of the other sailors. It might have been decided to kill the captain instead, since as captain, he might be held to have been responsible for the shipwreck. Alternatively, one of the older sailors might have been sacrificed on the grounds that a "fair inning" argument demanded that an older person be killed.[39] Or the unlucky person might have been chosen on the utilitarian grounds that the best person to kill was the one who was least socially useful. Would these sailors have hypothetically consented to any one of these reasons, as if they were behind a veil of ignorance? I very much doubt it because they were not actually behind any veil of ignorance when deliberating the options. Each of them knew very well what each principle would mean for them and their likelihood of survival.

The relevance of this case is that various interest groups might object to the liberal position if there were a moratorium on the controversial issue because (like the sailors) they would see how the liberal position would affect them. So those whose conceptions of the good would be disfavored under the liberal moratorium (e.g., the pro-lifers) will reject the liberal argument, whereas pro-choicers will welcome it.

---

[38] See: https://la.utexas.edu/users/jmciver/357L/QueenvDS.PDF.

[39] The classic paper is A. Williams, "Intergenerational Equity: An Exploration of the 'Fair Innings' Argument," *Health Economics* 6, no. 2 (Mar-Apr 1997): 117–132. According to this argument, a younger person should be favored on grounds of fairness.

The Criterion of Reciprocity

This brings me back to the criterion of reciprocity—the important idea that "when those terms are proposed as the most reasonable terms of fair cooperation, those proposing them must also think it at least reasonable for others to accept them, as free and equal citizens, and not as dominated or manipulated, or under the pressure of an inferior political or social position" (LP 137). Just as the cabin boy in *Dudley and Stephens* would not have accepted the proposal that the person without any dependents be sacrificed, religious groups who strongly oppose the legalization of abortion and MAID are unlikely to accept any balancing of political values, or any articulation of principles, if it leads to such legalization. Like the four sailors deliberating how to choose someone to sacrifice, religious groups strongly opposed to abortion and MAID would know in advance the possible implication of any political value or principle. In this sense, even if someone who puts forward a "balancing of political values," or who articulates a political principle, in justifying terms of agreement "think[s] it at least reasonable for others to accept them, as free and equal citizens," it does not follow that those adversely affected will think it reasonable for them to accept such terms. The only difference is that whereas the sailors were concerned with their own survival and hence their self-interest, the religious groups strive for their own conceptions of the good.

To recap, the criterion of reciprocity may fail at two levels. First, it may fail when and because some political values or principles relating to a specific social issue are not those that some religious groups could reasonably be expected to accept. Second, it may fail when and because the liberal political conception of justice is not one that some religious group could be reasonably expected to accept in the first place.

Note that there is an important difference between "reasonably accept" and "reasonably be expected to accept." It might be that a religious group should reasonably accept MAID, for instance, but it does not follow that it can be reasonably expected to accept MAID. Moreover, there is a difference between the notions (1) that a religious group could "reasonably accept" a proposal and (2) that those proposing terms of arrangement also think it at least reasonable for others to accept these terms. The captain and another sailor might have thought that it was reasonable for the cabin boy to accept that he should be sacrificed. But this does not mean that the cabin boy would agree, or that it was indeed reasonable to sacrifice the cabin boy.[40]

Yet, am I being too pessimistic or concerned about the nonfulfillment of the criterion of reciprocity? Or am I too concerned about people actually not accepting an arrangement even if the criterion of reciprocity is fulfilled? After all, Rawls restricts the parties for political agreement to be (religious and nonreligious) groups that are "reasonable" (or tolerant)—that is, those who are willing to cooperate with others on terms that others would also be willing to accept. But it should be observed that when the four sailors deliberated over the principles to follow in choosing the person to be sacrificed, they too might have been "reasonable" in the sense that there was an "overlapping consensus" that a solution be found. Yet it does not follow that they would have agreed on the principle which doomed one of their number to death. Similarly, citizens who subscribe to different comprehensive doctrines may not agree on the political values or principles by which they can proceed to deal with problems, such as abortion.

I believe this objection works more against Nagel's argument than against that of Rawls. This is because if, as Nagel says, there should be no law on any controversial private issue, it is too obvious which group would object, based on only the effect of the legal moratorium. But this might also present an objection to Rawls. Although Rawls would leave the resolution of a controversial issue to an enactment expressing the opinion of the majority, and hence there is no clear outcome to which a group could object in advance, some group might object to the very idea of political conception of justice. Recall that for Rawls the political conception of justice is liberal democracy. Some religious groups might be against liberal democracy, which they think might undermine the kind of

[40] After I wrote this, it has come to my attention that Steven Wall makes an analogous point in his "Is Public Justification Self-defeating?" *American Philosophical Quarterly* 39, no. 4 (October 2002): 385–394. Moreover, I have discussed consent—both hypothetical and actual— in "Contractualism and Punishment," *Criminal Justice Ethics* 34, no. 2 (2015), esp. 184–188, DOI: 10.1080/0731129X.2015.1067959

religious community they cherish. If a religious belief clashes with the law, do members of a religious sect have reason to follow the law, if they believe that their teaching is derived from divine revelation? (One example is faith-healing, which I shall discuss at the end of this chapter.) Rawls claims that political institutions in constitutional liberal democracy realize political values and ideals that normally outweigh opposing values (PL 209). But some religious groups might disagree.

Nevertheless, Rawls should be commended for putting forward a platform for people espoused to different reasonable doctrines to communicate and discourse in terms of public reasons. It cannot be predicted in advance that all sides would agree to the outcome of the discourse. Indeed, Rawls takes into account the possibility of disagreement among citizens, legislators, judges, and chief executives even after a lengthy process of deliberation and discourse in terms of public reasons. On his view, the final outcome should be decided by the majority in the legislature (or the majority in the Supreme Court). The right attitude toward the idea of public reason is the following. As Nagel points out:

> Rawls emphasizes that public reason is not to be thought of as an effective decision procedure, guaranteed to produce agreement, but rather as a special kind of disagreement, argument, and counterargument, which tries to use mutually recognized methods of evaluation and evidence, whether these produce consensus or not. Even if we are not convinced by an opponent's arguments about distributive justice, for example, we can recognize them as offering grounds that he thinks would be reasonable for us to accept, simply in virtue of the reasoning capacity that we all share. The same cannot be said for appeals to faith or revelation.[41]

The idea of public reason is not intended to solve all problems, just as the legal requirement of admissibility does not directly lead to a verdict. Rather, it is supposed to reorient the discussion or debate (regarding constitutional essentials and matters of basic justice, as well as issues that border on them and are politically divisive) by eliminating arguments based on comprehensive doctrines—just as the law disallows the general

[41] Thomas Nagel, "Rawls and Liberalism," in *Cambridge Companion to Rawls*, 76.

use of hearsay evidence and other types of evidence whose prejudicial effects outweigh their probative values. Even if various reasons are recognized to be public reasons, it does not mean that they will win the argument, since they might point at different directions and hence can be used by different or even rival positions.

How to Proceed?

With these caveats, we can proceed to see how Rawls' idea of public reasons—in particular political values and principles—can be brought to bear on bioethical issues. As I indicated, Rawls' use of "reasonable balance of political values" does not seem to be able to do the job. The basic problem is that his idea is too vague and undeveloped. The resolution of bioethical disagreement might require more than balancing some political values. In particular, it requires *arguments*, and particularly persuasive ones. We will have to invoke political principles as well as arguments to see how a position can be defended or denied. (I shall elaborate on this in Chap. 7.)

Apart from Rawls' (undeveloped) idea of reasonably balancing political values, another source of inspiration is his idea that political values and principles should be used by judges (especially judges in appellant and supreme courts) in legal judgment. As is widely known, Rawls has joined several other preeminent philosophers—namely, Ronald Dworkin, Thomas Nagel, Judith Thomson, Robert Nozick, and T. M. Scanlon—in submitting (as "amicus curiae") a brief to the Federal Supreme Court to argue that American citizens should have the constitutional right to assisted death (or MAID).[42] The arguments presented in this brief, since Rawls was a co-author, should satisfy his requirement that the arguments there consist of liberal political conceptions and values. In this brief, the arguments advanced consist of legal arguments referring to previous legal judgments by the Supreme Court, and also arguments as to why citizens should have the legal right to assisted death. None of these arguments are

[42] "Assisted Suicide: The Philosophers' Brief (March 27, 1997)," *The New York Review of Books*. Henceforth, "Dworkin et al. 1997."

built on religious or moral doctrines. On the contrary, they are self-standing, or else built on previous landmark cases.

Finally, an important source of inspiration as to what it is to use public reason is available in the work of T. M. Scanlon. To Scanlon's work, we now turn.

T. M. Scanlon

As is well known, Rawls intended to revise his political liberalism according to the ideas expressed in "The Idea of Public Reason Revisited." But this plan was never carried out because of his untimely death. Yet, I believe that Scanlon's contractualism provides inspiration for us to interpret and extrapolate Rawls' unfinished work, especially regarding how public reason is to be applied to concrete bioethical issues.

Rawls' political liberalism and Scanlon's contractualism bear a striking degree of congeniality or similarity.[43] To be sure, there are major differences between them. For one thing, Rawls is concerned with the political problem of stability and how public reasons should be used to address issues arising from constitutional essentials and matters of basic justice, whereas Scanlon deals with moral issues in general—covering personal and social ethical scenarios—as well as political issues.[44] For another, whereas Rawls hopes to resolve political problems by reasonably balancing political values vis-à-vis different stakeholders in society, Scanlon explains how principles are reasonably rejectable vis-à-vis "generic personal reasons" of different parties.

[43] Scanlon's contractualism is articulated and summarized in Hon-Lam Li, "Contractualism and the Death Penalty," *Criminal Justice Ethics* 36, no. 2 (2017), 152–182, DOI: 10.1080/0731129X.2017.1358912. I have employed contractualist reasoning in this article as well as in the following articles: Hon-Lam Li, "Contractualism and Punishment," *Criminal Justice Ethics* 34, no. 2 (2015), 177–209, DOI: 10.1080/0731129X.2015.1067959; Hon-Lam Li, Nancy S. Jecker & Roger Yat-Nork Chung, "Reopening Economies during the COVID-19 Pandemic: Reasoning about Value Tradeoffs," *The American Journal of Bioethics* 20, no. 7 (2020), 136–138, DOI: 10.1080/15265161.2020.1779406; Hon-Lam Li, "What We Owe to Terminally Ill Patients: The Option of Physician-Assisted Suicide," *Asian Bioethics Review* 8, no. 3 (September 2016), 224–243.

[44] Scanlon thinks that his contractualism could be used to social and political issues, such as criminal punishment and sentencing.

Nevertheless, Rawls' political liberalism and Scanlon's contractualism converge in the following ways. First, they endorse the important criterion of reciprocity. Scanlon's contractualism proposes that an act is impermissible if and only if "its performance under the circumstances would be disallowed by any set of principles for the general regulation of behavior that no one could reasonably reject as a basis of informed, unforced general agreement" (WWO 153). On Scanlon's view, moral reasons flow from "the reason we have to live with others on terms that they could not reasonably reject insofar as they are motivated by this ideal" (WWO 154). Rawls' criterion of reciprocity[45] bears a striking similarity to Scanlon's formulation of contractualism, which was already clearly formulated in the latter's "Contractualism and Utilitarianism." Apart from their use of "reasonably acceptable" and "reasonably rejectable," both of their views presuppose a moral community. Thus according to Rawls' criterion of reciprocity, citizens are reasonable when "they are prepared to offer one another fair terms of cooperation according to what they consider the most reasonable conception of political justice; and when they agree to act on those terms, even at the cost of their own interests in particular situations, provided that other citizens also accept those terms" (LP 136). For Scanlon, reason in the case of the morality of right and wrong is "the reason we have to live with others on terms that they could not reasonably reject insofar as they also are motivated by this ideal" (WWO 154).

Second, Rawls has been much influenced by Scanlon (and vice versa). Here are some philosophical grounds for my interpretation. For one thing, Scanlon says that his version of contractualism does not rely on the idea of "the veil of ignorance." As early as 1982, in "Contractualism and Utilitarianism," and later in *What We Owe to Each Other* (1998), Scanlon explained why the veil of ignorance is not necessary. In fact, Scanlon casts doubt on Rawls' "veil of ignorance" argument.[46] On Scanlon's view, to arrive at a reasonable conclusion, we only need to use reasons (or "per-

[45] Rawls refers to the principle of criterion first in *A Theory of Justice* (1971), but this principle plays a considerably more important role in *Political Liberalism* (1993) and "The Idea of Public Reason Revisited" (1997) included in *The Law of Peoples* (1999).

[46] See T. M. Scanlon, "Contractualism and Utilitarianism," in *Moral Discourse and Practice*, ed. Stephen Darwall, et al., (Oxford University Press, 1997), 279–282, esp. 282.

sonal reasons"[47]) to support or reject principles.[48] Importantly, in *Political Liberalism* (1993), Rawls abandoned any argument relying on the veil of ignorance, instead employing arguments that satisfy the reciprocity criterion as well as are intuitive and embedded in contemporary liberal democratic society. For another, Rawls also acknowledged in *Political Liberalism* (1993) the help rendered to him by Scanlon so that the "view as given in III is now more clear than first sketched in 1980," after the publication of Scanlon's "Contractualism and Utilitarianism" (1982).

Third, in articulating or defending what they see as the right conclusion on a normative issue, Rawls and Scanlon do not appeal to religious, philosophical, or moral doctrines. They appeal to principles or values that are self-standing.

Although some have criticized Rawls for having given up the notion of truth,[49] I believe Rawls rejects *labeling* certain comprehensive doctrines as true, and others as false. For, in the political realm, once a group of people declare their comprehensive doctrine as true, they must reject other reasonable doctrines as false. Yet there is no authoritative decision as to which reasonable doctrine is true and which is false. People espousing other reasonable doctrines could also declare theirs as true and reject other reasonable doctrines as false. But those who declare their comprehensive doctrine as true would be motivated to seek state power to enforce their doctrines and to prevent others from affirming their own views, even if these views are not unreasonable (PL 61). This consequence would not only be illiberal but also politically catastrophic because this would destabilize society and could only be maintained by force.

However, I agree with Zhuoyao Li's observation that Rawls' shift from "comprehensive doctrines" to "freestanding view[s]" together with Rawls' political conception of justice—or justice-as-fairness as presented in

[47] As Scanlon explains in WWO, a personal reason is a reason that concerns our claims in the broad sense. That is, if we are harmed or benefited by a principle, we have a personal reason to reject or favor it, but fairness also counts as a personal reason.

[48] Scanlon holds that a principle is a conclusion of a process of normative reasoning, and that there are an indefinite number of principles.

[49] See Zhuoyao Li, "The public conception of morality in John Rawls' political liberalism," *Ethics and Global Politics*. 9, no. 1 (2016), DOI: https://doi.org/10.3402/egp.v9.28679. However, according to Samuel Freeman, Rawls believes his comprehensive doctrines in *Theory* to be true. See Samuel Freeman, *Rawls*, 355–359.

Political Liberalism—are grounded in a "non-foundationalist" moral view that is "structurally justificatory." I believe that Scanlon's theory also has a similar structure, because the principles which Scanlon employs in any concrete scenario are also free-standing (or self-standing).[50]

It is true that political justice is different from the inner scope of morality known as "what we owe to each other." Li argues that the two, viz., political justice and morality on which Scanlon's contractualism focuses, are different only in scope. I agree that Rawls' and Scanlon's theories differ in scope, but I dispute the word "only." As I said, the structures of their theories are somewhat different. Vague as it is, Rawls' idea is to balance different political values relevant to a politically divisive issue, without giving much detail as to how this should be done. On the other hand, according to Scanlon, we should consider principles relevant to a moral issue by going through all relevant principles that we can think of (despite there being an indefinite number of them). Each principle is the conclusion of a process of reasoning. Opposing parties put forward reasons ("personal reasons") that concern their well-being in the broad sense that include deontological considerations, such as fairness. These reasons are put forward by competing parties to support or reject different principles that bear on a course of action, a policy, or a law. But these reasons must be general ("generic reasons"), so that deliberation on the right to privacy, say, can be carried out without naming any specific individuals. Unlike utilitarian reasons, the weights of *these generic personal reasons do not aggregate*. A utilitarian would either be hard-pressed to explain why minor headaches, if sufficiently numerous, should not outweigh a life, or otherwise have to bite the bullet and say that headaches can outweigh a life. Yet a Scanlonian contractualist can maintain that a life outweighs headaches (each suffered by a different individual), no matter how numerous these are. (See WWO, Chap. 5, esp. 229–241, although the literature on aggregation has quickly accumulated since the publication of WWO.)

For Scanlon, reason is not grounded in any metaphysical theory. Reason is basic and is a consideration in favor of something else. There is

[50] In a grant application submitted to Research Grants Council (Hong Kong) in October 2015, I already argued that there are important similarities between Rawls' theory in PL and Scanlon's contractualism, and that Scanlon's contractualist theory can be used to flesh out Rawls' idea of public reason. This happened before Zhuoyao Li's paper was published.

no interesting metaphysical account of reason. Nor can reason be explained in psychological terms. For Scanlon, reason is "irreducibly normative."[51]

The advantage of showing that discourse in public reason can be carried out in terms of Scanlonian principles and Scanlon's idea of reasonable rejectability is that his theory has been worked out in detail and can be used in resolving controversial issues. On the other hand, Rawls' idea of balancing political values has not been worked out and does not look promising.

(B) Bioethics

The Problem of Medical Assistance in Dying (MAID)

In accordance with the liberal conception of public reason, I shall argue that MAID should be allowed when three conditions are met. First, a patient has to be suffering from a terminal illness. Second, her condition is unbearable. Third, she is competent and has deliberated on this matter a number of times, and every time she concludes that she wants to die. If these conditions are met, I believe that a reasonable balance of political values—to use Rawls' term—would point to the conclusion that such a patient should be allowed to receive MAID.[52] In euthanasia a physician administers a fatal dose of a drug to the patient, whereas in MAID, the physician provides the drug and the patient herself takes it. The advantage of MAID is that a patient could change her mind up to the very last moment, but its disadvantage is that young infants and the paralyzed would not be able to receive MAID.

---

[51] See T. M. Scanlon, *Being Realistic about Reasons*, 2013, esp. chaps. 1 and 2. On Scanlon's view, a reason is just a consideration in favor of (or against) doing something (or omitting to do it). R is a reason for doing X if and only if there is a consideration in favor of doing X under the circumstances (*WWO*, 17–19). R may be a reason for doing X under circumstances C1, but may not be a reason for doing X under circumstances C2. Thus, Scanlon subscribes to moral contextualism, the plausible view that whether R is a reason for doing X depends on the context.

[52] Although I do not believe that Rawls' idea of balancing various values is usually helpful, the current case in favor of allowing MAID is so obvious that, even without an argument, such balancing should point to an obvious conclusion. But I shall provide an argument in the next section of this chapter.

This does not, however, follow that any physician is under a duty to provide the service to any patient. A physician may have a religious faith or a moral or personal view that is incompatible with administering MAID. Therefore, if any physician is to provide it, it must be on condition that she be entirely voluntary.

The political values in favor of MAID are those of personal autonomy, the badness of continual unbearable suffering, and the futility and meaninglessness of such suffering. Against these, there are also values operating in opposite directions. Those against MAID are usually motivated by (1) the religious view that only God can decide who should die, (2) the slippery slope argument, (3) the argument from abuse, (4) the argument from pressure on elderly patients to seek an early exit, and (5) the argument from the Hippocratic Oath (which I shall discuss in Chap. 4). David Velleman's argument against assisted suicide, grounded in his peculiar understanding of dignity, will be discussed in Chap. 7.

I shall now present an argument for the decriminalization for MAID.

The Case for Decriminalizing MAID[53]

The term "passive euthanasia"[54] used in philosophy covers several different cases in which the patient wants and seeks to die:

(1) Patients who want to die refuse treatment and are left at the mercy of nature running its course. (Refusal Case)
(2) Patients who want to die have their wishes honored and thus the life-saving equipment is withdrawn. (Withdrawal Case)
(3) Patients who want to die are given morphine with the result that the patient will die sooner than otherwise. These cases can further be distinguished into:

---

[53] This is a revised version of my argument in "What We Owe to Terminally Ill Patients—the Option of Physician-Assisted Suicide," in *Asian Bioethics Review* 8, no. 3 (September 2016): 225–230.

[54] The term "passive euthanasia" is used mostly by philosophers. When physicians use the term "euthanasia," they mean what philosophers refer to as "active euthanasia." Physicians generally avoid the term "passive euthanasia" and prefer refusal, withdrawal, and foreseeable death, and so on.

(3a) Patients who want to die are given morphine to alleviate pain with the *foreseeable* result that they will die. (The physician *foresees*—but does not intend—that her patient will die.) (Beneficial Foreseeable Death Case)

(3b) Patients who want to die are given morphine with the *intended* result that they die. (Beneficial Intended Death Case)

I take it that both Refusal and Withdrawal Cases are uncontroversial. In the United States, a patient (terminal or not) has the legal right to refuse treatment and, if he is already connected to life-saving equipment, the legal right to have it withdrawn.[55] *A fortiori*, a terminally ill patient should have these legal rights. It is widely thought that they have the same moral rights.

The Beneficial Foreseeable Death Case (3a) is legally permissible and regularly practiced, and is widely thought to be morally justified as well. This is because it can be in the interest of a terminally ill patient to have a shortened pain-free life than a longer life in unbearable pain.[56] Other equally pressing reasons include loss of autonomy (e.g., dementia), decreasing ability to engage in activities that make life enjoyable, and loss of dignity. In the scope of this chapter, however, I shall limit my focus to cases where unbearable pain or suffering is the reason for requesting MAID.

Opponents of MAID hold that offering a patient morphine with the foreseen consequence that he will die, as in (3a), is justifiable because there is no intention to kill or cause death. They distinguish this from MAID in (3b), where the physician *intends* that the patient will die.[57] Their view gains support from the Doctrine of Double Effect (henceforth "DDE"). DDE has various versions,[58] but the common feature among all versions is that there is an important moral difference between

[55] See *Cruzan v. Missouri Department of Health*, 497 U.S. 261 (1990).

[56] F.M. Kamm makes this point in her "Four-Step Arguments for Physician-Assisted Suicide and Euthanasia," in her *Bioethical Prescriptions*.

[57] See John Finnis, "A Philosophical Case against Euthanasia," in *Euthanasia Re-examined*, ed. J. Keown, (Cambridge, UK: Cambridge University Press, 1997): 23–35. It is more correct to describe a physician who performs active euthanasia as *intending* her patient to die, than one who assists in suicide, because in most cases of MAID, patients who obtain lethal medication choose not to take it (Ganzini 2000; Ganzini 2001).

[58] Most versions have a requirement that the good effect (which the agent intends) be proportionately good enough compared with the bad effect (which he does not intend).

an action intended to cause death and one that is foreseen to cause death but is not intended to do so: the former is impermissible, whereas the latter may be permissible. Thus DDE is thought to permit foreseeable death from, say, using morphine, but it does not permit MAID. At the minimum, DDE entails that at least in some cases, *everything else being equal*, an action intended to cause death is impermissible, whereas an action merely foreseen to cause death is permissible.

Judith Thomson and T. M. Scanlon have independently argued that DDE is untenable. They hold that an agent's intention is irrelevant to whether an action is *morally permissible*.[59] For instance, Scanlon argues that in an imaginary example of voluntary (active) euthanasia (which Scanlon assumes to be permissible), it is equally permissible, whether the physician intends the best interest of her patient or whether the physician turns out to be the patient's enemy and relishes the opportunity to put the patient away (T. M. Scanlon, *Moral Dimensions*, henceforth "Dimensions," 20–21). In these two cases, the *permissibility* of the action—based on patient's self-determination and his reasonable desire to die—remains the same, and it would seem ridiculous to demand that the patient must wait for a different physician with better intentions (*Dimensions* 21). However, the *moral characters* of the agents (i.e., the physicians) are vastly different (*Dimensions* 21).[60]

One reason in favor of the Thomson/Scanlon thesis is that whether an act performed under the circumstances is permissible has to do with reasons that arise from the objective factors in the circumstances only, and consequently what the agent thinks—including her intentions—is irrelevant to whether the act is *permissible*.[61] Thus in wartime, whether it is permissible for a pilot to drop a bomb that will not only destroy a military target but also kill civilians, cannot depend on what he thinks or what his intention is, but only on the reasons for or against dropping the bomb—reasons arising from objective circumstances.

[59] See Judith Thomson, "Self-defense" and T. M. Scanlon, *Moral Dimensions*, chaps. 1–2.

[60] As Scanlon holds, the physician (who is the enemy of the patient) should, given his bad intention, himself regard his action as impermissible (*Dimensions*, 16). But it does not follow that his action is indeed impermissible. It only reveals a fault in him (*Dimensions*, 58).

[61] See *Dimensions*, chaps. 1–2.

Scanlon is careful to qualify his claim in two ways. First, an intended act to cause X is usually more likely to succeed than an act that is merely foreseen to cause X, even though this is not necessarily so. Scanlon calls this the "predictive significance" of intent (*Dimensions,* chaps. 1–2). The predictive significance of intention, though important, is derivative and depends on the significance of the effect of an action on others. Thus, *everything being equal*, the thesis that intention is irrelevant to permissibility seems intact.

Second, there are cases in which intention seems to make a difference regarding permissibility. Consider these two cases. Jane and Mary visit their hospitalized grandfather. Jane visits her grandfather because she cares for him and wants to cheer him up. Mary, on the other hand, visits her grandfather only because she cares about what she will inherit under his will and wants to give him the false impression that she cares for him.

It is obviously permissible for Jane to visit her grandfather. Mary's intention to deceive her grandfather counts against her visit and, in the absence of countervailing considerations (such as cheering him up), may count decisively against her visiting her grandfather. Scanlon argues that in the second case, it may be impermissible for one to visit one's grandfather because such an act offends the principle that disallows one to act deceitfully (*Dimensions*, 38–39). If so, the evil intention matters only in a *derivative* way, because it violates the moral requirement not to mislead others or take advantage of their mistaken beliefs about one's intention. What differentiates the two cases is not merely intention, but also the fact that one case violates another principle, whereas another case does not violate any principle. Hence, the "everything else being equal" condition is not met. Is intention relevant to permissibility in a more fundamental way? Scanlon thinks not.

One might, however, object. Consider these cases:

Case 1: Someone drives a car and intentionally runs over a pedestrian, who dies shortly afterward. (Premeditated homicide)

Case 2: Someone is provoked by another person ("the victim") and drives to run her over. The victim dies shortly afterward. (Manslaughter)

Case 3: Someone drives carelessly and runs over a pedestrian, who dies shortly afterward. (Careless driving causing death)

I assume that the moral seriousness of these acts corresponds to their legal culpability.[62] These cases are distinguished on grounds that have to do with intention. In Case 1, the driver acts most wrongfully, because he intended to kill the pedestrian. In Case 2, the driver acts less wrongfully because his wrongful act was caused by a provocation. In Case 3, the driver does not intend to kill but nevertheless causes death by his own carelessness.

If the wrongfulness of these three cases is distinguished by the intention (or its absence) on the part of the perpetrator, one might ask, how can the Thomson/Scanlon thesis be correct? Thomson and Scanlon seem not to have anticipated this objection. This is, I believe, the reason for which philosophers do not take the Thomson/Scanlon thesis seriously enough, as Thomson herself acknowledges.[63]

I believe that we should admit that intention *does* affect how wrongful an act is, even though it does not affect the permissibility (or impermissibility) of the act. Therefore, the only way to defend the Thomson/Scanlon thesis is not to interpret it as the (false) view that intention is totally morally irrelevant to the act. Instead, I hold that although intention is morally irrelevant to the *permissibility* (or impermissibility) of an act, it is relevant to the degree of *wrongfulness* of the act and not merely the moral character of the agent. Let me explain.

To begin with, we must realize that the permissibility of an act is *all-or-nothing*, because it would not make sense to say that an act is permissible to some extent. Thus, intentional homicide and careless driving causing death are both impermissible, period. But it does not follow that they are both wrongful to the same degree because the gravity of an act—or how wrongful it is—is a matter of *degree* and is a function of the harm

[62] In other words, the case of murder is morally the worst, whereas the case of careless driving causing death is the least bad of the three cases.

[63] "It is not clear to me why people fail to take this seriously enough" (Judith Thomson, "Physician-Assisted Suicide: Two Arguments, *Ethics* 109, no. 3, (April 1999), 497–518 at 517). On her view, it is because they believe that if they do not accept DDE, then they must subscribe to utilitarianism. I think that her diagnosis is misplaced.

the perpetrator has caused as well as his intention. Thus, although premeditated homicide and careless driving causing death are both morally impermissible, they are wrongful to different degrees because the murderer intends for his victim to die, whereas the careless driver does not (even though the harm these two acts have caused are the same). Those who reject the Thomson/Scanlon thesis probably fail to notice this intricacy. Scanlon has reiterated that intention is relevant to the moral character of the agent. Although I agree with him, I have argued that intention is also relevant to how *wrongful* an act is.

In any case, the Thomson/Scanlon thesis is defensible insofar as it claims that intention is irrelevant to the *permissibility* (or impermissibility) of an act. It is, however, silent on whether intention is relevant to the wrongfulness of an act.

What is the relevance of this to MAID? The Thomson/Scanlon thesis says that two acts are equally permissible (or impermissible) if the only difference between them is the intention, *everything else being equal*. The relevance to MAID is as follows. If the foreseeable causing of death is permissible (Beneficial Foreseeable Death Case), where the patient is terminally ill and suffering excruciating pain—as this is uncontroversial—then, *everything else being equal*, the Beneficial Intended Death Case should also be permissible.[64] It does not mean, however, that these two cases are equally right (or equally wrongful), because that would depend on intention.

If what I have said so far is correct, then the following seems to follow. If the Beneficial Foreseeable Death Case is morally permissible, so should be the Beneficial Intended Death Case. This is what the Thomson/Scanlon thesis entails, and I think that this thesis is correct and defensible, if charitably interpreted. However, it does not mean that both cases are equally right (or wrongful) to the same extent. To fix ideas, consider:

> Abel operates a workshop that produces certain products. A by-product that Abel knows of is a toxic gas that can be lethal. Bernie lives near the

[64] After drafting this argument, I noticed that Judith Thomson also employs the same argument for MAID in Thomson, "Physician-Assisted Suicide," op. cit., 514–518. However, as I tried to show, her case is less tenable than mine because of her (and Scanlon's) failure to concede that intention may be relevant to the *wrongfulness* of an act itself.

workshop and has complained of headaches and allergies since the workshop's first day of operation. Abel foresees that the toxic gas will cause Bernie to die, and eventually Bernie dies. (Harmful Foreseeable Death Case)

Same as the earlier case, except that Abel hates Bernie and wishes Bernie dead. Abel builds the workshop (that produces the lethal gas) next to Bernie's home with the intention to kill Bernie. The toxic gas kills Bernie. (Harmful Intended Death Case)

Both cases are "harmful" for Bernie because death is bad for him. While Abel's actions in both cases are impermissible, he acts worse in "Harmful Intended Death Case" than in "Harmful Foreseeable Death Case." We can generalize that if Z is a bad outcome, then intentionally causing Z is worse than merely foreseeing and letting it happen.[65] (This point is different from the point about "predictive significance" of intention.)

We can also generalize where the outcome is a good one. Where Y is a good outcome, bringing about Y intentionally and actively is usually better than merely foreseeing and letting Y happen. Recall Jane who visits her grandfather in order to cheer him up because she cares for him. Consider Amanda, who visits her grandfather and merely foresees that her visit will cheer him up. Unlike Jane, Amanda does not care about her grandfather. And unlike Mary, she does not care about whether or how much she will inherit under his will. She visits her grandfather because she has nothing better to do. Unlike Mary's visit (which may be disallowed by a principle that we should not mislead people of our true intention), Amanda's visit is not disallowed by any principle. Everything else being equal, Jane's visit is morally better than Amanda's visit.

When a terminally ill patient suffers unbearable pain that is unavoidable and if he has made repeated requests to have his life terminated, his termination of life is—or can be—a rational and beneficial thing for him to do. In this case, helping him achieve his aim (that is, to end his life) is better than letting him achieve it himself (that is, letting nature takes its course in Refusal/Withdrawal cases), not because killing and letting die are morally equivalent, but because of the "predictive significance" of

[65] Letting Z happen passively could include either the case of foreseeing the death of some innocent person or the case of letting nature take its course.

intent. Intention being held constant, letting nature take its course would likely be more painful and hence more cruel.[66]

Furthermore, it is also morally worse to merely *foresee* that a patient will die than to intentionally *assist* him to die. As argued earlier, if death is a relief and hence beneficial for a patient suffering unbearable pain (and if he wants to have such relief), intentionally bringing about such relief is morally better than foreseeing that he will have such relief. In addition, intentionally bringing about such relief has "predictive significance," compared with merely foreseeing that the patient will die. For instance, she is more likely to have such benefit if the relief is intentionally made available than if such benefit is merely foreseen.

To recap, I have adopted Frances Kamm's view that it can be in the interest of a terminally ill patient to have a shortened but pain-free life than to have a longer life suffering in pain. I then argue that on these grounds, it is permissible for a physician to give morphine to a terminally ill patient to alleviate his pain, even if it is foreseeable that his life will be shortened. Finally, relying on the Thomson/Scanlon thesis that intention is irrelevant to permissibility, I argue that *everything else being equal*, it is permissible for a physician to give morphine to the terminally ill patient with the intention that the latter may take it to kill himself. Of course, I assume that in this last case, the patient is competent and has exercised his choice of self-determination to want to die.

A Religious View

Some people are against MAID because they hold that determining when to die would be playing God. This objection can take a religious form, as when some Christians argue that only God can decide when one should die, and thus MAID does violence to their religious faith. This objection proceeds from a religious comprehensive doctrine, and therefore

[66] Consider these cases. Case 1: In the battlefield, soldier Ryan stepped on a mine. He was very badly hurt and will certainly die. Ryan begged comrade Frank to kill him and thereby relieve him of the pain. Frank shot Ryan. Case 2: Same as Case 1, except that Ryan begged comrade George to relieve him of his pain by shooting, but George refused and chose to let Ryan die. Ryan was left on the battlefield and died 24 hours later. Clearly, it is crueler to let Ryan die.

(according to the idea of public reason) would be inadmissible in the public forum regarding constitutional essentials and matter of basic justice.

Slippery Slope Arguments: The Theoretical Version

An important objection to legalizing MAID is the so-called slippery slope argument. This argument has a theoretical and a practical version. The theoretical argument goes as follows. If there is a right to MAID, why limit it to patients dying in pointless suffering? Why not extend active euthanasia to dying patients who are so feeble or paralyzed that they cannot take the pills themselves but who beg a doctor to kill them by, say, injecting a lethal drug into them? And why not extend MAID to patients who are not dying but face years of intolerable physical or emotional pain, or crippling paralysis or dependence? Finally, why not extend it to anyone who has formed a desire to die—such as a 17-year-old suffering from a severe case of unrequited love?[67] This argument, which was raised by some judges of the Federal Supreme Court,[68] seeks to infer that if the law allows MAID, there is no natural or non-arbitrary line to draw between cases where MAID should be legalized and those where MAID should not be legalized.

The argument could be understood as follows. First, there is no clear line between cases where MAID is morally right and those where MAID is morally wrong. Yet, the law must draw fairly clear lines between what it permits and what it forbids—otherwise people would not know how to obey the law, in which case it would be unjust to enforce it against them. So, if there is no clear line to be drawn between cases where MAID is to be permitted and cases where it is not, the law must either forbid or permit assistance to a person seeking or committing MAID. And if these are the only choices, it is evident that MAID should always be forbidden.

However, this is not a good argument. It assumes that if there is no clear line between the cases where MAID is morally right and those where

[67] Ronald Dworkin et al., "Assisted Suicide: The Philosophers' Brief," *New York Review of Books*, 44 (27 March 1997): 41–47.

[68] Ibid.

it is morally wrong, the law may not draw a sharp and clear line between the cases it permits and those it forbids. This is not so. The law does not and cannot make all and only moral wrongs crimes. For a variety of perfectly good reasons, the law cannot criminalize all moral wrongs; similarly, the law must forbid some classes of acts some of whose members are not wrong at all. In the case of MAID, the court should try to identify a range of cases where MAID is not wrong. Although this might be time-consuming, it is preferable that the court does this than outlaws MAID in a blanket sort of way.[69] It is enough that the cases are clearly marked out, and that the cases marked out are clearly cases where medical assistance in dying is desirable. It is not necessary that every case where assistance is morally desirable should be included.

Slippery Slope Arguments: The Practical Version

The practical version of the slippery slope argument goes as follows. In a community, if a certain practice, A1, is allowed, A2 will as a matter of fact happen, which will then lead to A3, and perhaps A4 as well. Assuming that A2 and especially A3 and A4 are highly undesirable consequences, this argument—if supported by empirical evidence—would undermine the attempt to legalize A1 in this particular community. Proponents for this argument claim that if MAID is legalized (A1), it will be only a matter of time before voluntary euthanasia will be legalized (A2) as well. Non-voluntary euthanasia (A3) and perhaps involuntary euthanasia (A4) will eventually be legalized. They claim that voluntary euthanasia is bad, but non-voluntary and especially involuntary euthanasia are much worse. So, they claim that MAID should not be legalized.[70]

Unlike the theoretical version of the slippery slope argument, the practical version is not an *a priori* argument but instead relies on empirical data. Unlike the theoretical version (which is a *non sequitur*), the practical version in the abstract might present a genuine issue in the context of a particular community. For instance, it is understandable that most

[69] Ibid.

[70] Div. of Medical Ethics (University of South Florida College of Med.). 2001. "Physician-Assisted Suicide: The Legal Slippery Slope," *Cancer Control* 8 (1).

Germans object to the practice of voluntary euthanasia because of the Holocaust, but they have less of an objection to MAID.[71] Moreover, even if voluntary euthanasia is to be prohibited in Germany, it does not follow that it should be disallowed in the Netherlands, since the history, cultures, and traditions in of the two different communities may differ.

However, whether the practical slippery slope argument is persuasive depends on whether there is any empirical evidence to support the conclusion that MAID will lead to bad consequences in a particular community. Stephen Smith has looked at the empirical evidence in the Netherlands and in the state of Oregon. He concludes that there is insufficient evidence to support the practical slippery slope argument in these places.[72]

Abuse and Other Bad Consequences

Although we have no *a priori* solution to the practical slippery slope argument, it does not mean that we must give up any attempt to arrive at a reasonable conclusion. We should entertain various speculative objections to MAID.

In some jurisdictions where MAID is legalized (such as Oregon), a patient must be "terminally ill" (defined as having less than six months to live) in order to be eligible for MAID. According to one study, some 38% of "willing physicians" were "not at all" or "only a little confident" about their ability to "determine when a patient has less than 6 months to live."[73] However, just because there is such a risk of misdiagnosis, it does not follow that we should ban MAID, if only because these identical risks are present in a case where a patient decides to terminate life support.[74]

One very common objection to the legalization of MAID is that although MAID is morally permissible in some cases, there are certain

---

[71] See M. Battin, "Euthanasia: The Way We Do It, the Way They Do It: End-of-Life Practices in the Developed World," *Journal of Pain and Symptom Management* 65, no. 5 (1991): 298–305.

[72] S. W. Smith, "Evidence for the Practical Slippery Slope in the Debate of Physician-Assisted Suicide and Euthanasia," *Medical Law Review* 13, no. 1 (Spring 2005): 17–44.

[73] L. Ganzini, "Oregon Physicians' Attitudes about and Experiences with End-of-Life Care since Passage of the Oregon Death with Dignity Act," *JAMA* 285, no. 18 (2001):2363–2369.

[74] Ronald Dworkin, op. cit.

risks in legalizing it, and these seem to outweigh the possible gains.[75] The gain referred to here is the benefit a terminally ill patient suffering unbearable pain (which is not sufficiently alleviated by palliative care) would have should MAID be legalized. Ezekiel J. Emanuel (1999) lists several risks. First, MAID or euthanasia would undermine the medical profession because 25% of those physicians who have performed MAID or euthanasia regret having done so, as they feel like an "executioner." This suggests that 75% of those who have performed these procedures have no psychological or emotional problems afterward. The findings must also be compared with how withdrawing life-saving equipment would bother the physicians who honor the request made by the patients (or their families). In any case, the problem can be alleviated if physicians who are willing to perform MAID (or euthanasia) are cautioned or counseled by a psychologist before finally deciding whether to perform such procedures.

Another problem is "psychological anxiety" on the part of the patients if their physician raises the topic of MAID and euthanasia.[76] I think that we can resolve this problem in the following way. If MAID is legalized, a hospital would need to let eligible patients know of its availability—perhaps in the form of information sessions—because patients have the right to know about MAID, and importantly also the right not to opt for it. Anyone attempting to pressure a patient into MAID should commit an offense. Apart from information sessions, the law or hospital guidelines should disallow physicians to raise the topic of MAID with individual patients, unless a patient raises it first. The law should criminalize any attempt to pressure any patient to die, whether such pressure stems from the families or from the hospitals. Finally, there should be a law that requires two non-attending physicians (i.e., not attending the case) to authorize MAID.[77] Part of the function is to prevent abuse, whereas

[75] See J. D. Arrars, "Physician-Assisted Suicide: A Tragic View," *Journal of Contemporary Health Law and Policy* 13, no. 2 (1997): 361–389; E. J. Emmanuel, "What is the Great Benefit of Legalizing Euthanasia or Physician-Assisted Suicide?" *Ethics* 109, no. 3 (April 1999): 629–642, henceforth "Great Benefit;" Bonnie Steinbock, "The Case for Physician Assisted Suicide: Not (Yet) Proven," *Journal of Medical Ethics* 31 (2005): 235–241.

[76] See E. J. Emmanuel, "Great Benefit."

[77] Ronald Dworkin, "Assisted Suicide."

another is to screen off those who do not genuinely want to die from those who do.

The third problem has to do with the suspicion that although MAID is supposed to be a "last-ditch" intervention, which can be justified only after appropriate palliative options are attempted, MAID might be used before all palliative measures are utilized.[78] This fear has not materialized in the Netherlands, where two-thirds of requests for MAID or voluntary euthanasia have been rescinded, often as the result of palliative interventions[79] Ganzini also found little evidence in Oregon to suggest that vulnerable groups there had been given prescriptions for lethal medication in lieu of palliative care.[80] On the contrary, the physicians' experience with the Oregon Death with Dignity Act (1997) is that physicians are generally more aware of the patients' need to receive palliative care and consequently that patients tend to receive better palliative care. According to one study, "physicians grant 1 in 6 requests for a prescription for a lethal medication, and that 1 in 10 requests actually results in suicide. Substantive palliative interventions lead some—but not all—patients to change their minds about assisted death."[81]

Finally, a reason often used to object to MAID is that terminally ill patients, especially those from poor families, will be under pressure to opt for MAID so that their families will be relieved of the financial burden as well as the burden to visit them. By Emanuel's own admission, there are no accurate studies of the number of those who would opt for MAID because of family pressure (Emanuel, "Great Benefit," 637–638). Moreover, one might wonder why, if such pressure exists, should a patient not seek alternative ways to end one's life. One such option is the refusal to be saved at all. Another is the option to withdraw from life-saving treatment. Both of these options are much more cruel than MAID, because the pain associated with the dying process is dragged out. Finally,

---

[78] E. J. Emmanuel, "Great Benefit."

[79] L. Ganzini, et al., "Physicians' Experiences with the Oregon Death with Dignity Act," *New England Journal of Medicine* 342 (2000): 562.

[80] Ibid., 563.

[81] See both ibid. (2000) and L. Ganzini (2001), "Oregon Physicians' Attitudes about Experiences with End-of-Life Care since Passage of the Oregon Death with Dignity Act," *JAMA*, 285 (18): 2363–2369.

if MAID is not legalized, one could opt for terminal sedation, which is legal (in many jurisdictions). However, many patients would not prefer this to MAID because one's life is dragged on for a longer time, unconsciously, even if painlessly, and also because one loses dignity in the process.

However, it seems that while such a risk exists, the best way to tackle it is not by banning MAID. Instead, we can first distinguish patients who are not only terminally ill but are in unbearable pain from (a) those who are neither, and (b) those who are terminally ill but not in unbearable pain, and (c) those who are not terminally ill but are in unbearable pain.[82] Since MAID became legal in Oregon, only one in six patients who requested MAID has his request granted.[83] The first step is to grant the option of MAID to terminally ill patients suffering unbearable pain who desperately want to die.

Although the law must be general, it does not mean that it cannot distinguish between different kinds of cases in an intricate sort of way. For instance, the law of evidence generally rules out hearsay evidence as being inadmissible in a criminal trial. But this legal rule admits of exceptions, such as confession. Confession, in turn, admits of exceptions, as when it is acquired by force or inducement on the part of the law enforcers. Another example is the law on abortion. In various jurisdictions, abortion is permissible if the pregnant mother's life is endangered, but late abortions are generally disallowed.

Utilitarianism

Utilitarianism has dominated the philosophical scene for the past century and might make a difference to the argument that some elderly patients would feel obligated to opt for MAID. I shall argue that Scanlonian contractualism will provide a more plausible view than utilitarianism as

[82] I agree with Kamm that some patients experiencing unbearable pain should be allowed to receive MAID, even if one's illness is not terminal. See F. M. Kamm, *Bioethical Prescriptions* (New York: Oxford University Press, 2013).

[83] L. Ganzini, et al., "Physicians' Experience with the Oregon Death with Dignity Act," *New England Journal of Medicine* (2000), 342: 557–563.

regards the conflict of claims between (a) those who might be pressured to opt for MAID (but otherwise would not do so) and (b) terminally ill patients in unbearable suffering who desperately want to die.

Although it is Rawls' intention to rule out utilitarianism as inadmissible on the basis that it is grounded in a comprehensive doctrine, I wish to point out that this move might be unsuccessful.[84] Let me explain.

For one thing, Kant's and Mill's liberalisms are not political liberalisms, because they are grounded in comprehensive doctrines. Kant's liberalism was grounded in his doctrine of moral autonomy,[85] whereas Mill's liberalism was parasitic on his view on individuality. But it does not mean that there cannot be a self-standing "political liberalism" that does not rely on any doctrine of autonomy or individuality. Similarly, even if utilitarianism is so far founded on comprehensive doctrines, it is possible to have "political utilitarianism" that is not grounded in any comprehensive doctrine. Political utilitarianism can be based on the simple idea that the subject matter of morality or public policy is ultimately individual well-being,[86] and consequently that a policy that promotes better consequences is always to be preferred, compared with alternatives that do not do so. In fact, economists who are also utilitarians typically subscribe to a self-standing form of utilitarianism, according to which all that matters in morality is individual well-being and consequently that maximizing utility is what we should do.[87] Whether or not political utilitarianism is defensible is another question. Here, I am trying to show that not all forms of utilitarianism can be ruled out on the grounds that they are comprehensive doctrines or grounded in some such doctrines.[88]

---

[84] According to Samuel Freeman, "traditional utilitarianism is unreasonable as a political conception since it rejects the equality and priority of basic liberties" (*Rawls*, 396).

[85] Rawls holds that moral autonomy is not a political value (but is a doctrine), but political autonomy is a political value.

[86] Scanlon points out that the plausibility of normative utilitarianism is grounded in philosophical utilitarianism, the metaethical idea that the subject matter of morality is nothing but individual well-being. See his "Contractualism and Utilitarianism."

[87] I thank Tom Nagel for supplying this example in support of my contention that political utilitarianism can be a political conception because it need not rely on any comprehensive doctrine. I am aware that Rawls would hold that utilitarianism cannot be public reason because utilitarianism could justify slavery or some very unequal socio-economic system that maximizes collective utility. See, for instance, *LP* 50, 65, 79, 174.

[88] To be sure, Rawls argues that utilitarianism is excluded by the criterion of reciprocity (*LP* 40). But his argument seems to me too brief, or else the criterion seems too heavy-duty.

Political Utilitarianism vs. "Public Reason" Contractualism

Emanuel argues that when we look at the number of those who will benefit from MAID and those who will be harmed by it, it is not clear that more people will benefit from MAID.[89] He estimates that 2.3 million people die in the USA each year and that the maximum number of people who will benefit from MAID would be 25,000. According to him, such benefit is likely to be offset by the psychological distress of other patients, as well as an indeterminate number of terminally ill patients who might experience coercion to request MAID.[90] Finally, he thinks that too much time and energy has been wasted on the MAID debate, and that it is time to set aside the problem in order to focus our energy on things more useful.

There are two questions arising from Emanuel's paper. First, how does Emanuel know that the benefit of MAID will be offset by the harm, if we cannot even approximately determine how many people will be pressured into asking for MAID? Second, Emanuel's way of thinking seems to be a consequentialist or utilitarian view. Obviously, consequences are important. But to say that consequences are the only moral consideration is implausible. I shall offer an alternative way to think about this issue, based on contractualism.

Contractualism offers an attractive moral vision that is different from utilitarianism or other consequentialist theories, and from purely deontological theories. According to contractualism, an act, policy, or law is permissible if and only if it can be justified to everyone affected by it. Contractualism offers an interpretation as to what it is to count as a justification. Treating someone as a mere means is *ipso facto* not to justify the act, policy, or law to her. Contractualism demands that the justification be based on grounds that no one could reasonably reject, and that utilitarian maximization cannot (always) be counted as a justification.

The principal target of contractualism is utilitarianism because utilitarianism is an aggregative theory. Suppose we can save a life or

[89] Emmanuel, "Great Benefit."

[90] Ibid., 640.

alternatively alleviate a huge number (say, 10 million) of headaches, but not both. Whom should we help? I assume that the answer is to save a life. If this is correct, it is a mistake to maximize utility by aggregating relatively trivial utilities and then outweighing a life. Indeed, a utilitarian is committed to the repellant conclusion that we should alleviate headaches, if they are sufficiently numerous, rather than to save a life.

A real-life example that supports contractualism rather than utilitarianism is that the fact that a school or university in Hong Kong—and I take it the same situation would hold in Western countries as well—must be accessible to students with various kinds of disabilities, unless this would cause a school "unjustifiable hardship." Suppose that in a secondary school, out of 800 students there is one student with a physical disability and needs to use the wheelchair, or another student who needs regular access to psychological counseling. Accommodating such a student would cost the school resources that could have been used for upgrading the school library instead. A utilitarian would say that to maximize utility, not admitting such a student would be permissible, perhaps even mandatory. A utilitarian would care only about the collective utility, and the plight of the disabled student can be outweighed by the gain of other students (in the form of a better library). Yet, if we agree that this would violate the right of this student, we need an account as to why this is so. According to contractualism, ruling such a student out would be unjustifiable to her, whereas a utilitarian would not be able to account for this policy.

What is the relevance of this to MAID? A utilitarian would argue that we should compare (1) the collective utility of those who would be adversely affected by the legalization of MAID (for example, those who would be pressured into accepting MAID) and (2) the aggregated utility of those who would benefit from MAID. What is wrong with that?

First, simply counting the number of people adversely affected by the availability of MAID, and those who would benefit from it, is certainly the wrong way to go.[91] Even a utilitarian would not simply count the numbers. Second, if the "headaches" example is any guide, even summing

[91] Emmanuel argues that the number of patients who would benefit from MAID is small, as if the numbers were all that matter.

up relatively trivial utilities in order to outweigh fewer but more substantial benefits is problematic. We do not know how many patients would be pressured into accepting MAID, as opposed to those who would benefit from it. If the numbers are equal, surely the benefit of not having to drag through unbearable, meaningless, and seemingly endless pain and suffering is much more important than those who would succumb to family or hospital pressure into accepting MAID, or those who would opt for MAID only because of a sense of duty. After all, the latter two groups should be able to withstand such pressure. With the legal measures against undue influence on the elderly to choose MAID, educational sessions on patients' rights, and screening measures to distinguish between those in genuine need of MAID and those who simply "feel the heat" from their families or hospitals, the number of patients who involuntarily choose MAID should be quite small.

Suppose, for the sake of argument, patients who would be pressured to choose MAID outnumber those who would benefit from it by a factor of five to one. Would a ban on MAID be justified? Let us assume that the legal measures, educational sessions, and screening procedures are in place. The question will be whether (1) those who are pressured into choosing MAID and (2) those who opt for MAID out of a sense of duty could reasonably reject the legalization of MAID, as well as whether (3) terminally ill patients suffering in unbearable pain could reasonably reject the ban on MAID. If there are many patients being pressured into choosing MAID, then the legal, educational, and screening measures are probably inadequate. On the other hand, if these measures are effectively implemented, can a very "kind" or "considerate" patient who does not want to burden his family and who pretends to be entirely willing to choose MAID, reasonably reject the legalization of MAID? I would think not, because her decision is, at least to some extent, voluntary, and hence she is responsible for her own opting for MAID. On the other hand, terminally ill patients suffering in unbearable pain can reasonably reject a ban on MAID, if only because their suffering is unbearable, unavoidable, and pointless. In sum, if the various measures mentioned are in place, a ban on MAID cannot be justified to terminally ill patients suffering in unbearable pain and who have repeatedly asked for their lives to be terminated. We owe it them to have MAID legalized and make MAID available as an option.

I shall examine the argument that MAID clashes with the Hippocratic Oath in Chap. 4, and Velleman's view in Chap. 7.

(C) Concluding Remarks

In the first half of this chapter, I have articulated Rawls' political liberalism and his idea of public reason. I pointed out various concerns that this idea might entail. But as I argued, a case can be made to defend it. This is so especially if we bring in the resources provided by Scanlon's contractualism, which is congenial to Rawls' idea of public reason.

In the second half, I argued for the case that MAID should be legalized or decriminalized. Those who are against MAID object either on religious grounds, or on grounds having to do with the slippery slope, or due to concern over possible abuses. I replied to these arguments. Finally, I examined the argument that legalizing MAID would increase the number of people who would opt for it, because they worry about the pressure they have created for their relatives and about a drain on the social resources. I explained why this was not be a good argument against MAID being made permissible by law.

Although there seem to be no good *a priori* arguments for banning MAID, I admit that it is possible that in some particular community, MAID might be grossly abused and that in this community MAID should probably be banned. But an absolute ban would require strong empirical evidence for support. In all probability, however, if adequate measures are in place against abuse, I believe that it would be very difficult to justify an absolute ban on MAID, and even more difficult to justify such a ban on *a priori* grounds. For all citizens, the option to use MAID offers peace of mind. If the pain ever becomes unbearable, an individual can choose to leave the world without pain and with dignity.

Even though I have not used the kind of sweeping, macro argument employed by Nagel, I think that the ideas of public reason and public justification do indeed have an impact on the public discourse of various bioethical topics. The condition of public justification, as Nagel observes, "reverses the relative importance possessed by different values in the private domain. To a believer, salvation is more important than liberty, yet in political justification he may not appeal to the importance of salvation to justify the restriction of liberty, because liberty is a publicly admissible value and salvation is not."[92]

Finally, let me note the kind of tension experienced by individuals whose religion is against MAID, but who at the same time recognize that their agreement with the idea of public reason should require them to suspend one's religious view. As Scanlon points out:

> Rawls' requirement of public reason … involves a kind of division within the outlook of any person who accepts it. On the one hand, such a person accepts that, when argument about the basic institutions of society is concerned, no comprehensive doctrine, religious or otherwise, has greater standing than any other. What matters for settling these questions is just the reasons that individuals have for wanting to be able to adopt and live in accordance with their own particular view, whatever this view may be.[93]

Although the tension generated by the division can occur between public reason and any comprehensive doctrine, it is especially serious when the conflict occurs between public reason and one's religious faith. An individual who subscribes to a moral doctrine can in principle give it up in light of a better argument to the contrary. Religious faith is a different story, because it is to a considerable extent immune from any evidence and arguments to the contrary. In other words, the kind of division generated between public reason and religious faith can be considerable. Many persons of religious faith not only believe in their God but also think that they *know* God and that they are in constant communication with Him. In such a case, they may have little reason to allow their religious faith to yield to public reason in case of conflict. At least, if and when their religious faith requires them to practice something outside the constraints of the law, they may have no reason to yield to the demand of public reason. In the American state of Oregon, parents who practice

[92] Thomas Nagel, "Moral Conflict and Political Legitimacy," *Philosophy & Public Affairs* 16, no. 3 (1987): 215–240 at 237.

[93] T. M. Scanlon, "Some Main Points in Rawls' Theory of Justice."

faith-healing choose to pray instead of taking their sick children to a clinic run the risk of being prosecuted for not taking care of their children properly. In the past few years, parents who prefer faith-healing to modern medicine have been convicted of criminal mistreatment, criminally negligent homicide, second-degree manslaughter, and even first-degree manslaughter.[94]

In spite of these risks, Rawls' theory of political liberalism and the idea of public reason will play an important role in ensuring that different groups do not simply throw irreconcilable comprehensive doctrines—religious and nonreligious—at each other in the public political forum. Doing so would seem like one side seeking to impose their will on its opponents.[95] The idea of public reason, then, is to reorient the debates regarding constitutional essentials, or controversial cases that border on such essentials. Even if at the end there is no consensus among different groups in society, this is probably how public debates should go.[96]

[94] The case of the Bellows is a classic case of public reason clashing with nonpublic reason. The Bellows are members of Church of the First Born and practice faith-healing religiously. They believe that God will listen to their prayer. Given their religious belief, they have a nonpublic, religious reason to practice faith-healing. In other words, they have such a reason for not treating their sick children with modern medicine (and likewise when they themselves have fallen ill). Yet the state of Oregon has a public reason for demanding that parents take their sick children to receive medical treatment, namely, to avoid the preventable deaths that might otherwise result. According to *The Oregonian* (May 15, 2011), Oregon is the longtime home of the Followers of Christ church, a sect based in Oregon City whose members believe in treating sickness with prayer and oil rather than medicine. Getting glasses or dental care is sometimes considered acceptable. Seeking medical help for yourself or your sick child is not. This helps explain the church's high child mortality rate, its long history of children dying from treatable conditions, and the rows of children' headstones in the church cemetery.

[95] In "Engelhardt on the Family" (2013), *International Journal of Chinese and Comparative Philosophy of Medicine*, 153–160, I argue that the Conservatives who put forward their view on the family are in effect imposing their taste to those who are homosexuals and lesbians, as well as those who do not want to get married and those who want to get married but do not want to have children.

[96] As pointed out above, even if there is no consensus after discussion in terms of public reason, the majority's view in the legislature (whose members are elected into their offices) will be legitimate.

2

The Natural Law Tradition, Public Reason, and Bioethics

Dominic Farrell LC and Joseph Tham LC

Over the last fifty years, many countries have passed legislation to permit a series of controversial biomedical practices such as in vitro fertilization, surrogate pregnancies, abortion, and euthanasia. In each case, the change in legislation has enjoyed a certain amount of public support and has been defended as bringing both moral progress and social benefits. Normally, however, a significant sector of the public has opposed it on moral grounds. Consequently, there is generally a heated debate surrounding such legislation, and rightly so. What is at stake in these debates over first-order ethical issues is not just another matter of public policy, but conflicting views on important underlying second-order questions, such as personal identity and human dignity. But by what standard can our political and legal institutions settle first-order bioethical issues and the underlying second-order questions legitimately?

This is where theories of "public reason" are promising. In contemporary philosophical parlance, this term of art has gained currency, thanks

---

D. Farrell LC (✉) • J. Tham LC
Pontifical Athenaeum Regina Apostolorum, Rome, Italy
e-mail: Dfarrell@upra.org; jtham@legionaries.org

H.-L. Li, M. Campbell (eds.), *Public Reason and Bioethics*,
https://doi.org/10.1007/978-3-030-61170-5_2

to the influential work of John Rawls. In his anti-perfectionist liberalism, the political and legal rules of a society are just and legitimate if and only if they pass the test of public reason: if they are acceptable to all reasonable persons.

While Rawls's proposal remains the most influential, alternative accounts of public reason have been developed. It has been argued, for example, that natural law theory (NLT) in the Thomist tradition, a variety of perfectionist moral and political philosophy, yields a better account of public reason.[1] In this chapter, we shall outline NLT's conception of public reason and argue that it is more consistent than that of political liberalism, especially when it comes to bioethical legislation. After offering a précis of NLT (Sect. 1), we shall turn to Alasdair MacIntyre's treatment of the rational resolution of moral disagreements and argue that public reason should be construed more broadly, as is the case in NLT, as a shared political deliberation that is rooted in truth-directed inquiry (Sect. 2). Furthermore, such a truth-directed inquiry can only be carried out via a commitment to the primary precepts of natural law. A just political order should pursue and promote this sort of public reasoning, even in a pluralistic society, where there is systematic and apparently intractable moral disagreement. However, whereas MacIntyre believes that such shared deliberation is no longer possible outside local communities, we shall reflect on how his proposals might carry over to the full-fledged public reasoning of political decision-making. To do so, we shall examine NLT's conception of legislation and how it works for bioethical issues (Sect. 3).

---

[1] John Finnis, "Public Reason, Abortion, and Cloning", *Valparaiso University Law Review* 32 (1998): 361–82; Robert P. George and Christopher Wolfe, "Natural Law and Liberal Public Reason", *The American Journal of Jurisprudence* 42 (1997): 31–49; Robert P. George and Christopher Wolfe, *Natural Law and Public Reason* (Washington, DC: Georgetown University Press, 2000); Robert P. George, "Public Morality, Public Reason", *First Things* 167 (2006): 21–26; John Haldane, "Public Reason, Truth, and Human Fellowship: Going Beyond Rawls", *Journal of Law, Philosophy and Culture* 1 (2007): 175–90; Martin Rhonheimer, "The Political Ethos of Constitutional Democracy and the Place of Natural Law in Public Reason: Rawls's '*Political Liberalism*' Revisited", in *The Common Good of Constitutional Democracy: Essays in Political Philosophy and on Catholic Social Teaching*, ed. Martin Rhonheimer (Washington, DC: Catholic University of America Press, 2013), 191–264; Ibid., "Rawlsian Public Reason, Natural Law, and the Foundation of Justice: A Repsonse to David Crawford", in *The Common Good of Constitutional Democracy: Essays in Political Philosophy and on Catholic Social Teaching*, ed. Martin Rhonheimer (Washington, DC: Catholic University of America Press, 2013), 265–91.

Although this chapter is a contribution to a work on public reason and bioethics, it will focus first on public reason in general, and then on its specific application to bioethical issues, such as abortion and medically assisted suicide (henceforth MAS). Reasonable disagreement is just as likely to beset public debates on any issue falling under "the burdens of judgment" (e.g. economic justice). For this reason, success at public reasoning, whether in bioethics or any other issue, depends first and foremost on settling upon the most consistent theory of public reason.

1 A Précis of Natural Law

The substantive conclusions that a bioethicist reaches will depend, as in any area of applied ethics, on which tradition (i.e. a historically developing conceptual framework) of normative ethics he or she deems more compelling and worthy of endorsement.[2] As should be expected, some bioethicists endorse the natural law tradition as the proper framework for their field.[3]

Bioethicists who defend a theory of natural law are often Catholics and sometimes even describe their standpoint as that of "Catholic bioethics".[4] This is understandable. The Catholic Church's official teaching, especially since Leo XIII, defends a doctrine of natural law and has appealed to it consistently in addressing most ethical issues, bioethical ones included.[5]

[2] This definition of tradition is taken from Alasdair C. MacIntyre, *Three Rival Versions of Moral Enquiry: Encyclopedia, Genealogy, and Tradition* (Notre Dame, Ind.: University of Notre Dame Press, 1990), 371.

[3] Alfonso Gómez-Lobo and John Keown, *Bioethics and the Human Goods: An Introduction to Natural Law Bioethics* (Washington, DC: Georgetown University Press, 2015); E. Christian Brugger, "The First Principles of the Natural Law and Bioethics", *Christian Bioethics* 22 (2016): 88–103.

[4] William E. May, *Catholic Bioethics and the Gift of Human Life* (Huntington, IN: Our Sunday Visitor, 2000); Anthony Fisher, *Catholic Bioethics for a New Millennium* (Cambridge: Cambridge University Press, 2012).

However, one of the main Catholic bioethicists, Cardinal Elio Sgreccia, proposes a metaphysically grounded personalism, that transcends any specific religion, as the proper framework for bioethics. See, Elio Sgreccia, *Personalist Bioethics* (Philadelphia: The National Catholic Bioethics Center, 2012).

[5] For the historical background to Leo's use of natural law doctrine, cf. Thomas C. Behr, "Luigi Taparelli D'Azeglio, SJ (1793–1862) and the Development of Scholastic Natural-Law Thought as

Nevertheless, although an integral part of Catholic bioethics, natural law is not a specifically Catholic doctrine.

True, given their belief in the truth of Catholic doctrine, Catholics have reason to accept their church's teaching on the existence and content of natural law.[6] However, by proposing a doctrine of natural law, the Catholic Church teaches that they have another reason to do so.[7] Whenever it proposes a moral teaching as belonging to "natural law", it is claiming that there are reasons, external to Christian revelation, and accessible to anyone who gives due consideration to the relevant facts, for accepting it. Indeed, magisterial documents often attempt to outline those reasons.[8]

From here on, therefore, natural law theory will be treated as a tradition of normative ethics rather than a specifically Catholic doctrine, and for the following reason. It is as a tradition of moral and political philosophy that natural law can constitute a framework for public reason. Treating it as such is to claim that its substantive conclusions are accessible, rationally justifiable, and acceptable to all duly informed and right-thinking persons.

---

a Science of Society and Politics", *Journal of Markets and Morality* 6 (2003): 99–115. Russell Hittinger, "Two Modernisms, Two Thomisms: Reflections on the Centenary of Pius X's Letter against the Modernists", *Nova et Vetera* [English Edition] 5 (2007): 843–80. For an overview of natural law in Catholic moral theology, see Hittinger, "The Situation of Natural Law in Catholic Theology", *Nova et Vetera* [English Edition] 9 (2011): 657–70.

The International Theological Commission's 2009 document *In Search of a Universal Ethics: A New Look at Natural Law*, provides an overview of the Catholic doctrine on natural law. For the document and accompanying essays, see William C. Mattison III and John Berkman, *Searching for a Universal Ethic: Multidisciplinary, Ecumenical, and Interfaith Responses to the Catholic Natural Law Tradition* (Grand Rapids, Michigan: William B. Eerdmans Publishing Company, 2014).

[6] On the varying degrees of authoritativeness of the Church's Magisterium and the degree of assent which the faithful should give it, see Congregation for the Doctrine of the Faith, Instruction on the Ecclesial Vocation of the Theologian *Donum Veritatis* (24 May 1990), *Acta Apostolicae Sedis* 82 (1990), 1559–61.

[7] To this effect, *Catechism of the Catholic Church*, 2nd ed. (Vatican City – Washington, DC: Libreria Editrice Vaticana – United States Catholic Conference, 1997), nn. 1954–56.

[8] In the case of bioethics, see Congregation for the Doctrine of the Faith, *Declaration on Procured Abortion* (18 November 1974), *Acta Apostolicae Sedis* 66 (1974): 730–47; *Declaration on Euthanasia* Iura et bona (5 May 1980), *Acta Apostolicae Sedis* 72 (1980): 542–82; *Instruction* Donum Vitae *on Respect for Human Life at Its Origins and for the Dignity of Procreation* (22 February 1987), *Acta Apostolicae Sedis* 80 (1988): 70–102; *Instruction* Dignitas Personae *on Certain Bioethical Questions* (8 September 2008), *Acta Apostolicae Sedis* 100 (2008): 858–87.

One way of getting a handle on the natural law tradition is to study the author who is generally considered to have given its canonical statement: Thomas Aquinas.[9] However, such a study is out of the question here, even though his conception of natural law does inform this chapter. Instead, it is necessary to cut to the chase and outline the central theses of this tradition.

The natural law tradition (NLT) grows out of certain analysis of action and practical reason. This analysis leads it to hold both robust moral cognitivism and moral realism. Furthermore, as its name suggests, NLT proposes a variety of naturalism.

It defends naturalism insofar as it holds that "good" is, from a logical standpoint, an attributive rather than a predicative adjective.[10] On these grounds, it argues that moral judgments have the same form as judgments regarding natural goodness. In the latter case, the member of a kind, whether natural or artifactual, is good if and only if (i) it has the qualities that render a member of that kind good, (ii) it is a member of a goodness-fixing kind, and (iii) it is possible for some members of that kind to not possess those qualities and for at least one to possess them.[11] Moral judgments have to do with ways in which a human is good, not qualifiedly, but unqualifiedly: as a member of the human species. Consequently, moral judgments have the same form as judgments regarding natural goodness. They differ from judgments regarding other goodness-fixing kinds, however, in that only humans are capable of reason and action. A good human, therefore, is one who embodies excellence of reason and action, namely, the virtues. Furthermore, since the virtues are produced by, and produce in turn, the various kinds of action and affective response that characterize a good human, moral judgments can be duly expanded upon to describe virtues and vices, the various associated

[9] See Michael Bertram Crowe, *The Changing Profile of the Natural Law* (The Hague: M. Nijhoff, 1977), xii; John Finnis, *Natural Law and Natural Rights*, 2nd ed. (Oxford: Clarendon Press, 1980, 2011), 28.

[10] These terms come from Peter Thomas Geach, "Good and Evil", *Analysis* 17 (1956): 23–42. However, the status of the good as an attributive adjective has its roots in Aristotle's proper function argument and in Aquinas (*Summa theologiae* I, q. 5).

[11] This is outlined in Judith Jarvis Thomson, *Normativity* (Chicago, IL: Open Court, 2008).

kinds of action, and the goods that are constitutive of human excellence and fulfillment.[12]

By working out the implications of this conception of normativity, NLT comes to defend a non-consequentialist and naturalist version of ethical perfectionism. This distinguishes it, on the one hand, from consequentialism, and, on the other, from those varieties of normative ethics that reject either natural goodness itself or its normativity: Kantians, divine command ethics, or non-Aristotelian (e.g. Humean, Nietzschean) virtue ethics. This conception of ethical normativity can be spelt out as follows.

1. There exists, independently from the stance of individual humans, an objective, categorically, and universally normative order of human goods.
2. A voluntary action is (morally) good if and only if it is not contrary to the objective order of human goods.
3. The objective order of human goods is founded upon natural teleology.
4. The objective order of human goods can be discovered by considering the teleology of human nature (or practical reason).[13]

---

[12] This conception of normativity implies the existence of real essentialism and natural teleology and is a variant of Aristotelian ethical naturalism.

On the metaphysical and philosophical underpinnings of NLT, see David Oderberg, "The Metaphysical Foundations of Natural Law", in *Natural Moral Law in Contemporary Society*, ed. Holger Zaborowski, *Studies in Philosophy and the History of Philosophy* (Washington, DC: Catholic University of America Press, 2010), 44–75, 45. Anthony J. Lisska, *Aquinas's Theory of Natural Law: An Analytic Reconstruction* (New York: Oxford University Press, 1996), 85.

The natural law tradition's relation to Aristotelian ethical naturalism is considered throughout Terence Irwin, *The Development of Ethics: A Historical and Critical Study*, 3 vols. (Oxford: Oxford University Press, 2007–2009). For another statement of Aristotelian ethical naturalism, see Philippa Foot, *Natural Goodness* (Oxford: Clarendon Press, 2001).

Most proponents of Aristotelian ethical naturalism accept, on the other hand, evolutionary biology but see it as addressing a different question: the diachronic development of a species rather than the synchronic consideration of its life-form.

[13] Proponents of the "new-natural law theory", such as Germain Grisez, Joseph Boyle, John Finnis, Robert P. George, endorse the law of Hume and hold that we come to know these goods through the inherent directedness of practical reason itself rather than through a third-person consideration of the teleology of human nature.

5. The fundamental goods and their immediate requirements are manifest to any person with use of reason. More reflection is needed to discover their further requirements.

Traditionally, these theses have been packaged under the label of "*natural* law" to distinguish the deliverances of practical reason, working on its own lights, from "*divine* law", namely, divinely revealed morality. The epithet "law", on the other hand, is used to indicate that the deliverances of right practical reason are prescriptive and thereby have the force of a law. Practical reasoning, in fact, begins with the awareness that the good is to be done and pursued, and what is bad avoided. This principle is its first. However, since we are dealing with practical reason, this principle is not only grasped, but also prescribed. Consequently, its first principle is also a first precept. Moreover, whenever we understand a kind of act as good or bad, we view it as falling under this first precept and giving expression to it. Our practical reasoning thereby either prescribes or proscribes that kind of act. Hence, natural law corresponds to the normative deliverances of right practical reason regarding action types.

The precepts of natural law are either primary or secondary. The former regard a fundamental (i.e. non-derived) good to which we are directed in virtue of our physical, animal, and rational nature. They are known non-inferentially because they are not derived from a more ultimate precept. Secondary precepts, on the other hand, regard the application of primary precepts to a certain kind of situation and are worked out inferentially, with a greater or lesser degree of difficulty, depending on the complexity of the matter. Furthermore, the precepts of natural law regard what is a good for each human being as a member of a community, and so only achievable through communal interaction with others. In this sense, they regard common goods.[14]

Natural law counts as a "law", in the strict sense of the term, because it is of divine origin, though not because it is divinely revealed. Revelation may support practical reasoning and set it straight; it does not substitute it.

---

[14] Alasdair C. MacIntyre, "Intractable Moral Disagreements", in *Intractable Disputes About the Natural Law*, ed. Lawrence S. Cunningham (Notre Dame, IN: University of Notre Dame Press, 2009), 1–52, 4–6, henceforth "*IMD*".

Consequently, proponents of NLT often adhere to further theses that account for the range and strict legal character of natural law.[15] These theses are theological in character, although it may be possible to defend them on purely philosophical grounds. Nevertheless, we shall restrict our attention to the five abovementioned theses. We do so to stress that most normative conclusions of the natural law tradition do not depend on a commitment to theism in general and to Christianity in particular.

So far, NLT has been considered somewhat narrowly as a variety of normative ethics, but it can also be construed more broadly as a framework for politics, indeed, as a tradition of political and legal philosophy.

As such, it is ostensibly committed to political perfectionism rather than anti-perfectionism.[16] The former is the view that the state, in its various actions, should deliberately pursue the perfection of its constituents. Anti-perfectionist political philosophies, on the other hand, hold that the state should not employ coercive means to promote a certain conception of the good life because, to do so would impinge illegitimately upon the personal autonomy of citizens. Addressing the merits, viability, actual range, and implications of political perfectionism falls outside the scope of this chapter. For present purposes, it suffices to note that a commitment to political perfectionism does not necessarily entail that the state be the primary promoter of the well-being of individuals or that it uses coercion to deprive individuals of their capacity of self-determination and impose a set way of life. As Joseph Raz notes, political perfectionism does not necessarily require that there be legal means in place for the pursuit of every moral objective. Deciding which kinds and cases of moral objective be so pursued is a distinct issue.[17]

No matter what such a political and legal philosophy looks like, it could only belong to NLT by holding the following thesis.

---

[15] These can be formulated as follows. (7) By grasping and following the normative requirements of the objective order of goods, right practical reason expresses the natural moral law, established in creation, by which God directs human beings to their common good. (8) Perfect conformity to natural law does not suffice for attaining complete fulfillment as a human being (because grace, the central aspect of divine law, is even more necessary).

[16] A commitment to ethical perfectionism does not necessarily entail a commitment to political perfectionism. Cf. Douglas B. Rasmussen and Douglas J. Den Uyl, *Norms of Liberty: A Perfectionist Basis for Non-Perfectionist Politics* (University Park, PA: Pennsylvania State University Press, 2005).

[17] Joseph Raz, "Facing Up: A Reply", *Southern California Law Review* 62 (1988): 1153–235, 1231.

6. The man-made laws of any political society, which direct human beings toward the good that they can attain during their life and through their own efforts, are just and legitimate if and only if they enjoin actions which satisfy (2).

In fact, one of the main reasons why the normative requirements of human goods have been characterized as "natural law" is to indicate that they constitute pre-political moral constraints on legislation.

At present, many countries have legalized medical practices, such as abortion or euthanasia, which NLT deems wrong. Moreover, the legalization of these practices often enjoys widespread public support. In this case, NLT appears to be out of touch rather than a plausible framework for public reason, especially when it comes to bioethical issues. It should not be written off prematurely, though. Its approach to public reason in a pluralist society has certain strengths.

2 Public Reason as Shared Deliberation and Truth-Directed Inquiry

A certain conception of public reason is implicit in the abovementioned theses. According to those theses, the substantive conclusions of NLT are accessible, rationally justifiable, and so acceptable, *de jure*, to any duly informed and right-thinking person. In this regard, natural law reasons are potentially public in character. As such, not only should they be introduced into shared deliberation about the common good of a political society but also defended and, if justified, should constitute the standard for evaluating of laws. Such deliberation can be called "public reasoning" because it regards the good of the people and can be either conducted publicly or at least questioned publicly. For NLT, therefore, "public reason" would be an inquiry shared by the members of a political community. Such inquiry is directed toward discovering the objective order of goods and what the political community should do to promote it.

Whereas political liberalism defends a narrow conception of "public reason", that of NLT is broad.[18] Political liberalism construes "reasonableness" narrowly as "a commitment to political liberalism". However, NLT construes it more broadly as a disposition to revise, whenever there are compelling reasons, the grounds upon which one bases a certain policy and, if necessary, the policy itself. In other words, NLT construes public reason as a variety of reasoning in general—a form of truth-directed inquiry—and believes that its legitimacy depends upon it being conducted according to the criteria of rational inquiry. Moreover, this broader conception of "reasonableness" is more consistent than that of political liberalism. Unlike the latter, it is not circular.

For political liberalism, the reasonable is distinct from the rational and consists primarily of two social virtues.[19] These two virtues are "the willingness to propose fair terms of cooperation and to abide by them provided others do" and "the willingness to recognize the burdens of judgment and to accept their consequences for the use of public reason in directing the legitimate exercise of political power in a constitutional regime".[20] These social virtues act as aretaic constraints rather than epistemological ones and derive from one's commitment to a particular conception of society.

> Observe that here being reasonable is not an epistemological idea (though it has epistemological elements). Rather, it is part of a political ideal of democratic citizenship that includes the idea of public reason. The content of this ideal includes what free and equal citizens as reasonable can require of each other with respect to their reasonable comprehensive views.[21]

The argument, then, is that, society, as a system of cooperation between free and equal citizens, is possible under certain conditions. First, all must be willing to abide by fair terms of cooperation that are acceptable to all who are committed to such to society, irrespective of their comprehensive

[18] George and Wolfe, *Natural Law and Public Reason*: 2, 70.

[19] John Rawls, *Political Liberalism* (New York: Columbia University Press, 1993), 48, henceforth "*PL*".

[20] Rawls, *PL*, 54.

[21] Rawls, *PL*, 62.

doctrines. Second, precisely because these terms of cooperation constitute the principles on which such cooperation is based, all must be committed to conducting debates on basic constitutional matters and policies within the constraints of these principles.

For political liberalism, therefore, the content of public reason derives from the principles of a liberal political conception of justice. Such principles regard the basic structure of society and express the political values of a constitutional democracy (e.g. freedom, equality, and fairness [FEF]). They are presented as self-standing, depending not on any one comprehensive doctrine but supported by an overlapping consensus between parties holding different comprehensive doctrines.[22] They are construed, therefore, in a thin, political rather than in a thick, "metaphysical" sense. There is no agreement over the deeper grounds for them. Rather, the only relevant grounds for endorsing them is that all reasonable (i.e. politically liberal) people do.

> The primary grounds for accepting Rawls's theory is that it is constructed from the important values of freedom, equality, and fairness… They are accepted by all reasonable persons, and so Rawls can make use of them without invoking any particular comprehensive doctrine over which reasonable people disagree.[23]

Here "reasonable person" is taken in the idealized Rawlsian sense: a person who accepts FEF, construed politically rather than metaphysically. If so, the proposition which Jonathan Quong proposes as grounds for the validity of political liberalism, "All reasonable persons accept the values of freedom, equality and fairness", is tautological and uninformative. It works fine for an internal conception of political liberalism.[24]

---

[22] John Rawls, "The Idea of Public Reason Revisited", *The University of Chicago Law Review* 64 (1997): 765–807, 776.

[23] Jonathan Quong, *Liberalism without Perfection* (Oxford: Oxford University Press, 2011), 230.

[24] The general argument of Quong's *Liberalism without Perfection* is that only an internal conception of political liberalism is tenable, and that external conceptions are not. In other words, political liberalism cannot justify itself to non-liberals. It is meant to resolve an internal problem instead. It is meant to explain to liberals why liberalism gives rise inevitably to reasonable disagreements and how to deal with them.

There are problems, though, with adopting epistemic abstinence toward FEF and treating these as principles that are not to be asserted as true, but as the actual or potential object of an overlapping consensus in a well-ordered society. Critics object that any political theory must propose external reasons for its fundamental claims and that it is illegitimate to sidestep the demands of justification by falling back on an internal conception of the theory in question. If political liberalism is to characterize as "unreasonable" anyone who does not endorse it, it needs to do so on independent, non-circular grounds. By failing to do so, it leaves the notion of "reasonability" ungrounded.[25]

It cannot excuse itself by appealing to epistemic abstinence. Its status as a philosophical theory means that the fundamental values it bases itself on (whether (i) FEF or (ii) consensus-based social stability and cohesion) are only valid insofar as these are proposed as fundamental values on grounds *G*, *H*, *I*, and so forth. Theoretical justification is an internal and non-delegable task of philosophical theories.[26]

Rawls's definition of reasonableness is circular, therefore. It equates the reasonable with a commitment to political liberalism but denies the possibility of assessing this commitment on grounds that are both external and self-standing. In this case, political liberalism, despite its vaunted neutrality, is just another comprehensive liberalism.[27] At best, it is simply a comprehensive liberalism of a pragmatist bent.[28]

---

[25] Stephen Mulhall and Adam Swift, *Liberals and Communitarians*, 2nd ed. (Oxford: Blackwell, 1996), 238.

A more detailed argument on this point is Joseph Raz, "Facing Diversity: The Case of Epistemic Abstinence", *Philosophy & Public Affairs* (1990): 3–46, 15.

[26] Quong argues that political liberalism can endorse FEF on shallow foundations and epistemic abstinence because it leaves the task of providing deeper foundations to citizens. Quong, *Liberalism without Perfection*: 227–31.

[27] These points are made in Stephen Mulhall and Adam Swift, "Rawls and Communitarianism", in *The Cambridge Companion to Rawls*, ed. Samuel Freeman (Cambridge: Cambridge University Press, 2002), 481–84.

[28] This is how Richard Rorty understood Rawls's turn toward a political conception of justice. Richard Rorty, "The Priority of Democracy to Philosophy", in *The Virginia Statute for Religious Freedom: Its Evolution and Consequences in American History*, ed. Merrill D. Peterson and Robert C. Vaughn (New York: Cambridge University Press, 1991), 262–65. This essay has been brought to my attention by Michael J. Sandel, *Liberalism and the Limits of Justice*, 2nd ed. (Cambridge: Cambridge University Press, 1998), 194.

Political liberalism's self-referential definition of reasonableness is inconsistent in another regard. It risks counting as reasonable beliefs which are held on unreasonable grounds. Let us suppose that there is a group that holds FEF because, according to their religious leader, all men and woman, thanks to a scientific experiment which an advanced alien race carried out hundreds of thousands of years ago, are born with nanotechnology, transmitted via procreation, which endows them with Rawlsian moral powers, makes them satisfy the political conception of persons, and thereby renders them free and equal. Whereas political liberalism must deem this group reasonable because it holds FEF, surely their belief in freedom and equality is reasonable only insofar as it rests upon grounds which are external to their belief in implanted alien nanotechnology. This last belief is unreasonable because, ostensibly, there are not any plausible historical, metaphysical, or scientific grounds for it. Hence, reasonableness cannot be construed self-referentially. We deem a belief reasonable not only because of its content but also because of its grounds. It is reasonable for *A* to believe *X* if *A* has good grounds *G* for believing *X*. This is something that political liberalism itself acknowledges by recognizing that there are "burdens of judgment". In this case, there is an internal inconsistency within political liberalism. It recognizes that a belief is only reasonable if held on good grounds, but brackets questions about the reasonability of the grounds underlying its own principles of FEF.

It is not viable, then, to construe "reasonableness" narrowly, as political liberalism does. This is already a point in favor of NLT's broader conception of public reason. However, can NLT yield a plausible account of public reason, one capable of rationally resolving "the fact of reasonable disagreement".

This is where the work of Alasdair MacIntyre proves useful. He is a proponent of NLT whose work has centered on how to rationally resolve the intractable moral disagreements that arise in modernity with the breakdown of a shared moral culture. Furthermore, he has argued that the exercise of reason in a pluralistic society requires full-scale dialectical engagement with rival traditions, and that such engagement is only possible by committing to the fundamental precepts of natural law.[29] He thereby provides a philosophical argument for making NLT the framework for public reasoning.

His parting point is the acknowledgment of the seeming implausibility of NLT, given the "fact of reasonable disagreement". The precepts of natural law are meant to be universally binding principles of reason recognized by anyone with use of reason. In this case, there should be virtually universal agreement on fundamental moral issues. What we find instead is widespread reasonable yet intractable disagreement with NLT's position on abortion, torture for public security, extra-marital sexual activity, just wages, and, in some cultures, honor killings. These are not merely cases of inevitable and reasonable justificatory disagreements. Rather, they are the result of foundational disagreements. They stem from conflicting views on the principles (or primary precepts) of practical reason and, ultimately, from irreconcilably different underlying accounts, whether explicit or implicit, of human nature, action, and reasons for action. They derive from conflicting metaethical presuppositions. Furthermore, each rival conceptual framework has its own standards and mode of justification. As a result, the disagreement cannot be resolved by appealing to common criteria. This leads to a deadlock between the various parties. However, if NLT is correct, it should not be possible for there to be foundational disagreements that are both reasonable and irresolvable. According to NLT, there are first principles (or primary precepts) that are recognized by all, whereas the positing of alternative principles fails to satisfy the requirements of practical reasoning. In this case, disagreements should be easily resolvable through practical inquiry and debate. But this is not at all what we find.

Faced with this situation, the proponent of NLT appears to have only two options: either abandon ship or salvage it. Salvaging it means

[29] Alasdair C. MacIntyre, "Intractable Moral Disagreements". The first six sections of the paper are a rewrite of MacIntyre, "Aquinas and the Extent of Moral Disagreement", in *Ethics and Politics*, ed. Alasdair MacIntyre, *Selected Essays* (2006), 64–82. The next four sections provide a version of arguments that MacIntyre presents in various writings, including "Epistemological Crises, Dramatic Narrative and the Philosophy of Science", *The Monist* 60 (1977): 453–72; *Whose Justice? Which Rationality?* (Notre Dame, Ind.: University of Notre Dame Press, 1988), 349–69.

In the paper, MacIntyre claims to set out for the first time a systematic statement of both the nature of practical disagreement and his own proposed solution.

providing an alternative account of the nature and extent of moral disagreement. That is unfeasible. Nevertheless, instead of abandoning ship, MacIntyre identifies and defends a third option: show that the claims of NLT are compatible with the current widespread, irremediable moral disagreement. To this end, he distinguishes between the philosophical problem of intractable moral disagreement and the practical one.

Intractable moral disagreement calls for the rational justification of moral rules and principles. This is a philosophical problem. Dealing with the current radical moral and political disagreement, however, is a practical problem. We discover and must resolve radical moral disagreement through practical reasoning: by deliberating on what we should do, here and now. This discovery is especially acute when we are presented with deeply conflicting answers. In this case, "what rationality requires is that we deliberate further with others about how such disagreement should be resolved, including among those others with whom we most deeply disagree".[30]

In other words, deliberation should be construed as social in character rather than a merely individual affair. In this case, the proper question to be posed in deliberation is not, "How should *I* act, here and now?", but "How should *we* act, here and now?" For this reason, Aristotle and Aquinas, in their account of practical deliberation, point to how the one-sidedness of one's own standpoint can be overcome by learning to view a matter from the vantage point of several others. Doing so is important in theoretical inquiry, but even more so in practical deliberation, where we are apt to be deceived by our desires and good intentions. Solitary deliberation is prone to such self-deception and its consequent errors, but communal deliberation is particularly good at exposing it. Nevertheless, even social deliberation is not exempt from error. It too is insufficient unless it is regulated by "norms of objectivity".[31]

Establishing such norms of objectivity is, however, a seemingly impossible task. Among other things, moral disagreements are fundamentally about the nature of our ultimate end, an issue that falls outside the scope of deliberation. Deliberation presupposes and parts from a certain

---

[30] MacIntyre, *IMD*, 15.

[31] MacIntyre, *IMD*, 15–17.

conception of our end. It regards means *qua* means rather that ends *qua* ends. Hence, it cannot establish which conception of our end is correct. What it can do, however, is prompt us to reflect upon the actual nature of our ultimate end. In other words, we can pass from a practical knowledge of some conception of the good to theoretical inquiry into it.

MacIntyre's point is that moral disagreement has a positive role to play. It prompts us to question our immediate ends. This can lead, in turn, to the identification of the further chains of reasoning that underlie our moral beliefs and, if all goes well, to the identification of rival underlying conceptions of human good, then to inquiry into them. Such inquiry is theoretical in kind, but practical in its origin and end. It arises from deliberation and is aimed at setting it straight.[32] Indeed, theoretical inquiry into human good is unavoidable.

> But when disagreements turn out to be systematic and irresolvable in the context of immediate deliberation, then the identification of their character, let alone any attempt to resolve them, has to involve a resort to theoretical enquiry. Practice itself now requires us to engage with theory.[33]

A proponent of the anti-perfectionist conception of public reason could object that it is unrealistic to expect shared theoretical inquiry to establish the fundamental principles and institutions of a just social order. The anti-perfectionist conception is superior in this regard, the objection goes. It can yield fundamental social principles from a political conception of justice, without broaching the theoretical issues which underlie the divergent doctrines.

MacIntyre, however, is just as aware as the anti-perfectionist liberal of the actual difficulty of getting all parts of society on board for the sort of theoretical inquiry which he envisages. Some, if not most, will withdraw from shared deliberation when it comes to addressing the question of ultimate human good. The problem is that, in such a fractured society, communal decision-making will be based on some authority or a kind of

[32] MacIntyre, *IMD*, 18–19.

[33] MacIntyre, *IMD*, 19.

social contract. In either case, power, rather than right practical reason, is decisive.[34] Hence, MacIntyre's objection to the anti-perfectionist conception of public reason is that, faced with radical moral disagreement, it fails to satisfy the requirements of practical reason. These requirements are as follows.

R1 Any party facing systematic disagreement may not assume that its own account of human nature is right but must enter into inquiry on the matter with those who hold opposing views.
R2 An account of human good is adequate if and only if it is sustained by an ongoing inquiry which has truth, construed as the correspondence between thought and thing, as its end and good.
R3 The good of truth is a constitutive part of human good [from R2].
R4 Due accord must be given to the good of truth (i.e. it must not be overridden by arbitrary, inappropriate preferences) [from R3].
R5 Theoretical inquiry must be given some continuing and significant place in one's life by adopting as one's own the rules and virtues which such inquiry requires, when conducted according to its own inherent rules and in a virtuous manner.

R5 requires not only that one overcome the distortive influence of one's prejudices and selfish preferences, but also that one abstain from manipulative rhetoric in discussions with one's fellow enquirers. The inquirer needs to cultivate intellectual and moral virtues, especially in dealing with fellow inquirers.

Now, rational inquiry is only possible if, among other things, the parties be free from victimization and injustice. To engage in inquiry, I need to be assured that fellow enquirers will not endanger my physical integrity, freedom, or property. I also require some assurance that I will not be deceived and that promises will be honored. Furthermore, I need these rules to be universally and permanently binding. Now, these requirements of rational inquiry correspond materially to primary precepts of natural law. They correspond respectively to the fifth, seventh, and eighth of the Ten Commandments, which NLT has always viewed as a divinely

[34] MacIntyre, *IMD*, 19–20.

revealed statement of the fundamental precepts of natural law.[35] They also have the same formal characteristics of the natural law.[36]

FC1 *Universality.* They are universal because any one is a potential partner in deliberation.

FC2 *They do not admit exceptions.* Allowing exceptions to them would open the door to the victimization of one's partners in rational inquiry, thereby undermining the possibility of cooperative rational inquiry.

FC3 *Immutability.* They are the same for all.

FC4 *They are first principles of practical reason.* They are conditions of rational inquiry, which one, by pursuing such inquiry, has implicitly assumed and accorded authority to rather than worked out inferentially.

According to MacIntyre, these precepts do not emerge from a theoretical argument. An argument can only show that they are conditions of rational inquiry and that practices which do not presuppose them fail to be rational. MacIntyre concludes, therefore, that the existence and characteristics of moral disagreement do not offer any reason for rejecting the Thomistic understanding of the precepts of natural law. Rather, that conception of the precepts of natural law is not only consistent with the phenomenon of moral disagreement, insofar it corresponds to the requirements of shared deliberation, but also provides the best starting point for understanding it. Shared deliberation comprises different stages that follow from the identification of moral disagreement. However, that deliberation can break down at any point because of a refusal to do what reason requires. The refusal to do what reason requires amounts to a violation of some precept of natural law. As MacIntyre notes, Aquinas's account of law is a sequel to his treatment of sin, where sin is classified as a "transgression of reason".[37]

Summing up, deliberation is social rather than individual in character; shared deliberation requires rational inquiry into human good; rational

[35] On this aspect of NLT, see Dominic Farrell, *The Ends of the Moral Virtues and the First Principles of Practical Reason in Thomas Aquinas*, Analecta Gregoriana 318 (Roma: Gregorian & Biblical Press, 2012), 263–308.

[36] MacIntyre, *IMD*, 20–24.

[37] MacIntyre, *IMD*, 24–27.

inquiry requires the fundamental precepts of natural law. The primary precepts of natural law, therefore, are requirements of public reasoning.[38]

It is worth noting that MacIntyre does not infer the precepts of natural law from a comprehensive doctrine (a thick conception of human good) but establishes them as requirements of shared deliberation. Of course, an anti-perfectionist liberal is likely to endorse the primary precepts of natural law, but not any underlying comprehensive doctrine. The problem is that the political liberal's epistemic abstinence is incompatible with the abovementioned requirements of rational inquiry (R1 and R2).

Furthermore, MacIntyre does not believe that his own defense of NLT will convince adherents of other traditions of normative ethics in well-conducted rational public debate, because the debate is over deeper meta-ethical and philosophical issues. What is required is a more "radical philosophical, moral, and cultural critique of rival standpoints".[39] In this case, the sort of inquiry in which public reason must engage must not be restricted to a debate between rival political philosophies, but must extend to their deeper, underlying philosophical and cultural differences.

At this point, however, proponents of political liberalism could object that a truth-directed inquiry between adherents of different comprehensive doctrines is neither a desirable nor a viable basis for public reasoning. On the one hand, it would not be desirable for the state to coerce individuals to engage in such an inquiry. On the other hand, the need for social cohesion trumps a project with such low prospects of success. On these grounds, the former moral and the latter pragmatic, political liberalism simply requires that citizens, for the sake of the public good, do not appeal to their own comprehensive doctrines in public reason. It does not allow that the abovementioned conditions hold of public reasoning.

[38] Aware that his account of the development of moral disagreements is contentious, MacIntyre proceeds to discuss the utilitarian tradition as a test-case of a rival account of practical reason. What this brings to light is that utilitarians and Thomists espouse different, incompatible standards of practical reason. In his view, this is not a genuine case of incommensurable disagreement, but is an irremediable one. Nevertheless, it is possible to show that one of the traditions is rationally superior to the other, while being unacceptable to a member of the rival tradition.

[39] MacIntyre, *IMD*, 50–52.

These objections, whether they are sound or not, call our attention to an important point. Public reasoning of the political kind takes place against the backdrop of political and legal institutions with the power to coerce. The conditions of shared deliberation outlined above may work in civic society. There the members of a group can engage in shared deliberation out of their own accord. The political liberal, however, objects that these conditions of shared deliberation cannot work for public reasoning of the political kind.

To some extent, MacIntyre concurs. He too believes that the sort of shared political deliberation which he advocates is not possible in the modern state. This is not to say that the shared deliberation which he describes is apolitical. Any individual's effective practical reasoning ends up being political to some extent. First, we can only identify our individual goods through shared deliberation on common goods. This amounts to political reasoning. Given the constitutive interdependence of human beings, practical reasoning is embedded in networks of giving and receiving. Consequently, deliberation about one's good as an individual is inseparable from deliberation about a common good, and the latter is necessarily a shared endeavor. In second place, many goods are shared. As a result, one cannot determine what place that good is to play in one's own life independently from the place which the community gives to it in its political decisions. We need to be able to participate in political deliberation to determine what place many individual goods have in our life.[40]

The problem, in MacIntyre's view, is that such shared deliberation cannot take place within the politics of the modern state. First, the politics of the modern state prohibits philosophical debate on first, principles since these call into question the tacitly presupposed boundaries of political discourse. Second, it does not allow for debates over substantive issues connected with ways of life (e.g. whether current agricultural policy should be abandoned to preserve the way of life proper to farming). Third, political debate is not conducted systematically according to the

[40] Alasdair C. MacIntyre, *Dependent Rational Animals: Why Human Beings Need the Virtues* (Chicago: Open Court, 1999), 140–41. See also, MacIntyre, "Politics, Philosophy and the Common Good", in *The MacIntyre Reader*, ed. Kelvin Knight (Notre Dame, Ind.: University of Notre Dame Press, 1998), 235–52, 240–41.

standards of rational inquiry and is often unduly influenced by lobbies and elites.[41] Fourth, shared deliberation is not possible in the liberal pluralist state, but only possible in the Aristotelian polis, where there is a shared conception of the common good and accepted modes of rational inquiry.[42] Fifth, the modern state aims at ensuring public goods rather than the common good. Both public and common goods can only be attained through cooperative activity. However, public goods belong to the individual *qua* individual; common goods to the individual as a member of community.[43] The pursuit of a common good can only take place in a community which has a shared end. This is not possible in the modern state, with its bureaucratic institutions and liberal pluralism. Consequently, modern politics lacks an institutional forum in which ordinary people "are able to engage together in systematic reasoned debate, designed to arrive at a rationally well-founded common mind on how to answer questions about the relationship of politics to the claims of rival and alternative ways of life, each with its own conception of the virtues and of the common good".[44] In MacIntyre's view, the modern state is irremediably based on compromise between power interests rather than truth, and only intermediate societies—civic society—offer the possibility for shared, practice-based, and virtue-forming rational deliberation and decision-making.[45]

This does not amount to an abdication of political action. Groups that are concerned with sectorial common goods—such as that of the family, school, and workplace—are driven to engage in political action because their sectorial common goods are often interconnected. A concern for the good of one's family is bound up with deliberation on the common good of one's workplace and on that of the school where one's children study.

[41] MacIntyre, "Politics, Philosophy and the Common Good", 236–39. MacIntyre identifies a third characteristic of modern politics: it effectively promotes some ways of life and harms others. This undermines it purported neutrality but does not exclude shared rational inquiry in and of itself.

[42] MacIntyre, "Politics, Philosophy and the Common Good", 241.

[43] Alasdair C. MacIntyre, *Ethics in the Conflicts of Modernity: An Essay on Desire, Practical Reasoning, and Narrative* (New York: Cambridge University Press, 2016), 168–69.

[44] MacIntyre, "Politics, Philosophy and the Common Good", 239.

[45] MacIntyre, *Dependent Rational Animals*, 129–46.

By addressing these concerns, albeit on a local level, individuals pursue a common political good.[46]

MacIntyre therefore advocates a politics of making and sustaining local communities. It may lead its participants to engage directly or indirectly in the politics of the modern state, but it differs from the latter in two respects. It requires (i) actual shared deliberation, rather than hypothetical public reasons which pass the "acceptability" test; and (ii) a shared conception of just procedure, namely, a shared recognition of the precepts of natural law.[47]

It could be objected that this politics of communities is compatible with political liberalism. Political liberalism only requires the state to be neutral in its justification of policy but not that the outcomes of its policies be neutral for all parties. It only rejects state-perfectionism but allows for civic society to pursue social perfectionism.[48] It can give civic society all the space that it supposedly needs. However, MacIntyre does not propose the politics of community as a substitute for the modern state. He proposes it as the sole way of reviving, to some extent, authentic political deliberation (i.e. authentic public reasoning) within the modern state.

In MacIntyre's political philosophy, therefore, public reasoning is shared deliberation concerning common goods and can only take place within a politics of local communities rather than within the modern state's public good-oriented politics. Consequently, to promote public reasoning, the state should apply the principle of subsidiarity, enabling rather than restricting the legitimate activity of civic society.

The problem, though, is that full-fledged public reasoning may not be possible within local communities. Unlike political society, these communities are incomplete.[49] Following Aristotle, NLT regards the political

---

[46] MacIntyre, *Ethics in the Conflicts of Modernity*, 177.

[47] MacIntyre, *Ethics in the Conflicts of Modernity*, 177–78. MacIntyre presents the Guild of Thorupstrand Coastal Fishermen and the Associatilo Comunitària Monte Azul as examples of this politics of community: MacIntyre, *Ethics in the Conflicts of Modernity*, 178–83.

[48] Will Kymlicka, "Liberal Individualism and Liberal Neutrality", *Ethics* 99 (1989): 883–905, 883–84.

[49] For this objection, see Thomas Osborne, "Macintyre, Thomism and the Contemporary Common Good", *Analyse & Kritik* 30 (2008): 75–90. Similar objections are brought forward in Thomas S. Hibbs, "Macintyre, Aquinas, and Politics", *The Review of Politics* 66 (2004): 357–83; V. Bradley Lewis, "The Common Good against the Modern State? On Macintyre's Political Philosophy", *Josephinum Journal of Theology* 16 (2009): 357–78.

community to be a complete or perfect one because it has the means to meet all life's needs.[50] It is arguable, therefore, whether MacIntyre's politics of local communities can give rise to full-fledged public reasoning. Public reason is the deliberation of a political community. While aware of the need for a complete community, such as the Greek *polis*, MacIntyre believes that its functions cannot be discharged by the modern nation-state but only by local communities. The problem with this view, however, is that no local community possesses the political authority to arbitrate over other intermediate societies, coordinate them, and order the goods of political society. Full-fledged public reasoning cannot take place in any of them. As Thomas Osborne points out, even if MacIntyre's diagnosis of the defectiveness and incoherence of the modern state is correct, it could still be argued, as natural law theorists have traditionally done, that a complete community is one which exerts coercive authority.[51] This conception allows for the modern state to count as a complete community, despite its defects, because of its legal system and authorities. Its institutions remain worthy of acceptance, even when they are not entirely justified and legitimate, in virtue of the indispensable contribution that they make to fostering justice and social order.[52] Indeed, to a large extent, NLT articulates its conception of public reason in its anti-positivist, naturalist philosophy of law. MacIntyre does not give sufficient attention to this aspect of NLT.

He presents an account of how practical reasoning is both public and political in character, and of how public reason, like any form of practical reasoning, needs to be rooted in a broader truth-directed inquiry. He shows compellingly how, even in a pluralist society, shared deliberation is inescapable and must follow certain requirements of rational inquiry, which are none other than the precepts of natural law. What he fails to explain is how this shared deliberation should take place on the part of political and legislative authorities. Nevertheless, some standards of public reasoning can be adduced from his account of shared deliberation and its requirements.

---

[50] Aristotle, *Politica* I, 2, 1252b27–1253a1. Thomas Aquinas, *Political Writings* (Cambridge: Cambridge University Press, 2002), 9–10 [*De regno* I, c. 2].

[51] Osborne cites Francisco de Vitoria, *Political Writings* (Cambridge: Cambridge University Press, 1991), 42.

[52] Cf. Osborne, "Macintyre, Thomism and the Contemporary Common Good", 89.

First, if MacIntyre is right, our political institutions and laws should be formed to allow for ongoing shared deliberation and, whenever necessary, the revision of those laws and institutions. Since laws are meant to be rational ordinances directed to the common good of a political society, they need to be based on truth-directed inquiry rather than on an uncritical acceptance of certain beliefs or positions.

For example, should it be shown that the reasoning on which the legalization of abortion (e.g. *Roe vs. Wade*) was based is defective and that abortion should be proscribed, the law should be changed accordingly. The same applies to the legalization to MAS (e.g. *Gonzales v. Oregon*). Ruling out in advance the possibility of such change amounts to a refusal to support public reason.

Second, public reason should be limited in scope. Since political decisions should respect the subsidiarity requirement, public reason should limit itself to establishing what is necessary for the functioning of society and maximizing the sphere of action of intermediate communities. In legislating over controversial biomedical issues, therefore, governments and superior courts should do all that is possible to guarantee the freedom of intermediate communities and avoid coercing them to facilitate medical practices which they object to on moral grounds. There is a considerable difference, for example, between liberalizing abortion or euthanasia laws and making it mandatory for all health care services to provide abortion upon request from eligible applicants. By refraining from constricting health care agents and institutions to provide upon demand medical practices, such as abortion and euthanasia, to which they object on moral grounds, the state would show a greater attention to subsidiarity and resolve at least one category of bioethical dispute.

3 Natural Law, Legislation, and Public Reason

In the previous section, we pointed out that NLT may constitute an adequate framework for full-fledged public reasoning only if it does likewise for political decision-making and legislation.

The central aspect of NLT's conception of political decision-making and legislation is its anti-positivism. It holds that (6) laws must be compatible with the normative requirements of objective human goods. This stipulation is essential but insufficient. It does not specify whether there are any other constraints peculiar to practical reasoning of the political variety. If there are, these would constrain full-fledged public reasoning as well. Significantly, though, Aquinas, the canonical natural law theorist, identifies certain essential characteristics of the law.[53] These amount to conditions that public reason must satisfy.

First, since the law is meant to direct all members of society to cooperate willingly for the common good, laws need to be public, clear, general, stable, and practicable.[54] So does public reason. Just as the law regards all members of society, so too must all adult members of the political community be treated as partners of public reason. This is only possible if there are public standards which are clear, general, stable, and practicable.

Second, in line with (6), laws need to be based on grounds accessible to all reasoning members of society (i.e. public reasons) if they are to promote its common good. Hence, when another party of public reasoning does not accept the authoritative reasons of one's own tradition (e.g. religion), one should recur to rational argument.[55] Political liberalism makes a similar demand. The difference is that, for NLT, the reasons adduced do not have to be derivable from a liberal political conception of justice.

Third, a polity's laws, and public reason with them, should be restricted to that which is necessary for promoting and safeguarding the common good of political society. The law of a polity is narrower in scope than natural law. The latter is coextensive with the full range of virtuous action.[56] A polity's laws, on the other hand, should be limited to prescribing those actions which are necessary for the conservation of political society and safeguarding its common good. This means that they should

[53] This classification of the features of law is based on Finnis, "Public Reason, Abortion, and Cloning", 363–64. In his view these features amount to a theory of public reason.

[54] Thomas Aquinas, *Summa theologiae* I–II, q. 90, a. 4; qq. 96–97.

[55] Thomas Aquinas, *Quaestiones de quolibet* IV, q. 9, a. 3, co.

[56] Thomas Aquinas, *Summa theologiae* I–II, q. 94, a. 3.

prohibit only the gravest offenses, above all those that harm others.[57] Consequently, the fact that natural law requires φ-ing does not necessarily mean that a polity's law should also require it. Rather, a polity's law prescribes or prohibits φ-ing insofar as social cohesion, peace, and justice require it. For example, adultery, though morally objectionable, is not a crime in most countries. Presumably, this is because legislators believe that its criminalization makes no real contribution to the peace and cohesion of society, but would probably only be counterproductive. This is not to say that sexual mores fall outside the scope of the law. Most countries criminalize certain sexual misdemeanors, such as rape and intercourse with minors, which are injurious and socially harmful. On the other hand, by setting minimal ages and certain conditions for marriage, legislators attempt to promote stable, loving marriages. They set certain conditions to help ensure that couples getting married assume their commitment freely and maturely. They are thereby legislating to promote virtuous conduct (i.e. natural law). All that (6) requires, therefore, is that a polity's law should not contravene fundamental requirements of natural law, such as the human good of marriage.[58]

Public reason, therefore, is more limited in scope than natural law reasoning. It is coextensive with practical reasoning of the political kind, which is restricted to determining that which promotes the common good of political society. Public reasons, therefore, are not identical to natural law reasons. Rather natural law reasons constitute the normative framework of public reason and only become public reasons if they are required by practical reasoning of the political kind.[59]

As noted earlier, the existence of a law that goes against fundamental human goods does not necessarily delegitimize political authorities entirely. In such a situation, one still has reasons to deem the political

[57] Thomas Aquinas, *Summa theologiae* I–II, q. 96, a. 2.

[58] For example, some believe that the legalization of same-sex marriage contravenes the fundamental human good of marriage.

[59] On this point, see Rhonheimer, "The Political Ethos of Constitutional Democracy", 200, 42–44.

However, Rhonheimer's claim that natural law reasons are a source of social conflict and public reasons are acceptable to all is questionable. What the public accepts is not necessarily what is acceptable from the standpoint of right practical reason. Indeed, as we have argued in the previous section, public reasoning, even if it is narrower in scope than moral reasoning, has to be rooted in a truth-directed inquiry.

authorities and the legal system legitimate in virtue of their contribution to social cohesion, peace, and justice. In this regard, Martin Rhonheimer has argued that, if, as is widely believed, a constitutional democracy of some sort is the most appropriate political regime, then its fundamental political values (peaceful coexistence, political equality, and liberty) enjoy lexical priority over what one takes to be a morally better social ordering. Consequently, one should accept, within certain limits, the outcome of constitutional political and legislative procedures that run contrary to what one takes to be required by fundamental human goods.[60]

How, then, would bioethical issues be addressed within this conception of public reason?

First, many bioethical issues belong to the forum of public reason because of their political importance. Those in favor of the legalization of abortion, euthanasia, and in vitro fertilization believe that the law should permit each of these practices because a certain class of individual has a right to it. Opponents believe that these practices harm and undermine the social order because they violate fundamental rights of third parties or some fundamental human good. They believe that the law should prohibit these practices. No matter what side one takes, these bioethical debates revolve around questions of justice. They thereby fall under the purview of public reason and call for legislation of some sort.

The problem is that any legislation on these first-order matters is fraught with difficulties and controversial issues. It must take a stance on contentious second-order questions, such as personal identity and the grounds for a rights-claim. Political liberalism recognizes that reasonable disagreement on bioethical issues is inevitable, because these fall under the burdens of judgment. However, it places many of the underlying anthropological and metaphysical issues out of bounds. It thereby closes the door to a truly rational resolution of these issues and is continually exposed to the risk of making inconsistent and arbitrary decisions in these matters.[61] These deeper philosophical issues cannot be bracketed, therefore, as political liberalism proposes. By insisting upon the need for

[60] Ibid., 216–18.

[61] John Finnis, for example, consider how the U.S. Supreme Court and John Rawls justify pro-abortion legislation and argues that their justification is marked by inconsistencies and partiality. See Finnis, "Public Reason, Abortion, and Cloning".

a process of public reasoning that is rooted in truth-directed inquiry, NLT lays the premises for a more satisfactory and reliable resolution of these problems and an eventual rectification of eventual missteps. In this regard, its conception of public reason is superior to that of political liberalism.[62]

The legalization of abortion, for example, requires a moral justification. It thereby calls for bioethical reflection. However, such bioethical reflection cannot restrict itself to a consideration of the best available science. To reach a justified ethical conclusion, it needs to interpret that science through the prism of controversial philosophical concepts, such as those of "rights" and "personhood". Bioethical reflection on abortion, therefore, parts from biology's explanation of how a human being remains the same organism from the initial state as a zygote, through all successive stages of biological development, until death. Determining the ethical implications of this fact requires inquiry into when human beings acquire the "right to life": the right to be immune to physical aggression and for their life to be protected by the state and its coercive apparatus. Rawls claims that human beings enter society by birth.[63] This entails that it is only with birth that the members of society attain the right to life. It could be objected, though, that this is an arbitrary stipulation. There is no fundamental biological difference between the organism of an unborn infant and one that has been delivered, even if the former is not recognized as a citizen. Perhaps his stipulation presupposes that the mother should be free to terminate the pregnancy and that her freedom to do so rules out the possibility for the unborn to have a right to life. Alternatively, the stipulation may simply suppose that neither the unborn child nor one already born has any right to life, but that the state confers this right upon the latter in the interests of society. Nevertheless, any attempted justification of the stipulation must recur to a philosophical theory of rights or justice, a theory which needs to be justified in turn.

[62] As Martin Rhonheimer has noted, by basing public reason on a political conception of personhood, Rawls may be abstracting illegitimately from the naturally given aspects of personhood and society, such as procreation, marriage, and family. This may vitiate his views on issues such as abortion and marriage. See Rhonheimer, "The Political Ethos of Constitutional Democracy".

[63] Rawls, *PL*, 301.

Faced with these problems, one might conclude that the right to life can only be accounted for consistently if it is treated as a natural right: one which belongs to each member of the human species. Still, this does not put an end to inquiry. A theory of natural rights requires a philosophical account of human nature and personhood. Treating the right to life as a natural right of every human person raises the metaphysical question as to what constitutes a person. There are rival answers to this question. Some defend an animalist conception, while others defend a psychological notion of personhood. If the latter is correct, the fetus would not have a right to life; if the former is correct, as pro-lifers generally claim, it would.[64] At any rate, it is only possible for there to be morally justified laws on these issues if, as NLT claims, public reason is rooted in a broader truth-directed inquiry that ranges over second-order questions. Only in this way can legislation give non-arbitrary consideration to the underlying philosophical issues.

The same problem arises, though less clearly, with the legalization of MAS. In legalizing MAS, the state needs to ask what kind of action it is sanctioning and, in some cases, even obliging physicians to perform.

On the one hand, it needs to consider whether MAS falls under the description of a medical act. Medicine is generally understood as the science of the diagnosis, prognosis, treatment, and prevention of illness or disease. Medical acts and techniques, therefore, are directed toward the preservation and restoration of health.

Of course, there are times in which there is nothing a doctor and associated health care practitioners can do to cure a treat an illness or injury. There are also interventions, such as the amputation of a gangrenous member, which nonetheless constitute medical acts because, though they harm and impair the subject, they are directed toward the preservation of the patient's life and health. Can the same be said of MAS though?

It is an act in which the physician chooses means that are directed immediately toward ending the patient's life, even if it is performed at the

[64] For a fuller statement of the natural law argument against abortion, see Patrick Lee and Robert P. George, *Body-self Dualism in Contemporary Ethics and Politics* (Cambridge: Cambridge University Press, 2008), 118–50.

patient's duly deliberated request and to alleviate extreme suffering. As such, it has two characteristics that appear to qualify it as a medical act.

On the one hand, MAS is performed upon a consenting and duly informed patient by a medical professional, employing medical expertise. Any willing party could use brute force or a commonplace object to kill a patient. In MAS, however, the physician employs medical expertise to bring about the patient's death in the least painful and distressing way possible. On the other hand, there is reason to believe that MAS is either a form of palliative care or pain relief. As a deployment of medical expertise, it is directed toward alleviating the patient's extreme suffering. In this regard, it is akin to the administration of analgesics.

These features notwithstanding, there is reason to believe that MAS lacks the essential characteristic of a medical act and even constitutes a morally disordered use of medical expertise. If the constitutive feature of a medical act is its directedness toward the maintenance and restoration of health, then MAS lacks this essential characteristic of a medical act and, arguably, aims at the contrary end. Rather, MAS is directed immediately toward ending the life of the patient. Though the motive may be noble and praiseworthy, it is an act of killing. It consists of the use of medical expertise to help the patient commit suicide in the least painful and distressing way. As such, it is not directed to the maintenance and restoral of health. Health is a quality of the member of a biological species. By its very nature, therefore, a health-promoting act cannot aim at destroying the life that is its very support. As an act of killing, MAS is not a medical act. This, in turn, means that a physician's application of medical expertise to a suffering patient does not suffice to constitute a medical act. Rather, a medical act derives its essential identity from its inherent teleology. For this reason, MAS is essentially different from the medical act of administering an analgesic. Whereas MAS aims directly at ending the life of the patient as a means of relieving suffering, the latter aims directly at relieving pain while preserving the life of the patient. It is a medical act, whereas MAS uses medical expertise to go against the end of medicine. The medical act of administering analgesics differs from the use of drugs in MAS or a mere murder in two regards. First, it administers a dosage that will not kill the patient. Second, even if the dosage administered will shorten the life of the patient, this is foreseen rather than intended, allowed rather than done. It is a preterintentional effect rather than the direct end of the act. In MAS, on the other hand, ending the life of the patient is the intended deed.

This way of understanding the medical act has a long and distinguished pedigree. It is operative, for example, in the original Hippocratic Oath. That oath contains a pledge to never administer poison to a patient who asks for this and never to suggest its use to the person under one's care. Significantly, this pledge follows the one made to refrain from harming the patient and any wrongdoing. The Hippocratic Oath thereby treats the use of poison as a means of alleviating present or imminent suffering, not as a medical act, but as a form of wrongdoing and of doing harm to one's patient. This categorical condemnation of alleviative poisoning also extends implicitly to those cases which satisfy the conditions normally stipulated for MAS.

Defenders of MAS, on the other hand, may either accept the preceding definition of the medical act, but contest its moral significance, or argue for a more expansive one that allows for the killing with medical expertise as a means of pain relief. What is unavoidable in legislative debates, though, is the question of the nature of the medical act and its applicability to MAS. To be morally justified, the legislation of MAS must engage in a deeper inquiry into the essential characteristics of medicine and settle the ensuing debates over this issue.

Furthermore, MAS is indisputably a kind of killing. The legislation of MAS is justified, though, if and only if MAS is a morally justified kind of killing. Any such legislation, therefore, must establish that it is not a kind of murder.

According to a traditional definition, proper to the natural law tradition and Catholic moral theology, murder is the voluntary act of directly killing of an innocent human being. It consists of killing someone directly because that person's death is intended either as the end or a means to some other end. Furthermore, it consists of killing an innocent human being: one who is neither an unjust aggressor (whether a private person or an enemy combatant amid battle in a justified war) nor has been found guilty of a grave crime falling under the sanction of capital punishment.[65]

[65] The source of this understanding of murder is Augustine, *The City of God Against the Pagans* (Cambridge: Cambridge University Press, 1998), 32–40 (I, 20–26) and Thomas Aquinas, *Summa theologiae* II–II, q. 64.

Murder, construed thus, is considered wrong on roughly the following grounds. Each member of the human species possesses a singular and inherent dignity and so should be treated as an end and never as a mere means. Respecting the life of a human being is a fundamental requirement of such treatment. The direct killing of someone is incompatible with treating another person as an end because it intends the destruction of that person or treats that person as a mere means, namely, by intending the killing as a means to some other end. In certain circumstances, the direct killing of someone who is not innocent is compatible with treating human beings as ends and never as mere means. This is because the direct killing of the non-innocent is sometimes necessary to protect the lives of the innocent: to ensure that people are treated as ends rather than mere means. Furthermore, political authorities are responsible for the good of the innocent, namely, the common good. Consequently, only the state has the authority to directly kill the non-innocent. It exercises this authority through the legal system in the case of capital punishment, and through the security forces (e.g. soldiers, police) in the case of unjust aggressors.

By this account, therefore, neither killing in legitimate self-defense nor justified capital punishment falls under the description of murder or wrongful killing but constitutes distinct kinds of action. An act of legitimate, though lethal, self-defense is not a direct killing. If one chooses to deploy in self-defense only that measure of force that is needed to stave off the potential harm, then one acts with the intention of protecting the life of the person threatened, whether oneself or someone else, and not with that of bringing about the death of the unjust aggressor, whether as an end or means. Rather, the death of the unjust aggressor is foreseen as an unfortunate but preterintentional effect or consequence. Furthermore, killing in legitimate self-defense does not have the second essential characteristic of murder. It is not the killing of an innocent person but of an unjust aggressor. Capital punishment, on the other hand, does not qualify as murder because it too lacks the second but not the first essential characteristic. It is an act of direct killing. However, insofar as it is necessary for protecting the common good of society and carried out according to the rule of law by the public authorities, it does not qualify as the killing of an innocent person. In principle it is legitimate, although always open to abuse.

Like capital punishment, MAS is a case of direct killing. Unlike capital punishment, the suffering patient is an innocent person. For this reason, many consider it both wrongful killing and an action which the state does not have the authority to legalize.

A range of counterarguments are open to proponents of MAS. Besides contesting the preceding definition of murder, they might argue that respecting a patient's duly informed and autonomous request of MAS, to put an end to extreme suffering, trumps the duty to respect a patient's life. They could proceed to argue that, on these grounds, the state has the authority to legalize it under duly defined conditions. They need to explain, though, why MAS does not constitute a case of murder, an action which the state prohibits. What cannot be avoided, then, is a consideration of what constitutes wrongful killing, legitimate state-authorized killing, and of how the answers to these substantive second-order questions apply to MAS.

Furthermore, as with abortion, any legislative decision regarding MAS needs to consider what constitutes a person. Should the psychological criterion of personhood prove right, then human beings who have irremediably lost the capacity for exercising higher brain functions would no longer be persons and a case could be made that the medical termination of their lives is not murderous. However, proponents of MAS are generally concerned with justifying its use for conscious patients who continue to be persons under this description. They claim instead that one's biological life is of instrumental rather than intrinsic value for a person. According to this view, the value of one's functioning organism derives from its fitness for performing other, intrinsically valuable (generally conscious) activities. Hence, when one is no longer physically capable of performing intrinsically valuable activities adequately, life ceases to have any instrumental value and may be freely ended. However, there is a serious problem with this view of personhood and its implications for the value we should attach to life. It introduces a questionable separation of the psychological from the biological: of personhood from embodiment. However, the fact that these cannot be separated in practice renders dubious theoretical claims about their separability. All our interactions as persons, in fact, are embodied. This is a reason in favor of an animalist conception of personhood. Within the animalist conception of personhood, though, life is intrinsically rather than instrumentally valuable. It is a fundamental, constitutive, and intrinsically valuable good of the person. Any intentional action that is directed against a human being's life would amount to an act against the good of that person and would thereby be morally reprehensible.

With these considerations, we have not attempted to give a full statement and adequate defense of NLT's position on MAS and abortion.[66] We have merely tried to illustrate that, contrary to what political liberalism proposes, the legislation of MAS cannot escape complex second-order philosophical questions and may even end up endorsing questionable conceptions of personhood and medical care. In this regard, NLT's conception of public reason proves more consistent precisely because it requires that due consideration be given to these second-order questions.

Indeed, NLT's commitment to truth-directed inquiry protects it from accepting the moral positions of any given comprehensive doctrine uncritically, including its own. Rather, to be true to truth-directed inquiry, it must submit these positions to thoroughgoing scrutiny.

Hopefully, this chapter has managed to show that NLT not only yields an account of public reason but one which may be more sophisticated than is often assumed. Indeed, with its anti-positivist legal philosophy and constitutionalism, NLT yields, we submit, a conception of public reason which not only is valid but also avoids some serious deficiencies of political liberalism. It does so by rooting public reason in truth-directed inquiry. For the reasons already examined, side-stepping such inquiry undermines the possibility for authentic political discourse, especially when it comes to bioethics.

---

[66] For an analysis of MAS from the standpoint of NLT, see Luke Gormally (ed.), *Euthanasia, Clinical Practice and the Law* (London: Linacre Centre, 1994); Lee-George, *Body-Self*, 151–75; Craig Paterson, *Assisted Suicide and Euthanasia: A Natural Law Ethics Approach* (Abingdon: Routledge, 2017); Sgreccia, *Personalist Bioethics*: 663–710.

3

A Confucian Conception of Public Reason and Bioethics

Ruiping Fan

1 Introduction

Contemporary human lives are characterized by both moral agreement and disagreement. On the one hand, as H.T. Engelhardt points out, moral diversity is real. People hold moral disagreements not only on particular issues but also in principle. This diversity can be noticed obviously in the area of bioethics. Individuals make conflicting judgments on the morality of abortion, commercial surrogacy, genetic editing, health care allocation, organ sale, same-sex marriage, sex-robotics, head/whole-body transplants, euthanasia, medically assisted suicide (MAS), and so on.

The first draft of this paper was presented at the "International Workshop on Public Reason and Bioethics" organized by the Department of Philosophy and Centre for Bioethics, the Chinese University of Hong Kong, Hong Kong, 4–5 January 2018. I wish to thank the participants in the workshop for their comments on my draft.

R. Fan (✉)
Department of Public Policy, City University of Hong Kong, Kowloon, Hong Kong
e-mail: safan@cityu.edu.hk

H.-L. Li, M. Campbell (eds.), *Public Reason and Bioethics*,
https://doi.org/10.1007/978-3-030-61170-5_3

They also disagree about what type of normative ethical theory—teleological, deontological, or virtue-ethics-based—should be adopted to explore such issues.[1]

On the other hand, however, one cannot deny the existence of moral agreements, either. We hear same opinions, see common views, and find shared moral values in various human collectives, such as communities, nations, and even international organizations. The difficulty is how we can properly characterize such agreements and reasonably integrate them into a useful conception of shared values so as to legitimately apply the conception to direct political relations and justify public policy, without unreasonably ignoring or rejecting the existence of relevant disagreements.

John Rawls famously proposes the idea of public reason to achieve this purpose for contemporary Western democratic countries. In his words, "[t]he idea of public reason specifies at the deepest level the basic moral and political values that are to determine a constitutional democratic government's relation to its citizens and their relation to one another" (*Political Liberalism*, henceforth PL, 441–442). Can we similarly construct a particular idea of public reason and apply it to East Asian countries and regions which include both democratic and non-democratic regimes, but are all influenced by Confucian moral and political values? Significantly, as a liberal scholar, Rawls has offered a liberal conception of public reason for Western liberal democratic countries.[2] Is it possible for Confucian scholars to offer a Confucian conception of public reason for East Asian countries? In his work *The Law of Peoples* (2002) (henceforth LP), Rawls recognizes that there are decent people differing from liberal people in the world, and thereby explores general principles that can be accepted by both liberal and non-liberal peoples to regulate their behavior toward each other. Such general principles, acknowledged by Rawls, cannot be the full-fledged liberal principles adopted to direct liberal

[1] H. Tristram Engelhardt Jr., *The Foundations of Bioethics*, 2nd ed. (New York: Oxford University Press, 1996), 3.

[2] I distinguish particular conceptions of public reason from the concept of public reason, just as Rawls appeals to the distinction between the concept of justice and particular conceptions of justice (*A Theory of Justice*, henceforth *Theory* 5–6). Roughly, following Rawls' distinction, a concept is an abstract notion, whereas a conception contains specific or detailed ideas of the concept in it.

people. Along this line of thought, it would be reasonable to construct a non-liberal conception of public reason for decent, non-liberal people to adopt.

Confucian reasons are essentially communitarian reasons (originally held by Confucian people) which, for Rawls, cannot be called "public." Indeed, Rawls does not accept "communitarian" public reasons: "Not all reasons are public reasons, as there are the nonpublic reasons of churches and universities and of many other associations in civil society" (PL 213). He relegates such reasons to the "background culture" in contrast with "the public political culture," although he does not call the former private reasons, either, because "there is no such thing as private reason" (PL 220, note 7). However, I think it is only a matter of degree how "public" a reason is. It makes sense to talk about religious public reasons, such as Buddhist public reasons or Christian public reasons, if such reasons are shared by reasonable Buddhist or Christian believers and are used to guide their relations and provide justifications for solving their controversies when they belong to different sects or subcommunities in their respective religious communities. Indeed, each of such established large-scale religions not only exists internationally but also includes various internal orders and denominations. Within such a religious community there are not only conflicts of interests (that exist even in a small community such as a family or a university) but also different religious and moral visions, even if all members share certain religious beliefs in common. Thus, in such a large-scale religious community, there is not a unified, singular comprehensive doctrine held by all its subcommunities, even if they possess tremendous family resemblances in the eyes of other religious people. Their internal controversies, conflicts, and even warfare could be as severe and intense as those among different religions or countries (one can recall the bloody warfare between the Catholics and Protestants in the past, and those between different Muslim sects in the present). Moreover, the policies made by such large-scale, non-geographical religious communities vastly affect other people belonging to other communities of society. Accordingly, even if we do not want to count reasons shared by the fellows of a small church or temple as public, we can count shared Buddhist or Christian reasons as public, if they are reasonably worked out by and for these respective communities, to be neutral (in the sense of neutrality of aim in Rawls' terminology—see next section) to the different comprehensive doctrines held by their respective subcommunities.[3]

To emphasize the possibility of such communitarian public reasons is to indicate the inevitable culture-based and cross-national features that a meaningful public reason conception must carry. Just like Buddhist or Christian public reasons that will have to draw on their respective religious cultures, Rawls' liberal public reasons have been rooted in contemporary Western liberal culture. Moreover, just as Buddhist or Christian public reasons are shared by their respective international people, Rawls' liberal public reasons are held by liberal people in different Western nations. This is to say, public reason cannot be from nowhere but must originate from somewhere, and public reason should not be restricted to a particular nation but should be international.

Nevertheless, a notion of public reason for the citizens of a large-scale political nation would have less substantive content than a notion of public reason for the members of a large-scale international community. While the latter needs to accommodate the views of different sects under one culture or religion, the former needs to accommodate the views of different cultures or religions. For example, if an international Confucian family association is to be set up, its members can be expected to hold in common a series of Confucian public reasons embedded in profound Confucian beliefs, both religious and moral, such as the Mandate of Heaven (*tien ming*), the Dao of Heaven (*tien dao*), humanity (*ren*), filial piety (*xiao*), and harmony (*he*). In contrast, if one is to propose a Confucian conception of public reason for the citizens of mainland China, one needs at least to drop the idea of the Mandate of Heaven, because this idea carries a Confucian religious message that many Chinese citizens who are not religiously Confucian (but are morally and

[3] No public reason can be indefinitely broad. If one were to talk about a notion of public reason shared by all rational beings in the universe, wherever found, as what Kant seemed to have provided, that notion will definitely be short of sufficient substance in providing justification for public policy. That may be why Rawls wants to limit his notion of public reasons to the content of those reasons that are shared by liberal democratic peoples in modern Western countries, but not by other peoples (see next section). I will argue that it makes sense to talk about public reasons for non-democratic peoples as long as certain conditions are satisfied. It may also make sense to talk about public reasons shared by both democratic and non-democratic peoples.

politically Confucian) do not accept.[4] It is practically suitable for such a conception of public reason (to be adopted by religiously different citizens in a country rather than by Confucian believers in an international association) to focus on their common Confucian moral and political ideas, rather than on their religious controversies. This is in a way similar to the argumentative context of liberalism in which Rawls works: in attempting to establish a liberal conception of public reason for the Western democratic countries, Rawls cannot draw on the robust view of comprehensive liberalism, such as perfectionist liberalism; instead, as he argues, only political liberalism is the suitable version of liberalism for public reason in a modern Western country, in which the metaphysical loads of liberalism have to be dropped out (PL).

Since an important function of a conception of public reason is to provide a useful conceptual framework for citizens to use to formulate and justify public policies as they see proper, a conception of public reason **for national use** may be most significant. This is because in today's world, only sovereign states are the real authorities to make effective public policies and laws for their states. This is not to deny that a conception of public reason for national use may largely be shared by different sovereign states. Here I would like to echo my colleague Sungmoon Kim's standpoint that a notion of Confucian public reason is shared by the peoples in East Asian states and regions, such as mainland China, Hong Kong, Korea, and Taiwan,[5] although I disagree with him about the content of such a Confucian conception (see Sect. 4). Finally, to propose a conception of public reason for East Asian societies is by no means to affirm moral relativism in the sense that there is no universal moral truth. Rather, this is only to assume that pursuing universal moral truth cannot

---

[4] It is controversial, even among Confucian scholars, whether Confucianism is a religion in the strict (yet normally poorly defined) sense. I understand Confucianism as a religion in the sense that it includes not only moral and political instructions, but also the convictions of the ultimate reality (Heaven) and the necessity of the ritual practice (*li*) in communicating and interacting with supernatural beings, such as one's deceased ancestors, sages, and gods, in the authentic Confucian way of life. This is to say, even if the religious status of Confucianism is still disputed in academic circles, Confucianism can most reasonably be taken as a religion by looking at the substance of its comprehensive doctrine(s) actually held by its authentic believers in their way of life in such countries as China and Korea.

[5] Sungmoon Kim, *Public Reason Confucianism*, henceforth *PRC* (Cambridge: Cambridge University Press, 2016).

be the main purpose in establishing a conception of public reason. In addressing an appropriate conception of public reason, we attempt to find a suitable framework of values to formulate and defend public policy (such as a law regarding MAS, which I will address later in this chapter) for directing and regulating the people of diverse views and interests in a country.

Indeed, to set up a conception of public reason for national use is in particular helpful to deal with controversial bioethical issues through justifying proper policy and laws in particular countries. Kevin Wildes usefully provides a clear spectrum of moral agreement and disagreement regarding bioethical explorations, which includes three levels. The first level is "object-level" judgments, concerning what one ought or ought not to do in a given case or whether a particular course of action, such as MAS, should be permitted by law. The second level is "justificatory" judgments, which articulate the reasons, moral feelings, moral senses or moral values that apply to particular moral controversies or policy formulations. And the third level is "foundational" views, which define the basis and acceptability of justification.[6] For our main purpose of establishing a conception of public reason for national use (namely, to justify public policies and laws in general, and bioethical policies and laws in particular), I think we should discover such a conception from justificatory reasons available in a nation. Reasons existing on the "object-level" are too narrow to do the task, whereas reasons shared by people on the "foundational" level are too few to be sufficient. Of course, justificatory reasons cannot be entirely detached or isolated from foundational reasons, but people sharing justificatory reasons do not have to assume or accept a complete set of foundational reasons which can be offered only by a complete set of moral theories, principles, or rules from a moral tradition, such as Confucianism. That is, justificatory reasons may inevitably engage some foundational reasons so that they cannot be separated entirely from a comprehensive doctrine, but they do not need to engage all foundational reasons of the doctrine. All depend on what people have actually accepted and practiced in their particular way of life.

[6] Kevin Wm. Wildes, *Moral Acquaintances: Methodology in Bioethics* (Notre Dame: University of Notre Dame Press, 2000).

With this background information provided, in the next section I will lay out Rawls' liberal conception of public reason and explicate why it is not suitable for an East Asian country like China to adopt. Then the two Confucian conceptions of public reason proposed by Joseph Chan and Sungmoon Kim, respectively, will be introduced in the following sections. I will indicate how each conception faces dilemma in their arguments. In Sect. 5, I outline a different Confucian conception of public reason based on my view proposed in the *Reconstructionist Confucianism*,[7] which I believe overcomes the difficulties confronted by Chan's and Kim's conceptions. This is followed by a section reviewing the issue of MAS in terms of this conception of public reason, in which I argue why MAS should not be legalized in contemporary China. The final section covers concluding remarks.

2 Rawls' Liberal Conception of Public Reason

Rawls' liberal conception of public reason has been most influential in the contemporary academic world. He identifies public reason in the following way:

> Public reason is characteristic of a democratic people: it is the reason of its citizens, of those sharing the status of equal citizenship. … Public reason, then, is public in three ways: as the reason of citizens as such, it is the reason of the public; its subject is the good of the public and matters of fundamental justice; and its nature and content is public, being given by the ideals and principles expressed by society's conception of political justice, and conducted open to view on that basis. (PL 213)

Granted these three features for the concept of public reason, is it necessary to limit the concept only to a democratic people? Evidently, not every democratic people can construct a morally acceptable conception of public reason. For example, ancient Greek democratic polity excluded

[7] Ruiping Fan, *Reconstructionist Confucianism: Rethinking Morality after the West*, (Springer, 2010).

women from its citizenship and even argued for the legitimacy of slavery.[8] This example indicates that being simply democratic is not a sufficient condition for people to hold a proper idea of public reason. On the other hand, although most people do not think contemporary Chinese political regime is democratic, some Chinese authors claim that it is still a type of democracy, which is even more democratic than Western countries, because, according to their view, it represents the true interests of most Chinese people.[9] This controversy indicates that even if the current China can be taken as democratic, its type of democracy is not the liberal constitutional democracy that Rawls has in mind when he puts forward the above three features for the concept of public reason. Accordingly, no matter whether the current China is democratic or not, we may still have good reason to uphold the plausibility of conceiving a conception of public reason to meet the three Rawlsian features for China to adopt.

In fact, as Rawls specifies it, his conception of public reason is liberal (or liberal democratic) in the three following senses:

> first, it specifies certain basic rights, liberties, and opportunities …; second it assigns a special priority to these rights, liberties, and opportunities, especially with respect to claims of the general good and of perfectionist values; and third, it affirms measures assuring all citizens adequate all-purpose means to make effective use of their basic liberties and opportunities. (PL 223)

Rawls explains that in addition to its principles of justice, his conception of public reason also includes the guidelines of inquiry that specify rules of reasoning and evidence so as to direct citizens in applying principles to make laws and policies properly. Hence the liberal political values upheld under his conception are of two kinds: the values of political justice (such as equal political and civil liberties) and the values of public reason (such as the guidelines for public inquiry, the political virtue of reasonableness and the moral duty of civility) (PL 224).

---

[8] Aristotle, *The Politics of Aristotle*, ed. and trans. Ernest Barker, (London: Oxford University Press, 1958).

[9] Weiwei Zhang, *The China Wave: Rise of a Civilizational State*, (Singapore: World Scientific Press, 2012).

In this chapter I do not attempt to judge whether Rawls' liberal conception of public reason is suitable for contemporary Western societies. But I do not believe it is suitable for Chinese society. My first reason for this belief is that the social and historical conditions of modern China are essentially different from those of the West from which, as Rawls acknowledges, his liberal conception is constructed:

> The social and historical conditions of such a [Western] state have their origins in the Wars of Religion following the Reformation and the subsequent development of the principle of toleration, and in the growth of constitutional government and the institutions of large industrial market economies. These conditions profoundly affect the requirements of a workable conception of political justice: such a conception must allow for a diversity of doctrines and the plurality of conflicting, and indeed incommensurable, conceptions of the good affirmed by the members of existing democratic societies.[10]

In contrast, the social and historical conditions of China have their origins in quite different religious and historical factors from those of the West. First, the main culture or religion of China has been Confucianism, not Christianity. Although Confucianism has its religious or metaphysical teachings of Heaven (*tian*), the soul, and afterlife,[11] it has never formed a powerful religious organization like the Roman Catholic Church in the West. There is no controversy that the most important Confucian communities are families, not churches, but there has not been a Confucian association of families that has been placed in authority regarding Confucian beliefs as the Roman Catholic Church regarding the Christian faith. Accordingly, the Reformation could never have happened in China, and anything like the division of church and state, which has been enormously accentuated in the modern West, could not have been a significant matter in China.

---

[10] John Rawls, "Justice as fairness: Political not metaphysical," *Philosophy and Public Affairs* 14, no. 3 (1985), 225.

[11] Ruiping Fan, "DNA, Brain, Mind and Soul: A Confucian Perspective," in *Interreligious Perspectives on Mind, Genes and the Self*, ed. Joseph Tham, Chris Durante, and Alberto Garcia Gomez, (London: Routledge, 2019), 63–74.

Moreover, among the hundreds of the bloody wars in Chinese history, there was no war of religion like that between the Catholics and the Protestants in the West. Although there have certainly been social, ethical, and political conflicts among Chinese religions, especially among the so-called three religions (*sanjiao*), namely, Confucianism, Daoism, and Buddhism, they have never developed into a sanguinary religious battle. In fact, there has long been an attitude of toleration inherent in Confucianism toward other religious or metaphysical beliefs that are differentiated from Confucian convictions, as long as such beliefs can be interpreted as not including moral or political implications and requirements that violate generally accepted Confucian morality. In other words, compared to Christianity, Confucianism puts emphasis on moral virtues and does not demand exclusive religious allegiance and does not stress religious doctrinal orthodoxy. No doubt, this attitude of toleration was not always followed by the Chinese emperors or governors in the past. But political intolerance and persecution in traditional China rarely stemmed from genuine religious doctrinal controversies, in comparison with the conflicts of actual social and economic interests. In this sense, I may reasonably conclude that a Confucian principle of religious toleration had already been present in traditional China, and this principle is consistent with the modern Western principle of religious freedom developed from the Reformation and the Wars of Religion.

Of course, there have been great changes in the Chinese way of life ever since the East met the West in modern times. The twentieth century witnessed the gradual breach and fragmentation of the Confucian tradition in China through a series of radical Chinese revolutions and communist political movements, including the Republican Revolution (1911), the May Fourth Movement (1919), the Communist Revolution (1949), and the Cultural Revolution (1966–1976). The superstructures of Chinese politics and economics have ceased to be Confucian, and an officially promoted morality has been communist, which morality calls for sacrificing individual interests to seek an egalitarian society under the Communist Party's rule. However, Confucian morality still governs a wide range of contemporary Chinese life, and a substantive "Confucian personality" has continuously informed the social base of Chinese culture and morality and significantly accounted for the motivation behind the strategies adopted for the success of Chinese economic reforms in the recent four decades. There has been an increasingly remarkable disconnection between the officially announced communist morality and the actually operating, although damaged and even fragmented, Confucian morality in mainland China. If we need to characterize a Chinese way of life underlying the conception of contemporary Chinese public reason, it would still be more accurate to term it Confucian than communist or liberal or anything else.[12] This is the second reason why Rawls' liberal conception of public reason is not suitable for China.

Finally, Rawls' way of addressing the issue of neutrality as well as his discussion of the issue of abortion shows that a particular conception of public reason like his cannot stand totally independent of the individualistic feature of the comprehensive doctrines actually existing in contemporary Western societies. To be sure, under his liberal conception, Rawls argues only for the neutrality of aim, not that of procedure or effect: basic institutions and public policies are not to be designed to favor any particular comprehensive doctrine (PL 194), but are "to ensure for all citizens equal opportunity to advance any conception of the good they freely affirm" (PL 192). This kind of the neutrality of aim, he acknowledges, is not procedure neutrality, since the political values it appeals to are far more than procedure values only. Neither is it the neutrality of effect or influence, as he continues, since "[i]t may still affirm the superiority of certain forms of moral character and encourage certain moral virtues … such as the virtues of civility and tolerance" (PL 194). Rawls provides a long footnote discussing a particular application of the neutrality of aim in enacting a proper law regarding the issue of abortion (PL 243). He thinks that we need to consider the issue "in terms of these three important political values: the due respect for human life, the ordered reproduction of political society over time, including the family in some form, and finally the equality of women as equal citizens." His conclusion is that "any reasonable balance of these three values will give a woman a duly qualified right to decide whether or not to end her pregnancy during

[12] For a more detailed account of contemporary China's condition, see Ruiping Fan, "A Confucian notion of the common good for contemporary China," in *The Common Good: Chinese and American Perspectives*, ed. David Solomon and P.C. Lo, (Springer, 2014), 193–218.

the first trimester." He contends that "the reason for this is that at this early stage of pregnancy the political value of the equality of women is overriding" (PL 243). It is evident, in my view, that the three values as he formulates and the way of balancing them as he conducts are both characteristic of an individualistic moral feature which is embedded (and perhaps shared) in comprehensive Western doctrines, either religious or secular. They generally assume and indeed accentuate individual independence, authority, and autonomy rather than communitarian or relationalistic propriety as emphasized in Eastern comprehensive doctrines (such as Confucianism) regarding making a decision on abortion. While it emphasizes the equality of women in the issue, it overlooks the value of a united father-mother relation, the importance of the role of the father in the decision of abortion, as well as the enormous concern of the mother's parents as implied in certain Eastern non-individualistic moral views, such as the Confucian view.[13]

This is to say, even if Rawls is not disingenuous regarding the neutrality of aim toward different comprehensive doctrines in the West in designing his conception of public reason, such neutrality does not stand separable from a certain individualistic moral feature of Western comprehensive doctrines, whereas this feature may not be present in the comprehensive doctrines of non-Western societies. In this sense, his conception of public reason has inevitably been designed to favor Western comprehensive doctrines, not Eastern comprehensive doctrines, such as Confucianism, that do not carry this individualistic moral feature. For Rawls, "[t]he only comprehensive doctrines that run afoul of public reason are those that cannot support a reasonable balance of political values" (PL243). But our first question is where such political values come from. Presumably, Rawls wants to argue that even if some concepts of his proposed political values

[13] I understand that individualism vs. holism (or relationalism) involves highly sophisticated metaphysical and moral issues that cannot be addressed in this chapter. Here I only mean to point out that East Asian comprehensive doctrines share a family-oriented mode of decision-making in which the individual is not appreciated as possessing sole or exclusive decisional authority in biomedical matters independently of one's family, as in individual-oriented Western model (Fan 2015), and this familist or non-individualist moral feature of decision-making present in Confucian-influenced East Asian societies may constitute a reasonable challenge for the suitability of applying Rawls' liberal conception of public reason to these societies, because the requirement of the neutrality of aim cannot be met in such application.

(such as liberty and equality) are from Western comprehensive doctrines, they are self-standing and are not adopted to favor any **particular** Western comprehensive doctrine. Then our second question would be whether they are not adopted to favor Western (rather than non-Western) comprehensive doctrines. If my above observation of individualist versus non-individualist moral features differing between Western and Eastern comprehensive doctrines in relation to Rawls' discussion of abortion is correct, then it would be hard to conclude that Rawls' conception of public reason is entirely independent of the individualistic feature of Western comprehensive doctrines. Accordingly, if one simply imports Rawls' conception of public reason into East Asian societies, one would have intended to give a favor to those individualistic Western doctrines against those non-individualistic Eastern doctrines. This would violate the requirement of the neutrality of aim in constructing a conception of public reason as Rawls contends.

Does this comment on the unsuitability of Rawls' liberal **conception** of public reason to Confucian-influenced East Asian societies imply that the **concept** of public reason can be used to characterize and direct any kind of people, regardless of what comprehensive doctrines they hold or what political systems they accept? No! That is not what I have meant to argue for. I would stress one necessary condition for the applicability of the concept of public reason: it cannot apply to any totalitarian regime in which the people are thoroughly controlled by the government, without enjoying a basic sense of freedom for thinking, speaking, and acting. And neither can this concept make sense for a people who are, although democratic in a sense (e.g., ancient Greek democracy), subject to the tyranny of the majority whose constitution does not allow appropriating any private property nor protects a set of basic individual rights. In short, I would follow Rawls' basic thought developed in his late work, *The Law of Peoples* (henceforth LP), to uphold that in addition to liberal democratic people, the concept of public reason applies also to "decent" contemporary people enjoying a decent minimal list of human rights in Rawls' terminology, even if they are not liberal democratic people (LP). As long as this condition (of a minimal list of human rights) is met, it is up to particular people themselves to reasonably develop their own respective conceptions of public reason suitable to the particular historic-social-cultural situations of their societies, no matter whether their societies are liberal or democratic.[14]

Rawls provides two criteria for a decent society: first, it "does not have aggressive aims, and …. must gain its legitimate ends through diplomacy and trade and other ways of peace" (LP 64); second, its system of law secures for all its members a minimal list of human rights (which is only a subset of the liberal list of human rights, as I will address in Sect. 5), imposes legitimate moral duties and obligations on all its persons, and that there exists "a sincere and not unreasonable belief on the part of judges and other officials who administer the legal system that the law is indeed guided by a common good idea of justice" (LP 65–66). I think these two criteria are reasonable and should be generally accepted. I would assume that contemporary Chinese society, at least from the time it started reforms in the late 1970s, has gradually become such a decent society, although it has not become liberal or democratic. I am aware that this assumption is controversial. It is open to debate about how much the current Chinese society has met each element of the two criteria that Rawls has put forth. It may be more accurate to state that China is on the way to becoming such a decent society. In any case, I expect that the reformed situation of current China, along with the Rawlsian idea of the legitimacy of decent society in contrast with that of liberal democratic society, provides me a suitable context to reasonably explore the possibility of constructing a proper Confucian conception of public reason for China.

3 Joseph Chan's Confucian Conception of Public Reason

Strictly speaking, Joseph Chan does not propose a Confucian conception of public reason. He terms what he produces "moderate Confucian perfectionism" which "steers a middle path between comprehensive

[14] I do not attempt to offer more justification for this necessary condition for the concept of public reason than what Rawls has offered in his *The Law of Peoples*. Instead, the following sections of this essay will attempt to illustrate and defend a Confucian case of public reason for East Asian societies, including China, assuming that the necessary condition is met.

doctrines and what Rawls calls 'public reason'—a kind of shared political value within the political culture of a liberal society".[15] However, as I see it, what Chan actually offers is precisely a set of shared political values within the political culture of a Confucian-influenced society (not a liberal society), such as Hong Kong, so as to form a modern version of Confucian political philosophy to direct governmental policies and activities, such as moral education. This is done to suit the modern pluralist character of such a society. In this sense, there is no conceptual problem in taking his work as providing a Confucian conception of public reason for the society of Hong Kong, since the reasons he offers can reasonably be seen as public in the three ways of being public that Rawls portrays (see Sect. 2).

The main points of Chan's Confucian conception of public reason are as follows.

> I believe it is possible to appeal to judgments about the good life without basing them on any comprehensive doctrine .... Moderate perfectionism ... appeals to individual judgments about human goods and experiences in a piecemeal way ... I believe that traditional perfectionist philosophies such as Confucianism have rich insights and ethical resources to offer modern societies, but they took the form of a comprehensive doctrine that causes problems of legitimacy and harmony for modern-day politics. The best way to make these insights and resources relevant to a modern pluralist society is through moderate perfectionism. The moderate way of conducting moral education, for example, is to promote specific human virtues that constitute the good life, without basing them on a comprehensive doctrine. It is, I believe, possible to understand and appreciate the value of such Confucian virtues as respect, reverence, trustworthiness, sincerity, and benevolence without accepting the whole philosophy of Confucianism.[16]

It is crucial, so Chan insists, that one should drop any comprehensive Confucian doctrine but simply appeal to particular Confucian virtues

[15] Joseph Chan, *Confucian Perfectionism: A Political Philosophy for Modern Times*, (Princeton: Princeton University Press, 2014), 100.

[16] Ibid., 99–100.

(separated from any comprehensive Confucian doctrine) to constitute a system of political values. In this system, such Confucian virtues as respect and trustworthiness constitute an essential part of the content of public reason to guide contemporary Confucian-influenced societies, such as Hong Kong. Chan holds that other essential parts of his conception of public reason should stem from liberal values as Rawls provides. However, Chan will need to do a lot more work to bring these two types of values—liberal and Confucian—together to make a coherent Confucian conception of public reason, although the structure of his conception is already available in his argument. Now suppose his strategy is possible—namely, a framework of Confucian and liberal values can be set up in this way as public reason to do its task in a Confucian-influenced society—then two important questions arise. The first is why the separation between specific Confucian virtues and comprehensive Confucian doctrines is necessary for a Confucian conception of public reason. And the second is whether such a separation is possible.

The separation is necessary, for Chan, because it is the requirement of civility in modern society. Where does this requirement of civility come from? Chan admits that it does not come from Confucianism; instead, he imports it from Rawls into his argument. "Civility tries to diminish conflicts by seeking a *common ground* that underlies conflicting opinions and a *common good* that transcends partisan interests … to give reasons that others can share in justifying one's views."[17] Thus what he wants to promote is both "the rich insights and ethical resources" of Confucianism and the requirement of liberal "civility" in a modern Confucian-influenced pluralist society. What he has constructed is a combined liberal and Confucian conception of public reason. In this case, although his theory is called Confucian perfectionism, Confucian values can be promoted only in a piecemeal way, since "promoting Confucianism as a comprehensive doctrine is undesirable in the main because it damages civility".[18]

[17] Joseph Chan, "On the legitimacy of Confucian constitutionalism." In *A Confucian Constitutional Order*, trans. Edmund Ryden, eds. Daniel Bell and Ruiping Fan, (Princeton University Press, 2013), 102–103. Emphasis original.

[18] Ibid., 103.

I don't think this separation requirement is necessary because the liberal requirement of civility (at least under Chan's separation-requirement interpretation) may be too strong to allow any useful conception of public reason to be constructed for a contemporary Confucian society. When Chan requires that one has "to give reasons that others can share in justifying one's views," he needs to consider whether "others" include everyone in a contemporary Confucian society, or only the majority people who have been affected by Confucian values and have generally accepted such values. If it is the former, then no justification would be possible, because it requires every reason one offers to be shared by all other citizens. For example, if one offers a deductive justification for one's view, there should be no objection to one's major or minor premise by any other citizen. This is simply impossible, not only for a Confucian conception of public reason for East Asian societies but also for the liberal conception for the Western societies, because no society can expect all of its citizens to share the same starting premises. On the other hand, if it is the latter—namely, to give reasons that are shared or acceptable by most people in a society—then Chan's separation requirement would become unnecessary for a Confucian-influenced society. It may well be the case that not only some piecemeal Confucian specific virtues but also some comprehensive doctrines (such as the Confucian doctrine of filial piety in relation to the Confucian account of the family; see next section for details) are actually shared by most people in East Asian societies. Accordingly, it is unnecessary to require that such comprehensive doctrines not be drawn on as justificatory reasons to raise one's argument or even make public policy, as long as such Confucian comprehensive doctrines have been inherited and accepted spontaneously or voluntarily by most people, without being coerced or repressed by government.

For the second question, concerning whether the separation requirement is possible, Sungmoon Kim has offered a compelling argument to indicate that it is not. As Kim sees it, Chan's conception cannot avoid a seeming dilemma. On the one hand, Chan needs to demonstrate that the human virtues he enumerates are important Confucian virtues. Without such demonstration he would be unable to characterize his perfectionist view as Confucian rather than other, such as a liberal view: "it is far from clear how human virtues, such as respect, reverence, trustworthiness, and the like, can be called 'Confucian' when they are detached from the comprehensive doctrine of Confucianism" (PRC 181). On the other hand, however, if Chan highlights the Confucian roots of these human virtues, the Confucianism in question would ineluctably be a comprehensive doctrine, which he believes would violate the (liberal) value of civility (PRC 181). This is to say, in order to be a Confucian conception of public reason, the content of Chan's conception cannot be sharply divorced from a comprehensive Confucian doctrine as he intends. If he insists on the separation, his conception of public reason would no longer be so much Confucian as it is liberal.

4 Sungmoon Kim's Confucian Conception of Public Reason

Kim is the first Confucian scholar proposing a clear idea of the Confucian conception of public reason. He states that his conception has two following normative premises:

1. There is a valuable Confucian way of life distinct from a liberal way of life;
2. It is permissible for a (democratic) state to promote or discourage some activities, ideas, or ways of life based on the grounds of a constellation of Confucian values, such as filial piety, respect for elders, ancestor worship, ritual propriety, harmony within the family, and social harmony (PRC 87).

In this way, he explicitly distinguishes his Confucian conception of public reason from the Rawlsian liberal conception. At the same time, however, he emphasizes that the democratic values of modern (Western) society must be integrated into any possible conception of public reason, such as his own Confucian conception. For him, although democratic citizens may share general agreement about the political value of some core liberal liberties and rights, they can also differ radically as to the substantive moral content of each right and liberty. The source of such disagreement, in his view, must stem from the comprehensive doctrine that directs a person's moral life (PRC 80). Then from where should the normativity of public reason be? For Hobbes, it derives solely from the coerciveness of political power. For Rawls, it cannot and should not derive from such coerciveness, since Rawls does not take the Hobbesian approach to be appropriate. Given that Rawlsian public reason aims to achieve simultaneously restraint of and respect for diversity, Kim contends that Rawls' approach must respect value pluralism. This circumstance "renders the line between comprehensive doctrines and public reason far more porous than it is usually believed to be or even intended to be by Rawls himself" (PRC 80). "The result is acknowledgment of the possibility of competing interpretations of public reason" in relation to different comprehensive doctrines. Consequently, as Kim sees it, Rawls has to "open a backdoor allowing comprehensive doctrines to sneak into the domains of public reason," since they are needed for the legitimacy of certain differences regarding the substantive moral content of each right and liberty (PRC 80). Therefore, for Kim, a refined interpretation of Rawls' liberal conception of public reason can accommodate the true sense of respect for diversity, and this leads to the overlap or intertwinement between public reason and some comprehensive doctrines. As a result, Kim concludes, we are enabled to explore a new mode of normative theory, public reason perfectionism, of which public reason Confucianism is one type, for Confucian-influenced East Asian societies (PRC 81).

Moreover, Kim argues that the two normative premises of his Confucian conception are supplemented by six propositions that together render public reason Confucianism a kind of democratic perfectionism. Among the six propositions, P3 is a crucial one:

> P3: In a Confucian society, all citizens are equal to one another qua public citizens and together they exercise popular sovereignty. (PRC 88)

Kim admits that P3 is not an original Confucian value, but is imported into Confucianism from democratic values. As he asserts, his Confucian conception of public reason "must embrace as noninstrumentally valuable core values of democracy such as popular sovereignty and political equality" (PRC 27). This means that interestingly, while Chan inserts into his Confucian conception of public reason a core liberal value (civility), Kim inserts into his a core democratic value (equality). Thus, in my view, while Chan faces a dilemma between Confucian values and liberal values as Kim demonstrates, Kim faces a dilemma between Confucian values and democratic values (as I will show further in the chapter). If Chan cannot coherently go both ways, neither can Kim. I think they both have offered heuristic lessons for us to learn in order to characterize a genuinely Confucian conception of public reason.

Specifically, it seems to me that by sneaking the general democratic value of public equality into his Confucian conception of public reason, Kim overlooks and even rejects certain crucial nonegalitarian features of Confucianism:

1. Political inequality (due to the fact that the Confucian virtues are unequally possessed by individual citizens): for example, Confucian public reason may not accept that everyone has a moral right to vote.
2. Opportunity inequality (arising from the Confucian view of the family and the practice of the Confucian principle of differentiated and graded love): given that the family holds a fundamental place in Confucian society, opportunity equality must be appropriately qualified (family-based) equality, and certain reasonable inequalities generated by the existence of the family must be allowed and even respected.

The general democratic value of public equality emphasizes equal liberties and duties among citizens, whereas Confucianism advocates some unequal liberties and obligations,[19] although Confucianism should accept that certain liberties be asserted by everyone as basic rights (see next section). This is because, for Confucianism, the more virtuous and capable the individual citizens, the more political powers and obligations they should have. So Confucianism genuinely supports political meritocracy rather than democracy. Of course, it is not the case that those who espouse democratic equality do not know that some citizens have little intelligence or virtue. Neither is it that they trust that equal political

[19] Ruiping Fan, "A Confucian notion of the common good for contemporary China," in *The Common Good: Chinese and American Perspectives*, ed. David Solomon and P.C. Lo, (Springer, 2014), 193–218.

participation (such as one person one vote) would surely select out meritorious leaders. It is rather that they embrace an absolute egalitarian idea of the equal worth or dignity or status of human individuals so that everyone should have an equal right to political participation and accept the risk of likely bad consequences. In contrast, although Confucianism recognizes certain inherent moral equality among human beings (this is because, according to Confucianism, every human individual is endowed with virtue potentials so that everyone may become a sage through virtue cultivation), it also recognizes the moral inequality of individuals in their actual virtue cultivation, possession, and exercising. Accordingly, Confucian culture tends to reject an absolute egalitarian idea of moral status and political participation for every individual. There has been a series of intellectual papers illustrating and defending a legitimate type of political inequality for citizens from a Confucian meritocratic perspective.[20] It has also been recognized that egalitarian qualities sometimes conflict with good governance or the protection of liberties, so that democratic equalities should be restricted in order to defend liberal aspects from a Confucian view.[21]

A prominent example of political inequality espoused by Confucianism is that citizens have no human or moral right to vote. As Joseph Chan argues in an unpublished paper, whereas democratic egalitarians normally argue that everyone who is affected by the decisions of a government should have the right to participation in that government, including the moral right to vote, Confucianism sees that political participation involves not only the power of affecting one's own interests but also the other-regarding element of directing the lives of others. Thus, in Chan's view, Confucianism holds a "service conception of political authority": this authority is primarily responsibility, and it depends on one's capacity to govern and the cooperation of the governed.[22] Accordingly, the rights attached to this authority are justified by the contribution they make to the betterment of people's lives. There is no natural rulership just as there

[20] See, e.g., Daniel A. Bell and Chenyang Li (eds.) *The East Asian Challenge for Democracy: Political Meritocracy in Comparative Perspective*, (Cambridge University Press, 2013).

[21] Tongdong Bai, *Against Political Equality: The Confucian Case*, (Princeton: Princeton University Press, 2019).

[22] Joseph Chan, *Confucian Perfectionism.*

is no natural citizenship.[23] Thus, Confucianism can take universal suffrage as a useful instrumental/empirical mechanism: if this mechanism turns out to be helpful to the people's happiness or flourishing in a particular society at a particular time, it should be adopted; on the contrary, if it is not useful or is even detrimental to people's good lives, it should not be adopted. Accordingly, "one person one vote" may or may not become a political or legal right in a society, but it is not a fundamental human or moral right in the Confucian view. For a Confucian, the important thing is to pursue a good life as long as one's basic liberties and rights are protected, not to seek a one person one vote arrangment for casting a ballot at any cost.

Moreover, central to the Confucian way of life is the fundamental place of the family which is supported by an ontological-metaphysical account of the family and collides with many of the commitments of the liberal social-democratic state, particularly its egalitarian aspirations. In the Confucian understanding of the family, "the roles of men, women, and children cannot be freely created, but are held properly to reflect antecedent, non-socially-constructed moral norms, a normative reality that supports status obligations within the family".[24] However, the liberal social-democratic state is committed to the democratic equality of opportunity, and this commitment is in tension with the existence of such traditional families, because the latter are aimed at advantaging their own members in preference to others so as to prevent the full realization of fair equality of opportunity in society. Although most Western countries still tolerate such inequalities at present, this is not the direction of societal development guided by democratic justice. As Rawls acknowledges, although it is not urgent to abolish the family, yet the democratic idea of equal opportunity does incline in this direction (Theory 511). Western liberal democratic states have offered a web of publically funded

[23] Joseph Chan, "Is there a human right to political participation? A Contemporary Confucian perspective." (unpublished paper, 2015).

[24] H. Tristram Engelhardt Jr., "The Family: Crucial to and Divisive in Bioethics," *International Journal of Chinese & Comparative Philosophy of Medicine* XI.2 (2013), 114.

entitlements that make it easier for persons to exit the family and to have children outside of marriage.[25]

Confucian public reason must explore the challenge of pursuing opportunity equality that also supports the traditional family. The Confucian value of differentiated and graded love sets certain constraints on governmental policy and system in pursuing equality of opportunity in society. Such constraints are necessary in order to protect the integrity of the family and to preserve family-based (unequal) opportunities. In this regard, only some types of equality will be ensured in Confucian society, such as the formal equality before the law, the judicial independence and the rule of law that traditional Confucian society has failed to develop. But other types of equality, especially substantive or welfare equality such as a government-run, one-tier, robust egalitarian system of basic health care distribution as imposed by some Western countries like Canada, cannot be accepted by Confucian public reason. For this type of equality imposed by the state deprives citizens of useful ways to purchase a better basic tier of health care for themselves and their immediate family members, thereby failing to leave legitimate room for them to practice differentiated and graded love and obligations toward their family members as approved by Confucianism. Confucians should accept a general principle of equal opportunity in society, but it must be adequately qualified in accommodation with Confucian virtue-based familial moral sentiment so as to allow and even promote certain family-based opportunities, such as certain educational opportunities offered by parents to their children, even if these must lead to unequally distributed educational opportunities in society.[26] This is to say, in promoting fair equality of opportunity in a Confucian society, public policy and laws must leave a sufficient institutional space (such as the private sector of education, health care, and housing, in addition to a public system provided by the government)

[25] Charles Murry, *Coming Apart: The State of White America 1960–2010*, (New York: Crown Forum, 2012); H. Tristram Engelhardt Jr., "Fair Equality of Opportunity Critically Reexamined: The family and the Sustainability of Health Care Systems," *Journal of Medicine and Philosophy* 37, no. 6 (2012): 583–602.

[26] Ruiping Fan, Xiaoyang Chen and Yongfu Cao, "Family-oriented Health Savings Accounts: Facing the Challenge of Health Care Allocation," *Journal of Medicine and Philosophy* 37, no. 6 (2012): 507–512.

to individual citizens. In this space individuals can legitimately give unequal love or discharge special obligations to their family members, even if this would serve as a stumbling block to the full realization of equal opportunity for all.

In short, Kim has mistakenly sneaked the democratic value of public equality into his Confucian conception of public reason so as to clash with certain central Confucian values. Like Chan's, Kim's conception cannot avoid a dilemma. On the one hand, Kim needs seriously to embody the substance of the Confucian values (such as "filial piety, respect for elders, ancestor worship, ritual propriety, harmony within the family, and social harmony" as he enumerates) as he espouses in the content of his conception. Otherwise he would be unable to characterize his conception as Confucian rather than other, such as democratic. On the other hand, however, if Kim highlights the Confucian roots of these values as well as their entrenched connection with comprehensive Confucian doctrines, they would inevitably contain the nonegalitarian connotations and requirements that are ethically defensible. The importance of such requirements for Confucianism would not allow him to integrate the democratic values of "popular sovereignty and political equality" into his conception without losing its coherence as a Confucian conception. In short, it seems that, like Chan, Kim has not been sufficiently confident to propose a genuine Confucian conception of public reason for East Asian societies.

5 The Reconstructionist Confucian Conception of Public Reason

A more plausible Confucian conception of public reason can be worked out from the view I have developed in the *Reconstructionist Confucianism*.[27] In this section, I will first summarize the key points of this new conception. Then I will offer explanation and defense for it.

[27] Ruiping Fan, *Reconstructionist Confucianism*.

1. A series of fundamental Confucian virtues which are still generally accepted and practiced in contemporary Confucian-influenced East Asian societies must be contained in the Confucian conception of public reason as public values or reasons, along with other public reasons (especially a list of basic liberties and rights), to direct public discussions and the formulations of public policy in these societies.
2. A list of basic liberties and rights should be contained in the Confucian conception of public reason as public values or reasons, but such liberties and rights should be developed from fundamental Confucian virtues or values as stated in (1), or should at least stand no contradiction with their central requirements. This is to say, no fundamental reason or value contained in the Confucian conception of public reason should be directly inserted into it from another tradition (such as the liberal social-democratic tradition) even if it is in conflict or contradiction with some of the central requirements of fundamental Confucian virtues as stated in (1).[28]
3. The principles of political justice under this conception will include both the central requirements of fundamental Confucian virtues on the one hand, and the basic liberties and rights developed out on the other, to make a coherent constellation of public reasons for the regulation of contemporary Confucian-influenced societies.
4. Like any other conception of public reason, appropriate guidelines of inquiry (namely, a set of rules of reasoning and evidence) should be included in this conception to facilitate citizens to engage in public discussion and to make public policy.

Accordingly, the content of public reason under this conception is fourfold. First, it retains certain basic Confucian virtues, such as *ren* (humanity), *yi* (appropriateness), *li* (propriety), *xiao* (filial piety), and *he* (harmony), of which there are still general acceptance and practice in

[28] Following this line of argument, the fact that Chan inserts into his conception the liberal value of civility and that Kim into his, the democratic value of equality, makes their respective conceptions no longer genuinely Confucian, because the substance of their inserted values contradicts with some of the central requirements of fundamental Confucian virtues, as we discussed in the previous sections. In contrast, in my Confucian conception, although certain basic liberties and rights are necessary components, they are not so fundamental as the basic Confucian virtues, and neither are they in conflict with any of their central requirements.

contemporary Confucian-influenced East Asian societies, to continue to guide such societies. Here the point is not to offer an exhaustive list of basic Confucian virtues for the conception of public reason, neither is it to accept all the requirements of such virtues as they are traditionally understood or required. Rather, for the purpose of working out a suitable Confucian conception of public reason for application in contemporary East Asian societies, my focus is on the central requirements of such basic virtues that are still embodied in the way of life of the peoples in these societies. In short, it is such central requirements that should be constructed and formulated adequately to constitute an irreplaceable part of the Confucian conception of public reason to guide contemporary people. Second, this conception establishes a list of basic liberties and rights that are not only necessary for protecting legitimate individual interests in contemporary societies but are also not contradictory with any central requirement of the virtues. Third, the principles of political justice under this conception are inevitably twofold: they must integrate the central requirements of both fundamental Confucian virtues and the basic liberties and rights. In this system, no basic liberty or right may be sacrificed for the sake of general societal interests, and neither may any central requirement of the virtues be violated for promoting any (non-basic) liberal or democratic rights. Finally, it offers suitable guidelines of inquiry, such as rules of reasoning and evidence, for citizens to adopt to engage in public discussion and identify proper laws, policies, and decisions for contemporary society.

I think there are at least two central moral and political requirements of the Confucian virtues that are still accepted and practiced in contemporary Confucian-influenced societies. The first is that human lives are most noble (*gui*) or valuable among all types of lives in the world so that innocent human lives must be respected and protected. And the second is that human individuals should be encouraged and assisted to cultivate basic virtues to form appropriate relationships with one another, and that the moral propriety of such relationships is not subjectivist in the sense that it should not be determined solely by individuals' autonomous decisions or voluntary mutual consent. This second requirement is consistently correlated to a series of considered judgments that are still inherited or formed in East Asian societies. For example, the family is a deep metaphysical and normative reality for individuals to realize rather than a mere social structure that comes about through the agreement of the participants. Moreover, parent-child incest is morally evil, regardless of whether the child involved is already an adult and whether the incest performed is mutually voluntarily agreed and causes no psyche harm.[29] Furthermore, everyone is morally obliged to exercise differentiated and graded love and obligation toward other people depending on what kinds of relationships one stands with them.[30] Finally, contrary to the conclusion of typical liberal moral arguments offered by some leading liberal scholars,[31] East Asian people generally hold that adult children have a moral obligation to care for their elderly parents, even if they do not have any contract with their parents undergirding such an obligation.[32] It is necessary to accentuate that the general moral orientation of such central requirements from the fundamental Confucian virtues deviates from that of the full-fledged liberal democratic ideas of individual liberty, equality, and rights, even if it may be found that some contemporary Western people also share such central requirements. The Confucian conception of public reason must also contain a set of basic liberties, equality, and rights, and I will explain what I mean by "basic" later in this section.

Before I turn to the second component of my Confucian conception, let me digress to deal with the legitimacy issue first. Indeed, to establish such a conception of public reason, I have to face the challenge of pluralism. There must be non-Confucians in East Asian societies who do not accept such central requirements of the Confucian virtues. How can I justify grounding a conception of public reason in Confucian views without doing violence to the convictions of others who are not Confucian? I

---

[29] Ruiping Fan, "Is Confucian Harmony Foundationless? A Critical Question for Chenyang Li," *Philosophy East & West* 67, no.1 (2017): 246–256.

[30] Ruiping Fan, "Nonegalitarian Social Responsibility for Health: A Confucian Perspective on Article 14 of the UNESCO *Declaration in Bioethics and Human Rights*," *Kennedy Institute of Ethics Journal* 26, no.2 (2016): 105–218.

[31] E.g., Jane English, "What Do Grown Children Owe Their Parents?" in *Ethics in Practice: An Anthology*, ed. Hugh LaFollette (Oxford: Blackwell, 2002), 152–155; Norman Daniels *Am I My Parents' Keeper?: An Essay on Justice Between the Young and the Old*, (New York: Oxford University Press, 1988).

[32] Ruiping Fan, "Which Care, Whose Responsibility, and Why Family? A Confucian Account of Long Term Care for the Elderly," *Journal of Medicine and Philosophy* 32, no.5 (2007): 495–517.

can provide two points to deal with this challenge. First, this conception of public reason is based not only on Confucian views but also on the construction of a moral and political way of life manifested in East Asian societies. It is a sociological fact, for example, that most people in contemporary China still accept and practice fundamental Confucian virtues, especially their central requirements as summarized earlier, even if they do not identify themselves as Confucians. That is, religiously the people may be atheists, Buddhists, Confucians, Christians, Daoists, or others, but morally and politically they remain mostly affected and shaped by Confucian virtues. In Sungmoon Kim's terminology, their moral way of life is Confucian, not liberal or any other. This is to say, if we look at the moral and political way of life of the Chinese people, we can easily find Confucian public reasons such as the central requirements of the fundamental virtues as well as the related considered judgments to be entrenched or embedded in their practices. In the ordinary language they speak, we hear the concepts of virtues normally used and appealed to. We see the inklings of the Confucian public reasons that they share in their activities as well as in the common sense of the societies. In their conduct in treating themselves and other people, in their attitude to political and legal institutions and governments, and in their reactions to moral, social, and cultural events, we find no other ideas than Confucian that can more accurately characterize the nature of their reasons to be used. All these factors give me sufficient legitimacy to spell out the central requirements of the basic Confucian virtues and integrate them into the conception of public reason for these societies, even if there are individuals who may announce that they do not accept such central requirements or my ways of formulating them.

This is to say, my answer to the question of why Confucian values or reasons under my conception can be termed "public" in contemporary Confucian-influenced but pluralistic societies (such as mainland China) is primarily practical, not theoretical: such values and reasons are embedded in the people's moral and political way of life. In other words, these values cannot be an "overlapping consensus" of East Asian societies in the theoretical sense that the available doctrines, theories, and ideologies in these societies can be reconstructed to converge on such values or reasons. They simply cannot. For example, if we are serious about the substance of each of the various moral and ideological doctrines present in the current Chinese society, such as Marxism, liberalism, and feminism, as well as religious traditions, such as Confucianism, Buddhism, Daoism, Christianity, and Islam, it is hard to find any meaningful coherent overlap over them. Their starting points and core values are largely incommensurable with each other. Indeed, no one can discover an actual Chinese moral and political consensus by looking at the content of these doctrines, even if one can simply label and reject any part of such doctrines that one does not accept as "unreasonable." I am afraid no meaningful value can be left in this way. Accordingly, the force of the values in my conception of public reason cannot ground in any agreement of such underlying general or comprehensive doctrines, but only in the reasonable Confucian construction of the moral and political way of life lived by these people.

My second point to deal with the pluralist challenge to my Confucian conception is to draw on a comparative task. As I have indicated in the previous sections, each of the major alternative conceptions of public reason for East Asian societies, namely, Rawls' liberal conception, Chan's Confucian conception, and Kim's Confucian conception, has serious difficulties and does not suit these societies. In contrast, my conception does not suffer from the same difficulties as they do. Although this comparative point by no means constitutes a sufficient argument for my conception, it provides remarkable support based on a comparative advantage.

Finally, it is worth mentioning that some may think it more inclusive or popular to term my conception as "practical" or "pragmatic" than "Confucian," as it integrates into it other values (especially a universal concept of human rights as I am going to specify further) that are not originally Confucian. However, the terminology of "practical" or "pragmatic" is by no means more accurate than "Confucian" for my conception, but is a bit disingenuous, because I cannot deny that my conception carries a coherent feature that is categorically Confucian. On the one hand, it covers a series of fundamental virtues and their central requirements that are evidently Confucian in their moral origin and nature, although they are also accepted and practiced by peoples other than Confucians in East Asian societies. On the other hand it is required under my conception that non-Confucian values it integrates be either developed out of Confucian virtues or at least not contradictory with the central requirements of Confucian virtues. With this feature, it is pertinent to call it Confucian rather than "practical" or "pragmatic."

The second major component of my conception of public reason, namely, a list of basic liberties and rights, is needed for properly developing Confucian tradition and accommodating the pluralistic circumstances of contemporary East Asian societies. For instance, although basic Confucian virtues are still embedded in the contemporary Chinese way of life, some Chinese people no longer take these virtues to be their foundational moral precepts or their starting points for moral elaboration. Instead, they take some other moral precepts, such as Buddhist or liberal precepts, to be more fundamental than the Confucian virtues, although they also accept the Confucian virtues. Accordingly, it is reasonably necessary for the society to adopt the concept of rights to protect the people's interests in basic freedom of faith and conscience, let alone their material interests. On the side of Confucian citizens, it is true that they hold the concept of the Confucian virtues rather than the concept of individual rights, liberty, or equality to be the ground of morality. In other words, for Confucians, the ultimate nobility (*gui*) or dignity of the human life does not lie in enjoying rights, but in pursuing virtue. Nevertheless, virtue is not the only intrinsic value that they pursue. Confucian tradition has always proposed some other values, such as *datong* (grand union), *dashun* (great smoothness), and *zhishan* (the highest comprehensive good) for comprehensively good human life, that should be pursued along with the virtues. Indeed, in the Confucian view, *zhishan* should be sought by all individuals, families, and governments alike. The Confucian classics have shown it clearly that to pursue the highest comprehensive good for human beings, it is not enough for individuals to cultivate and exercise virtue, but it is also essential for government to respect and protect legitimate individual interests, such as their individual and familial property and wealth. In other words, Confucianism recognizes that even if a Confucian individual may still be virtuous when his or her legitimate self-interests are damaged, his or her life will not be able to reach the highest comprehensive good, *zhishan*, in this circumstance.[33] Accordingly, pursuing the highest comprehensive good implies not only that individuals cultivate and exercise virtue, but also that government respect and protect legitimate individual interests, so that the virtuous will lead good and happy lives. The way of accomplishing this comprehensive goal, to learn from classical liberals, is to grant individuals basic liberties and rights.

A list of basic liberties and rights to be covered by my Confucian conception of public reason includes two parts: one is universal human rights that should be granted by all societies, and another is special Confucian rights that should be enjoyed in East Asian societies. Universal human rights are a decent minimal list of basic liberties and moral rights that has been proposed by Rawls in his later work: it is "a special class of urgent rights" such as "liberty (but not equal liberty)", which does not represent a full-brown liberal democratic conception of rights (LP 79). Instead, it is "a proper subset of the rights" similarly possessed by citizens in a liberal constitutional democratic regime and the members of a decent hierarchical society (LP 81). This list is as follows:

> Among the human rights are the right to life (to the means of subsistence and security); to liberty (to freedom from slavery, serfdom, and forced occupation, and to a sufficient measure of liberty of conscience to ensure freedom of religion and thought); and to formal equality as expressed by the rules of natural justice (that is, that similar cases be treated similarly). (LP 65)

These rights, as Rawls stresses and I agree, "cannot be rejected as peculiarly liberal or special to the Western tradition" (LP 65).[34] They do not

[33] I take this Confucian idea of the highest comprehensive good (*zhishan*) to be similar to Aristotle's Eudaimonia, or human flourishing. It requires not only honoring human dignity by individuals through their virtue exercising, but also protecting legitimate individual interests by governments through safeguarding rights. In the Mencian terminology, a virtuous government must conduct "benevolent governance" (*renzheng*) through protecting legitimate individual interests, especially their material interests. Given the space limit of this paper, I cannot extend to a comparative study between Confucianism and Aristotelianism regarding this idea, but only suggest an inherent Confucian logical path to the concept of rights. For a detailed account of *zhishan*, see the Confucian classic of the *Great Learning* (*Daxue*).

[34] Rawls points out that among international human rights declarations, "some seem more aptly described as stating liberal aspirations, such as Article 1 of the Universal Declaration of Human

"belong to certain kinds of political institutions," liberal democratic, or other (LP 79). Instead, they set a limit to the pluralism among peoples and constitute a necessary condition of the decency of any political institution (LP 80). Although the Confucian tradition has never proposed this concept of universal human rights, I think such rights can be worked out from within the Confucian tradition because they are implicit, if not explicit, in the fundamental implications of the basic Confucian virtues. This is to say, they are not only needed to protect legitimate individual interests in Confucian-influenced societies, but can also be consistently developed out of the Confucian virtue-based tradition. In fact, in my previous work I have shown that such rights can be derived from the moral requirements of the basic Confucian virtues, such as *ren* (humanity) and *yi* (appropriateness), regarding how individuals should treat one another as well as how they should be treated by their government. Accordingly, they should be accepted and upheld by contemporary Confucian-influenced societies for pursuing comprehensively good life, such as *zhishan* I mentioned before.[35]

This is to say, although Confucian tradition has not proposed a concept of rights in the past, it should be and can be consistently developed and included in the Confucian conception of public reason for contemporary society to protect legitimate individual interests. However, whereas a thin concept of rights should be shared universally across traditions, the Confucian conception of rights must differ from the liberal conception of rights, because the latter includes certain individual-autonomy-oriented rights (such as a right to do morally wrong deeds) and carries excessive individualistic values that cannot meet the central requirements of the Confucian virtues.[36] For a Confucian conception of rights, individual interests will be seen as legitimate or illegitimate, depending on whether they are consistent with the central requirements of the virtues.

Rights of 1948: 'All human beings are born free and equal in dignity and rights. They are endowed with reason and conscience and should act towards one another in a spirit of brotherhood.' Others appear to presuppose specific kinds of institutions, such as the right to social security, in Article 22, and the right to equal pay for equal work, in Article 23" (LP 80, note 23).

[35] Ruiping Fan, *Reconstructionist Confucianism*, 56–61.

[36] Ruiping Fan and Wenqing Zhao, "Developing Confucian Virtue-based Rights," in *Religious Perspectives on Bioethics and Human Rights*, ed. Joseph Tham, Kai Man Kwan and Alberto Garcia (Dordrecht: Springer, 2017), 115–118.

Due to comprehensive considerations (such as for the peace of society), Confucian society may tolerate certain illegitimate interests (such as gambling or prostitution) in ways in which individual behavior for gratifying such interests may be seriously regulated but not be absolutely prohibited (or punished) by the law. But they should never be established as moral rights to be respected, because they are in conflict with the central requirements of the virtues. Thus the Confucian conception of rights is primarily virtue-based, whereas the full-blown liberal conception of rights is individual-autonomy-based. As a result, my Confucian conception of public reason cannot allow immoral or depraved conduct to become rights, even if such conduct does not directly harm unconsented others and thereby should be tolerated.

On the other hand, in addition to the universal human rights to be accepted, certain typical Confucian cultural rights should also be developed and included in the list of liberties and rights under the Confucian conception of public reason. Such cultural rights are morally and politically parochial, pertinent to Confucian-influenced East Asian societies only, because they are derivatively Confucian and may not be shared by other non-Confucian societies. A prominent example of such rights is one that, based on the Confucian virtue of filial piety (*xiao*), an elderly parent should have a right to care from one's adult children. Liberal scholars cannot even accept the rationale of adult children's moral obligation to care for their elderly parents,[37] much less the logic of such a parental right to be developed for the elderly. But this moral obligation is deeply grounded in the Confucian morality of filial piety and is still taken for granted by most people, both Confucians and non-Confucians, in contemporary Confucian-influenced societies.[38] In order to enhance the moral agency of elderly people as well as to protect their legitimate interests in contemporary society, this right should be developed on the basis

[37] English, 'What Do Grown Children Owe Their Parents?'; Daniel, *Am I My Parents' Keeper?*

[38] Qingjie Wang, 'The Confucian Filial Obligation and Care for Aged Parents,' in *Confucian Bioethics*, ed. Ruiping Fan (Dordrecht: Kluwer Academic Publishers, 1999), 235–256; Erika H. Y. Yu, 'Respect for the Elderly and Family Responsibility: Confucian Response to the Old Age Allowance Policy in Hong Kong,' in *The Family, Medical Decision-Making and Biotechnology: Critical Reflections on Asian Moral Perspectives*, ed. Lee Shui Chuen (Springer, 2007), 197–206.

of the Confucian filial obligation by drawing on the general logical correlation of rights and obligations.

Based on these components, the principles of justice covered in my conception of public reason would not only specify basic liberties and rights as well as their priority, but also the central requirements of the Confucian virtues with which the basic liberties and rights do not stand in contradiction. The principle of equal opportunity will be accommodated with the values of the family through which adequate policy and institutional rooms will be left for individuals to provide privileged opportunities for their family members in society.

Finally, what kind of guidelines of inquiry for public discussion and policy formulation should my Confucian conception of public reason offer? For his liberal conception of public reason, Rawls identifies three features that make a comprehensive doctrine "reasonable" so that these features, presumably, should also serve as guidelines of inquiry for public reasoning. The first is "an exercise of theoretical reason" that requires consistence and coherence to express "an intelligible view of the world." The second is an "exercise of practical reason" that gives instruction about how to weigh different values when they conflict. Finally, a reasonable comprehensive doctrine "normally belongs to, or draws upon, a tradition of thought and doctrine" and "it tends to evolve slowly in the light of what, from its points of view, it sees as good and sufficient reasons" (PL 59).

Martha Nussbaum convincingly shows that Rawls' criteria are too demanding to be held by reasonable citizens in contemporary Western societies. The problem comes from the first two criteria. As Nussbaum illustrates, if we check the comprehensive doctrines held by contemporary Western people against these two criteria, a number of world views based on Western traditional religions, astrology, and new age religion would become "unreasonable." The Christian doctrines of Trinity and grace, and traditional Judaist acceptance of mystery (that is, in principle, not graspable by reason) can by no means meet the Rawlsian theoretical requirement of consistence and coherence. The Judaist belief that there are no in-advance instructions for hard cases obviously runs afoul of

Rawls' second criterion for practical reason.[39] Similarly, as I see it, the Buddhist belief in reincarnation and the Daoist belief in bodily immortality held by a great number of people in Eastern societies can hardly meet the Rawlsian theoretical requirement of consistence and coherence, given today's basic scientific or biological knowledge generally shared by the people. Moreover, the Buddhist and Daoist experience of mystery and miracle (as well as their relevant ritual exercises leading to such mystery and miracle) may grotesquely go against the Rawlsian practical reason regarding how to weigh different values. Nussbaum's recommendation is that the public reason of political liberalism should give up any such requirements. Instead, she contends, "respect in political liberalism is, first and foremost, respect for persons, not respect for the doctrines they hold".[40] To stick with this ethical requirement of respect, she suggests that there is no need to hold any extra criteria as guidelines of inquiry. In her view, not only do we not need the first two Rawlsian requirements, there is also no need to contain the third requirement: any doctrine is "reasonable" as long as it is endorsed by a reasonable citizen (who respects other citizens), no matter whether it is "tradition-based, authority-based, argument-based, faith-based, or based in nothing but its allure."[41]

I think Nussbaum goes too far by rejecting all the Rawlsian criteria. I think she is right at dropping the first two criteria. But my Confucian conception of public reason would contend that we should maintain Rawls' third requirement, which contains, as I see it, two standards for a comprehensive doctrine to be reasonable for use in public discussion. The first is a traditional source standard: a reasonable doctrine stems from and belongs to an established tradition. The second is a gradual evolution standard: a reasonable doctrine evolves slowly according to its good and sufficient reasons. To hold this contention is not to hold a conservative attitude toward assessing comprehensive doctrines for public discussion, recommending no social change or even social stagnancy. Rather, it is to recognize the reasonable connection of a new position or proposal to

[39] Martha Nussbaum, "Perfectionist Liberalism and Political Liberalism," *Philosophy & Public Affairs* 39, no.1 (2011b), 25–29.

[40] Nussbaum, "Perfectionist Liberalism and Political Liberalism," 33.

[41] Nussbaum, "Perfectionist Liberalism and Political Liberalism," 33.

some values or reasons explicit or implicit in some comprehensive doctrine that has already been established or held by many citizens of a society, so as to facilitate adequate public discussion and to contribute to peaceful evolution as well as social and political stability. These two standards, as guidelines of inquiry, would be most suitable and beneficial for citizens in East Asian societies to adopt and engage in their public discussion and formulate their public policy.

6 The Bioethical Issue of Medically Assisted Suicide (MAS)

This chapter has no space to appeal to my Confucian conception of public reason to address political issues to illustrate its suitability and usefulness. Instead, I will only apply it to explore the bioethical issue of medically assisted suicide (MAS). Indeed, unlike Western countries, no East Asian society (such as Hong Kong, Taiwan, mainland China, South Korea, Japan, Singapore, and Vietnam) has legalized MAS or active euthanasia. Should MAS be legalized in such societies? What would be a proper law or policy about MAS in such societies? Based on the content of my Confucian conception of public reason proposed for these societies, I will need to elaborate and balance the following public values in order to pass a proper judgment on this issue: (1) the due respect for the patient's request for assistance for suicide; (2) the cherishing of the patient's life as demanded by the central requirements of basic virtues such as *ren* (humanity), *yi* (righteousness), *li* (ritual propriety), *xiao* (filial piety), and *he* (harmony); (3) the proper professional role of the physician as implied by the virtues of *ren, yi, li, xiao*, and *he*; and (4) the likely effects of such legalization on elderly people in these societies. I will show that my reasonable balance of these values leads to the conclusion that MAS should not be legalized in these societies, such as mainland China and Hong Kong.

It is true that patients in these societies sometimes voluntarily ask for MAS, and such request deserves certain respect. But what is the due respect for such request in the context? Following liberal ethical reasoning, some academic papers published in mainland China argue that patients have a right to end one's life or a right to death.[42] There are also scholars who vehemently object to granting such a right to patients.[43] Indeed, from the Confucian conception of public reason as I proposed, citizens (including patients) in such Confucian-influenced societies do not and should not have such a right, because this "right" would contradict the central requirement of basic Confucian virtues that an innocent human life must be honored and respected. Although in some contexts committing suicide is understandable and acceptable in Confucian tradition,[44] such cases are exceptional and are acceptable only under special circumstances and qualifications (see later). The idea of granting such a right to individuals for committing suicide in the medical context cannot be consistently understood as manifesting the honoring or respecting of an innocent human life, because this is to intentionally allow, affirm, and even will the provision of medical professional assistance to patients for destroying their lives. In Confucian view, medicine is, and ought to be, set up in society only for saving patients' lives and relieving their suffering. This is to say, under my Confucian conception of public reason, although society should give certain respect to a patient's request for MAS, the due respect for it cannot be granting him or her a right to MAS. There cannot be such a right in the Confucian list of individual rights.

---

[42] Z. Wang, "Ethical Controversy on Abandoning Treatment and Euthanasia" (in Chinese), *Chinese Medical Ethics* 18, no. 4 (2005): 61–63; M. Chen, (2014) "Debates and Analysis of Physician-assisted Suicide" (in Chinese), *Medicine and Philosophy* (*Yixue yu zhexue*) 35, no. 12 (2014): 15–19; J. Shuai, "Autonomous Choice of Right to Life" (in Chinese), *Jishou University Journal* (*Jishou daxue xuebao*) 39 (2018): 45–47.

[43] X. Liu, and J. Yang, "Right to Die: Myth of Autonomy and Limitation of Rights" (in Chinese), *Medicine and Philosophy* (*Yixue yu zhexue*) 36, no. 5 (2015): 22–24; H. Wu, "Challenge and criticism: Euthanasia should not be legalized" (in Chinese), *Medicine and Law* (*Yixue yu faxue*) 8, no. 4 (2016): 7–11; Y. Sun, "Terminally Ill Patients' Medical Autonomy: Focusing on Latest Legislation beyond Mainland China" (in Chinese). *Journal of Southwest University of Political Science & Law* (*Xinan zhengfa daxue xuebao*) 19, no. 5 (2017): 65–73.

[44] P.C. Lo offers examples in Confucian tradition of those who committed suicides in order to save others or prevent humiliation. See Ping-cheung Lo, "Confucian Ethic Death with Dignity and Its Contemporary Relevance," *The Journal of the Society of Christian Ethics* 19 (1999): 313–333. Such cases of suicide have been ethically acceptable and even admirable in certain circumstances in the tradition. However, our question here is whether the medical context constitutes such a legitimate circumstance. As Lo, I do not think it does.

What would be the due respect for such request, then? From the perspective of Hong Kong's Hospital Authority (which manages all public hospitals in Hong Kong), accepting such a request and performing MAS on the patient is conducting an unethical act—it is "direct intentional killing of a person as part of the medical care being offered." Instead, from their experience, a request for euthanasia or MAS by the patient is often a call for help because of uncontrolled physical symptoms, social problems, or psychological or spiritual distress. The due respect for such request is not to accede to it and perform MAS, but is to address the patient's problems properly. It requires, at least, careful communication with the patient and the family to understand their underlying concerns and help them deal with such concerns (Hospital Authority, 2002).[45]

Honoring innocent human life, as a central requirement of the virtues of *ren, yi, li, xiao*, and *he*, remains accepted and practiced generally in East Asian societies. On the part of the patient, cherishing one's life implies that, ideally, one should not ask for MAS. It is true that in Confucian tradition it is acceptable and even admirable for individuals to commit suicides in order to preserve their dignity in certain uncontrollable circumstances. However, as P.C. Lo indicates, such circumstances in traditional China were caused by inevitable external factors, such as the circumstances when an individual had to suffer humiliation or assault by the presence of hostile forces, such as a cruel and heartless emperor or enemy, that one had no way to avoid.[46] On the other hand, however, one's illness is an internal factor afflicted in one's body, and it can be dealt with by appropriate medical treatment, especially palliative care, even if it cannot be completely cured or relieved. As life is always accompanied with some suffering, one should not hold that one must be assisted to die immediately unless there is not any suffering at all. Of course, one does not have to accept any futile aggressive means for simply extending one's life. That is, withholding or withdrawing life-sustaining medical interventions would not contradict the virtuous requirement of cherishing one's life when such interventions are therapeutically futile. However,

---

[45] Hospital Authority, "HA Guidelines on Life-sustaining Treatment in the Terminally Ill," (Hong Kong, 2002): http://www.ha.org.hk/haho/ho/adm/124655e.pdf.

[46] Lo, "Confucian Ethic Death with Dignity and Its Contemporary Relevance."

asking for MAS is asking for directly killing or destroying one's life when it can go on under suitable medical care, which is hardly consistent with the virtuous requirement.[47]

On the part of family members, the patient as an ill family member should be humanely assisted and cared for. If a terminally ill patient is one's parent, the virtue of filial piety requires that the adult child must attend to the parent's physical, emotional, and spiritual needs and try every effort to enable the parent to spend a peaceful and comfortable final stage of life as far as possible. In the non-individualist culture of East Asian societies, such as mainland China, it is normally understood that even if the parent seems to voluntarily ask for MAS because of severe illness, what the parent really expects is to alleviate one's pain and reduce the children's burden rather than immediately to terminate one's life. Thus the genuine embodiment of filial piety in this context is not obeying the parent's request for MAS, but seeking more effective medical and nonmedical means to take care of the parent. Moreover, the conjugal relation is taken to be virtue-laden and to require taking care of each other in difficulty or illness. If the patient is a spouse, the other spouse should fulfill one's duty to accompany and take care of him or her. It would be unvirtuous to abandon the ill spouse by terminating his or her life through MAS.[48]

Under the influence of Confucian virtues, traditional Chinese medicine has been termed "the art of *ren* (benevolence)," requiring the physician to maintain a kind heart and sympathetic attitude toward the patient in practicing medicine.[49] This does not require that the physician always prolong the patient's life through aggressive treatments. If such treatments may not help the patient recover but simply prolong the patient's suffering, they may legitimately be withheld or withdrawn. However, the account of medicine under virtue does imply that the role of the

---

[47] Ping-cheung Lo, "Confucian Views on Suicide and Their Implications for Euthanasia," in *Confucian Bioethics*, ed. Ruiping Fan (Dordrecht: Kluwer Academic Publishers, 1999), 69–102.

[48] Ruiping Fan and Sihan Sun, "To Relieve or Terminate? A Confucian Ethical Reflection on the Use of Morphine for Late-stage Cancer Patients in China," *Developing World Bioethics* 20.3 (2020): 130–138.

[49] Ruiping Fan, "Discourses of Confucian Medical Ethics," *The Cambridge World History of Medical Ethics*, ed. Robert Baker and Laurence McCullough (Cambridge: Cambridge University Press, 2009), 195–201.

physician is using medical means to treat and care rather than terminate or kill the life of the patient. The physician must attempt to improve one's knowledge and practical techniques, including enhancing one's palliative skills, such as prescribing the right type and proper dose of painkillers like morphine—neither overdose nor underdose—to help the patient alleviate pain and have an as comfortable final stage of life as possible. In fact, the contemporary palliative medicine available to treat the patient under the reasonable guidelines of reasoning and evidence has been significantly improved. It is crucially up to the physician to learn more and practice more in this direction to assist the patient and family effectively.[50] In particular, a virtuous physician should also devote more time and energy to communicate with the patient and family to understand and sympathize with their concerns. This kind of sympathy is the concrete manifestation of "the art of *ren*" and indicates the proper role of the physician.[51] To provide MAS is not a moral choice for the physician to make in a Confucian culture.

Finally, to consider legalizing MAS, we should also pay attention to its likely effects on elderly people in East Asian societies, in consideration of their particular psychological features in the Confucian cultural context of these societies. This feature has been vividly summarized by Liang Shuming, a neo-Confucian scholar in the mid-twentieth century. In Liang's view, living in the ethical relation shaped by Confucian virtues, one tends to forget about one's own self-interests and primarily concentrates on the interests of one's counterpart in the family relation. For example, a good parent is one who gives weight to one's child's interests rather than one's own interests, and a good spouse is one who gives weight to one's spouse's good rather than one's own good. That is, one is normally willing to sacrifices one's self-interest in order to promote the counterpart's interests in the ethical relation.[52] Although this feature may have been diluted a bit in contemporary times, it is still evidently embodied in the lives of elderly people in these societies. As parents, they normally

[50] Atul Gawande, *Being Mortal: Medicine and What Matters in the End*, (New York: Metropolitan Books, 2014).

[51] Fan, "Discourses of Confucian Medical Ethics."

[52] Shuming Liang, *The Essentials of Chinese Culture* (zhongguo wenhua yaoyi), (Shanghai: Shanghai Renmin Press, 1949, 2003), 105.

consider their children's interests above their own. In the medical context, when they recognize that their illness cannot be cured and they will become a heavy burden on their children, they are willing to give up their lives for the sake of relieving their children's burden. This psychological feature of the elderly people would all the way be significantly strengthened if MAS becomes legalized. Like it not, law and morality are closely related to each other in East Asian societies. In addition to their regulative function, legal rules also play an educational function: if MAS is legalized in such a society, elderly people may not take it as something neutral for them to accept or refuse, but may take it as something "positive" or even "good" for them to conduct. In this case, a legal "right" for them to ask for MAS may overwhelmingly become a legal "obligation" for them to undertake, in order to relieve the burden of their children as well as the society. This undesirable effect of legalizing MAS should be seriously considered.

Taken all these values and considerations together, I think the bottom line is that MAS should not be legalized in East Asian societies.[53]

7 Concluding Remarks

This is a preliminary work for the possibility of a defensible Confucian conception of public reason for contemporary East Asian societies. This chapter shows that this conception must include a group of basic Confucian virtues, a list of basic liberties and rights that are not contradictory with any central requirement of the basic Confucian virtues, a

[53] It should be noted that although this conclusion has been made based on public values rather than any specific religious beliefs, such as those from Buddhism or Daoism, two major religions present in East Asian societies, it is a mistake to conjecture that these religions see suicide as a noble act. The truth of the matter is that suicide is frowned upon by both Buddhism and Daoism, although suicide committers are not condemned to Hell as the way Christianity has it. In committing suicide, one is building a circumstance that will make one's next birth more difficult than the previous life (for a Buddhist) or will deprive one of the opportunity of becoming an immortal (for a Daoist). In addition, perhaps under the influence of Confucianism, both Buddhists and Daoists note that a person's committing suicide causes grave pain and suffering to the family members and those who love the person, which makes suicide generally a bad act to perform, even if it is permissible under special circumstances. But it is usually believed that medical illness does not constitute such a circumstance.

coherent constellation of the principles of justice in terms of the central virtue requirements and the basic liberties and rights, and appropriate guidelines of inquiry (including the traditional source standard and the gradual evolution standard). My argument offered for this conception, as well as my argumentative illustration of applying this conception to the issue of MAS, has raised more questions than I can satisfactorily answer in the scope of this chapter. However, I hope I have at least demonstrated that this Confucian intellectual endeavor is worth attempting for both theoretical and practical purposes.

4

Replies to Farrell & Tham, and to Fan

Hon-Lam Li

1 Reply to Farrell and Tham

A law don at Oxford University, Martin Matthews, once said: "We lawyers have to resolve problems and get solutions. (You philosophers do not.)" There is some truth to this. Philosophers have been arguing about the nature of morality and ethical principles for thousands of years and more recently about practical ethics and bioethics. It is not foreseeable that many of these disagreements will be resolved soon. This is fine with philosophers. However, for the court of law and for the legislature, decisions have to be made urgently and cannot be put off.

I am grateful to Allen Wood, Bonnie Steinbock, Win-chiat Lee, Michael Campbell, Alastair Campbell, Sin-Yee Chan, and especially John G. Bennett for helpful comments on earlier drafts. I would also like to thank Joseph Chan, Jonathan Chan, Yong Huang, Chung-yi Cheng, Kwong-loi Shun, Yun-chak Chong, and Anthony Chong for discussions on the relationship between filial piety and medical assistance in dying. I thank David O. Brink and Joe Lau for the references to Peter Railton's article and Jean Davies' case respectively.

H.-L. Li (✉)
Department of Philosophy and CUHK Centre for Bioethics, The Chinese University of Hong Kong, Shatin, Hong Kong

H.-L. Li, M. Campbell (eds.), *Public Reason and Bioethics*,
https://doi.org/10.1007/978-3-030-61170-5_4

In a pluralist society where constitutional democracy is practiced, the first question we should ask is whether there is only one right comprehensive doctrine with its conception of the good? (If so, do we know what it is?) Or are there more than one reasonable comprehensive doctrines with their conceptions of the good? Plato and Aristotle, and the Christian tradition as represented by Augustine and Aquinas, hold that there is but one such doctrine that can be recognized by all citizens who are fully reasonable and rational (*PL* 134). The Natural Law Theory (NLT), which owes much to Aristotle and especially to Aquinas, is the theory espoused by Catholics.[1] Political liberalism, by contrast, maintains that because of the "burdens of judgment," there exists reasonable disagreement over comprehensive doctrines among people who are reasonable and rational (*PL* 135).

NLT, as articulated by Fr. Dominic Farrell and Fr. Joseph Tham, makes the following claims:

(1) "There exists … an objective, categorically, and universally normative order of human goods" which "is founded upon natural teleology."
(2) A voluntary action is (morally) good if and only if it is not contrary to the objective order of human goods.
(3) The objective order of human goods can be discovered by considering the teleology of human nature (or practical reason).
(4) The fundamental goods and their immediate requirements are manifest to any person with use of reason. More reflection is needed to discover their further requirements.
(5) The man-made laws of any political society, which direct human beings toward the good that they can attain during their life and through their own efforts, are just and legitimate if and only if they enjoin actions which satisfy (2).

Besides their claim that the objective order of human goods can be discovered by considering the teleology of human nature, Farrell and Tham also make a number of claims about the "common good" of a group or a

[1] Francis J. Beckwith, "Natural Law, Catholicism, and the Protestant Critique: Why We Are Really Not That Far Apart," *Christian bioethics: Non-Ecumenical Studies in Medical Morality* 25, no. 2 (August 2019): 154–168, https://doi.org/10.1093/cb/cbz001.

community. In particular, they claim that individual goods are inseparable from the common good, that deliberation of the common good of a community is "necessarily a shared endeavor," and that "[t]he pursuit of a common good can only take place in a community which has a shared end," but "[t]his is not possible in the modern state, with its bureaucratic institutions and liberal pluralism."

In the next section, I shall criticize the view that whether an act is morally good has to do with the human *telos*.[2] I shall also criticize the view that exaggerates the necessity of the common good.

Human *Telos* and Common Good

The term "*telos*" is often translated as "ultimate end." According to Farrell and Tham, "[a]mong other things, moral disagreements are fundamentally about the nature of our ultimate end, an issue that falls outside of the scope of deliberation. Deliberation presupposes and parts from a certain conception of our end." However, note that not all natural lawyers agree about the importance of the idea of human *telos*. John Finnis, whom Farrell and Tham cite approvingly,[3] is a prominent contemporary natural lawyer. Finnis rejects the idea that deliberation and choice should be based on facts about the human *telos* or human function (*ergon*):

> [Andrew Koppelman] is right in thinking that Grisez, George, Bradley and I reject as fallacious (and never argue on the basis of) any proposition like "natural functions or tendencies are moral standards and ought to guide deliberation and choice." But though this fallacy is certainly to be found from time to time in the tradition, Koppelman is mistaken in thinking that Aquinas's sex ethics depends upon it.[4]

[2] As Farrell and Tham claim, "a voluntary act is (morally) good if and only if it is not contrary to the objective order of human goods" as "founded upon natural teleology" and discoverable by "considering the teleology of human nature," as per (1) to (3) of their claims.

[3] See Chap. 2 of this volume, notes 1, 9, 13, 53, and 61.

[4] John Finnis, "Law, Morality, and 'Sexual Orientation'," in his *Human Rights and Common Good: Collected Essays*, Volume III (Oxford University Press, 2011).

Finnis does not rely on the argument that human beings have some natural functions or *telos*. He even thinks that this kind of argument is fallacious. The reason why Grisez, Boyle, and Finnis reject this argument is that, on their view, "the reality of free choice is incompatible with the supposition—for instance, of Aristotle—that there is a single natural end of human life."[5]

I believe there are further reasons for doubting the view that whether an act is morally good is based on facts about the human *telos*. In *Nicomachean Ethics*, Book 1, Aristotle suggests that all human beings want happiness (*eudaimonia*). But Aristotle argues (fallaciously) that the good for man is to be a good man.[6] Apart from this fallacious step, the most important problem is that human beings do not seem to have a *telos*—or inherent purpose, preordained goal, or ultimate end. (Aristotle argues that because all people want happiness, his project in *Nichomachean Ethics* is to explore what constitutes happiness.)

After Darwin, Aristotle is the second most important biologist in the long span of human history. Aristotle believes that the world has existed for an infinite period of time (*Physics* 1.7), and that all species have lived in this world for an infinite period of time (*Physics* 3.6, 206a25–27). He holds that each species has a unique *telos* or function (*ergon*). According to NLT, any act that goes contrary to the natural order of objective good is deemed morally impermissible. It is not uncommon for natural lawyers to argue that homosexual activity is morally impermissible on the grounds that it is unnatural or that it goes contrary to the human *telos*.[7]

The evolutionary theory of Charles Darwin, who is widely regarded as the greatest biologist in history, has become the new paradigm regarding

[5] German Grisez, Joseph Boyle, and John Finnis, "Practical Principles, Moral Truth, and Ultimate Ends," *American Journal of Jurisprudence*, 32:1, 99–151 at 101. For a contrary view, see Edward Feser, "Natural Law Ethics and the Revival of Aristotelian Metaphysics," in Tom Angier, ed., *Cambridge Companion to Natural Law Ethics*, 2019.

[6] Aristotle then claims that happiness consists in performing "activity of the soul in accordance with virtue," where virtue has to do with virtues of thought (or intellectual virtues) and virtues of character (or moral virtues). See Sarah Broadie, *Ethics with Aristotle* (Oxford: Oxford University Press, 1994). I realize that some Aristotelian scholars want to rescue Aristotle from his fallacious argument and propose the idea of an inclusive end. But I shall not delve into this matter here.

[7] Lo Ping-cheung has argued for this view in his *Starry Heaven and Morality: Ethics and Its Applications*. (in Chinese) (Hong Kong: Joint Publishing Co. Ltd., 1993).

how species come into existence and become extinct, depending on their adaptiveness to the changing environment. According to Darwin, all species exist on Earth as a result of "natural selection," and their continued existence is a contingent matter. Darwin's theory entails that there is no such thing as a human *telos*—the kind of inherent purpose or preordained function of human beings. As Marcus Hester says, "given Darwin's nonteleological concept of evolution, species do not have functions."[8] Anthony Kenny "believes evolutionary theory has made the function of man an unacceptable concept."[9]

Alasdair MacIntyre rejects Aristotle's metaphysical biology as untenable. He writes that on Aristotle's view:

> Human beings, like members of all other species, have a specific nature; and that nature is such that they have certain aims and goals, such that they move by nature towards a specific *telos*. The good is defined in terms of their specific characteristics. Hence Aristotle's ethics, expounded as he expounds it, presupposes his metaphysical biology. (*After Virtue*, 148)

> Aristotle's teleology presupposes his metaphysical biology. If we reject that biology, as we must, is there any way in which that teleology can be preserved? (ibid., 162)[10]

My focus here is on the fact that MacIntyre rejects Aristotle's metaphysical biology.

Can a natural lawyer claim that there exists a *telos* for human sexuality, and that homosexuality falls short of, or is contrary to, such a *telos*?[11] Farrell and Tham do not claim this. But if such a claim is put forward, it would have to be assessed according to whether there is only one kind of

[8] See Marcus Hester, "Aristotle on the Function of Man in Relation to Eudaimonia," *History of Philosophy Quarterly*, vol. 8, no. 1 (January 1991), 3.

[9] Ibid., 13.

[10] Furthermore, Martha Nussbaum and Anthony Kenny have both doubted that Aristotle really develops the concept of functions of species in his biological works. See Martha Nussbaum, *Aristotle's De Motu Animallum* (Princeton: Princeton University Press, 1978), 81, 93. See also Anthony Kenny, *The Aristotelian Ethics* (Oxford: Oxford University Press, 1978), 203.

[11] See Lisa Fullam, "Sex in 3-D: A Telos for a Virtue Ethics of Sexuality," in *Journal of the Society of Christian Ethics*, 27, 2 (2007): 151–170, esp. 163. Fullam claims that a virtuous sex life should have "a feel for incarnation," "an ability for intimacy," and "an eye for insight" (163–167).

sexual life for all people, or whether it would depend on whether some people are heterosexual or homosexual.

An alternative is to argue that acts against human nature are morally wrong. A natural lawyer might claim that certain acts, such as homosexual acts, are against human nature. Farrell and Tham claim that an act must be in accordance with the objective order of human goods in order to be morally permissible, and they *might* want to claim that homosexual copulation is against such an objective order of human goods.[12]

In addition to the argument from human nature, a natural lawyer may argue against homosexuality by relying on the argument that homosexual copulation does not promote any "common good." Thus, Finnis argues approvingly of Plato's view that homosexuality is unnatural: "Homosexual acts are radically and peculiarly non-marital, and for that reason intrinsically unreasonable and unnatural. Furthermore, according to Plato … homosexual acts have a special similarity to solitary masturbation, and both types of radical non-marital act are manifestly unworthy of the human being and immoral."[13]

Finnis further argues that sex within a marriage of the reproductive kind "can *actualize* and allow them to *experience* their *real common good*," which:

> is precisely their marriage with the two goods, parenthood and friendship … even if, independently of what the spouses will, their capacity for biological parenthood will not be fulfilled by that act of genital union. But the common good of friends who are not and cannot be married (for example, man and man, man and boy, woman and woman) has nothing to do with their having children by each other, and the reproductive organs

[12] Let me point out that, as Immanuel Kant argues, human nature (and by implication human *telos*) does not provide a sound metaethical foundation for morality. As Kant points out at the beginning of his *Groundwork for the Metaphysics of Morals*, human nature is not a good foundation for ethics [Ak:4: 389] because moral commands, such as "Do not lie," apply to all rational beings regardless of their nature. Rational beings who lack the desire to follow moral commands are not thereby exempted from morality because moral principles are categorical. More importantly (for our present purposes), if Kant is correct, it follows that the validity of moral commands, or precepts, should not depend on human nature at all (apart from the fact that an agent has rational capacity).

[13] John Finnis, "Law, Morality, and 'Sexual Orientation'," *Human Rights and Common Good: Collected Essays, vol. III*, 338–339.

> cannot make them a biological … unit. So their sexual acts together cannot do what they may hope and imagine. Because their activation of one or even each of their reproductive organs cannot be an actualizing and experiencing of the marital good…, it can *do* no more than provide each partner with an individual gratification. For want of a common good that could be actualized and experienced by and in this bodily union, that conduct involves the partners in treating their bodies as instruments to be used in the service of their consciously experiencing selves." (ibid., 340–341)

Furthermore, Finnis notes the "moral worthlessness" of solitary masturbation (ibid., 341). Moreover, any sexual act that does not actualize the common good between spouses is "unreasonable and wrong" (ibid., 348). This includes "[a]cts of the kind that same-sex partners engage in (intended to culminate in orgasmic satisfaction by finger in vagina, penis in mouth, etc., etc.) [which] remain non-marital, and so unreasonable and wrong, when performed in like manner by a married couple" (ibid., 347–348). This is because having sex merely for pleasure between a married couple, according to Finnis, is wrong unless it occurs within the boundary of a marriage where the good of friendship and *fides*, and the good of procreation and of children, can be realized. As Finnis approvingly expounds Aquinas' view: "if I choose this act of intercourse with my spouse, not for the sake of pleasurably actualizing and expressing our marital commitment, but 'solely for pleasure' … then my sex act with my spouse is non-marital and is in principle seriously wrong" (ibid., 345). Finnis also holds that "intentionally sterile intercourse" between a married couple (ibid., 346) is wrong because it negates the possibility of procreation. (This entails that Finnis is against the use of contraceptives as well.) Sexual intercourse between unmarried consenting adults is also wrong, in his view, because it does not occur within a marriage. On the contrary, those who perform these wrongful acts, as well as homosexual acts (which by their very nature must be outside of a marriage, for Finnis), treat themselves as instruments of pleasure. These claims by Finnis are highly implausible. His arguments would not convince anyone other than those who have already accepted his position.

Finnis makes three claims: (1) that sex for pleasure lacks any common good, (2) that it makes us become merely an instrument for pleasure, and

(3) that homosexual copulation is unnatural. Finnis' second claim seems to me too harsh. (Would massage for pleasure—not sexual pleasure—be wrong?[14]) But for my present purpose, I shall focus on (1) because NLT's linking the common good and an act's moral permissibility concerns us here.

I am not against the idea of the common good *per se*. For one thing, Rawls endorses the idea of the common good.[15] The view that an act is wrong *unless* it actualizes a common good is, however, highly implausible (and this view appears to be an implicit assumption in Finnis' argument). Suppose Johnny enjoys playing tennis by himself, knocking the ball against a wall. Does this mean that he fails to actualize any common good, and is therefore doing something "unreasonable and wrong" and his game "worthless"? Must Johnny play tennis with another person in order to actualize a common good, or else he will be deemed "unreasonable and wrong" and his game "worthless"? Surely not. Indeed, if an activity requires a common good, then many kinds of activities have to be deemed as wrong, including playing pinball by oneself, playing billiards by oneself, swimming by oneself, and exercising by oneself. Even drawing a sketch (if the artist does not intend to show it to others) must also be wrong.

For the sake of argument, let us assume that marital sex is better than non-marital sex because it is more conducive to creating a family and so on.[16] It does not follow that only marital sex performed not (primarily) for pleasure is morally permissible. To hold such a view is to be blind to the fact that there are in society homosexuals, lesbians, and couples (married or not) who do not want to have babies. Is there a moral obligation

[14] Would (2) entail that massage for pleasure is also wrong? Would (2) mean that playing any game or hobby for pleasure is wrong? Whether marriage is a good thing is contestable. Even if marriage is good and sex within a marriage is the best kind of sex, it does not follow that sex outside marriage, and sex for pleasure between spouses, are wrong.

[15] See *Theory*, 101, 123, 178, 232, 233, 243, 246, 425, 472, 478, 489. Note that the idea of the common good can be understood in different ways; see W. Hussain, "The Common Good," Edward N. Zalta, ed., *The Stanford Encyclopedia of Philosophy*, Spring 2018 edition.

[16] Finnis: "[M]arriage is rational and natural primarily because it is the institution which physically, biologically, emotionally, and in every other practical way is peculiarly apt to promote suitably the reproduction of the couple by the generation, nurture, and education of ultimately mature offspring" ("Law, Morality, and 'Sexual Orientation'," 349).

to have babies, even for married couples? It would be difficult for Finnis or NLT to show this.

Alasdair MacIntyre

In their explication of NLT, Farrell and Tham rely heavily on Alasdair MacIntyre's work. MacIntyre's moral heroes are Aristotle and especially Aquinas. Farrell and Tham also endorse Aquinas' view. Aquinas holds that there exist precepts of natural law, which are grounded in our knowledge of human goods, which are in turn based on human nature. Aquinas holds that these precepts are knowable by all rational human beings. The obvious question to ask, MacIntyre points out, is: why is there so much moral disagreement in contemporary society?

MacIntyre's solution to this question is inspired by Thomas Kuhn's theory of revolutionary science, which involves incommensurable paradigms and paradigm shifts, although there are important differences between these two theories.[17] In a nutshell, MacIntyre claims that where there is moral disagreement, people have to engage in shared practical deliberation ("Intractable Moral Disagreement," 19). The conditions for such shared deliberation to be possible at all are fundamental precepts of natural law that people must obey, which include the precept not to kill innocent people, to respect the property of others, and to speak the truth (ibid., 23). These precepts are necessary because without them, shared deliberation would be impossible. Together with the precept to cultivate understanding, they are also grounded in Aquinas' claim about the natural law that we should pursue human goods and avoid evils. Human goods are understood as (1) "goods for our physical nature," which have to do with our lives and health; (2) "goods of our animal nature," including sexuality and the goods to be achieved by educating our children; and (3) goods of "our nature as *rational* animals," namely, the goods of

[17]Alasdair MacIntyre, "Intractable Moral Disagreements," in L. S. Cunningham, ed., *Intractable Disputes about the Natural Law: Alasdair MacIntyre and Critics* (University of Notre Dame Press, 2009), 37. For Kuhn's important theory, see T. S. Kuhn, *The Structure of Scientific Revolutions*, 3rd ed. (Chicago, IL: University of Chicago Press, 1996).

knowledge, of nature, and of God, and of a social life "informed by the precepts of reason" (ibid., 5).

As Farrell and Tham remark, MacIntyre argues against utilitarianism by stressing that the goods of truth and justice, for instance, are not only means to ends but are constitutive of human goods (ibid., 20, 47). Thus, if I lie to others, I do wrong not only to them but also to myself as I fail to respect myself as a rational agent (ibid., 38). On the other hand, a utilitarian (whether an act- or rule-utilitarian) would decide what to do by reference ultimately to consequences alone.[18]

For MacIntyre, the disagreement between utilitarians and Thomists is intractable because both groups view the disagreement from the perspectives of their own traditions. This is analogous to the disagreement between Aristotelian defenders of pre-Galilean physics and Newtonians. In the case of physics, Aristotelians and Newtonians subscribed to incommensurable scientific paradigms through which the physical world is perceived. Hence, data about the world are "theory-laden." In the case of morality, utilitarians and Thomists belong to different traditions. What struck Thomists as a good argument against utilitarianism will not be seen in the same light for utilitarians, and vice versa in the case of a utilitarian objection to Thomism.

As regards revolutionary science, Kuhn holds that one paradigm is preferred to another (older) paradigm by the scientific community because the new paradigm has greater explanatory power, is more fruitful, or is simpler. But the new paradigm is not necessarily closer to truth compared with the old paradigm. Kuhn even holds that the concept of truth is unnecessary for explaining the way in which one paradigm replaces another because Kuhn does not believe that, for instance, Einsteinian physics is closer to truth than Newtonian physics. However, MacIntyre believes that one moral tradition (e.g., Thomism) can be seen closer to the truth than another tradition (e.g., utilitarianism) if and because the former can resolve an epistemic problem that the latter cannot handle, as when the latter is in an epistemic crisis, for instance.

[18] This is not to deny that Mill's claim that there are higher and lower pleasure is an Aristotelian one. See ibid., 49–50.

However, whereas Kuhn's ideas of paradigm and revolutionary science are highly plausible in portraying scientific revolutions, they *mutatis mutandis* seem less apt in accounting for a moral revolution (if there is such a thing as a moral revolution). This is because in scientific revolutions there are "crucial experiments," the result of which can settle issues between competing scientific paradigms.[19] It is doubtful that there exists any "crucial *moral* thought-experiment" whose outcome would conclusively favor one moral tradition over another one.

Aquinas, MacIntyre, as well as Farrell and Tham hold that there are moral truths. However, there is a need to consider morality at a certain time and place. Aquinas' position resembles virtue ethics to some extent: "[Aquinas] allows for the Aristotelian insight that the particulars of the situation always outstrip one's rules, so that one will always need the moral and intellectual virtues in order to act well (*Commentary on NE*, II, 2, 259)."[20] Similarly, MacIntyre's position, as well as that of Farrell and Tham's, is that to resolve moral disagreements we have to deliberate, *here and now*, and that the question we have to ask is "What should we do, *here and now*?"

According to Aquinas, MacIntyre, and Farrell and Tham, fundamental precepts of natural law are "universal," "immutable," and "do not admit exceptions."[21] As Mark C. Murphy points out, "[t]he natural law view rejects wholesale particularism."[22] How should we interpret these terms—universal, immutable, and without exception? There is one possible interpretation: if an A-type act is wrong in circumstance C, then another A-type act must be wrong in relevantly similar circumstances, but if circumstances change, then an A-type act might be permissible. This view, usually known as *moral contextualism* (or situational ethics), is highly

[19] One such crucial experiment was the Michelson-Morley experiment, which falsified the theory that light travels in the medium of ether, and which later led to Einstein's special theory of relativity.

[20] See Mark C. Murphy, "The Natural Law Tradition in Ethics," section 1.3, pp. 8–9, *Stanford Encyclopedia of Philosophy*.

[21] See Dominic Farrell & Joseph Tham, "The Natural Law Tradition, Public Reason, and Bioethics," Chap. 2 of this volume. This view is also held by MacIntyre, "Intractable Moral Disagreements," op. cit. However, Finnis holds that the idea that precepts of natural law are exceptionless applies to only negative precepts (e.g., do not kill innocent people), but not positive precepts. See John Finnis, *Natural Law and Natural Rights*, first edition, 1980.

[22] Mark C. Murphy, "The Natural Law Tradition in Ethics," op. cit.

plausible. If moral contextualism is accepted, it means that if circumstances C and C' are relevantly different, abortion under circumstances C might be impermissible whereas abortion under circumstances C' might be permissible. After all, the Roman Catholic Church approves of abortion if and when the fetus is implanted in the mother's fallopian tube,[23] or if the mother suffers from cervical cancer.[24]

Yet, I doubt that this is the correct interpretation of Aquinas' view. MacIntyre himself appears to endorse moral absolutism.[25] Both MacIntyre, as well as Farrell and Tham, believe that Aquinas is an absolutist at least with respect to fundamental precepts, such as "Never kill an innocent person," "Speak the truth," and "Respect the property of others" (see "Intractable Moral Disagreement," 24, 39). Following Aquinas and MacIntyre, Farrell and Tham hold a moral absolutist view about not taking the life of an innocent person, or else they could not have drawn the conclusion that abortion and medical assistance in dying (MAID) (or medically assisted suicide) are morally impermissible, regardless of the circumstances. At least, Farrell and Tham sound absolutist when they say: "At present, many countries have legalized medical practices, such as abortion or euthanasia, which NLT deems wrong."[26]

On the other hand, if Aquinas were a moral contextualist, then he could have accepted that abortion might be wrong in most cases, but not wrong if a mother's life is endangered by her fetus' continued growth in her fallopian tube. Given that the Roman Catholic Church allows abortion where (1) implantation of a fetus occurs in a fallopian tube, and where (2) the mother suffers from cervical cancer, by reference to the doctrine of double effect, moral contextualism rather than moral absolutism seems to be the Church's preferred view. (I have argued in Chap. 1

[23] Father Tadeusz Pacholczyk, "When Pregnancy Goes Awry: Ectopic Pregnancies," *Catholic Education Resource Center*, 2009. See: https://www.catholiceducation.org/en/science/ethical-issues/when-pregnancy-goes-awry-ectopic-pregnancies.html.

[24] "Roman Catholicism and Abortion Access: Possible Exceptions to the Ban on Abortion by the Roman Catholic Church," *Religious Tolerance*. See: http://www.religioustolerance.org/abo_hist_c1.htm.

[25] By "absolutism," I mean the view that if an act of type A is impermissible, then all acts of type A must also be impermissible even if the circumstances under which these acts are performed are vastly different. See his "Intractable Moral Disagreements," op. cit.

[26] Farrell and Tham, Chap. 2 of this volume.

that the doctrine of double effect is indefensible, and I shall not repeat the argument here.)

The consequence of the foregoing discussion is the following. If Farrell and Tham hold moral absolutism, this theory seems mistaken and can hardly be defended (because it would proscribe abortion even when the mother's life is endangered by the fetus). If, on the other hand, they follow the Church's view that abortion is permissible in cases where the mother's life is endangered by the fetus or by cancer, moral contextualism is called for, in which case it would be difficult to argue for a blanket proscription of abortion. (Note that there are two different senses of the term "abortion." As Thomson points out, a mother can permissibly remove the fetus from her body when, for instance, she is endangered by her fetus, even though this guarantees its death since a fetus, or at least a non-viable one, is not able to survive outside the mother's body. However, it is not permissible for the mother to secure the death of her fetus if a miracle occurs and the fetus survives after being removed from the mother's body. Both kinds of removal (i.e., where a fetus' death is to be secured, and where a fetus is foreseen to die) are, and deserve to be, regarded as abortion. See Judith Thomson, "A Defense of Abortion," *Philosophy and Public Affairs* 1:1, 66.)

Objections and Replies

Farrell and Tham raise several objections to political liberalism. First, they object that political liberalism construes "reasonableness" narrowly as "a commitment to political liberalism" and that "Rawls's definition of reasonableness is circular." This criticism is misguided. One way to understand this objection is that Rawls' definition is vacuous. Understood in this way, this objection is neutralized by the fact that there exists a diametrically different objection, raised by Martha Nussbaum, that Rawls' standard for reasonableness (as to when a comprehensive doctrine is reasonable) is too demanding.[27] Let me explain why I disagree with Farrell and Tham.

---

[27] Martha C. Nussbaum, "Perfectionist Liberalism and Political Liberalism," *Philosophy & Public Affairs*, vol. 39, no. 1 (Winter 2011), 3–45. Nussbaum's view is that Rawls is right to hold that reasonable people hold reasonable comprehensive doctrines. But Rawls' epistemic requirements (as to theoretical coherence and practical consistency), according to Nussbaum, are unnecessarily demanding because some reasonable people, who respect other people, believe in certain views that fail these requirements. She argues that some who believe in astrology, or others in some religious

According to Rawls, the reasonable and the rational are two perfectly ordinary ideas (*PL* 48). The reasonable is public in a way that the rational is not. They are independent ideas in that neither can be derived from the other. Rather than defining what "reasonable" means, Rawls has used the term "reasonable" in different contexts. First, the ideas of reasonableness and reciprocity are intertwined. Persons are reasonable when they are ready "to propose fair terms of cooperation and to abide by them provided others do" (*PL* 54). The second aspect of reasonableness is the willingness to recognize the "burdens of judgment" (i.e., the sources of disagreement) and to accept the consequences of using public reason in directing political power in a constitutional regime (*PL* 54). For citizens with (1) the capacity for a sense of justice and the capacity for a conception of the good and (2) powers of reason (the "two moral powers," *PL* 19), the sources of disagreement lie in "the correct (and conscientious) exercise of our powers of reason and judgment in the ordinary course of political life" (*PL* 56). Rational and reasonable citizens have to make judgments of various kinds, which involve difficulties in balancing various ends, making correct judgments of rationality, and also in taking an impartial view of other people's claims as well as their own (*PL* 56). Furthermore, they need to assess (empirical and scientific) beliefs and their evidence, and decide what weight to give them. They also have to think in terms of moral and political concepts, weigh moral and political values, and arrive at a judgment via inferences and arguments (*PL* 57). Because there are (at least) two sides to a debate, they have to make an overall assessment. Finally, any system of social institution is "limited in the values it can admit so that some selection must be made from the full range of moral and political values that might be realized" (*PL* 57). The burdens of judgment account for the fact of reasonable pluralism.

As Rawls points out, religious and philosophical doctrines express views of the world and of our life with one another (*PL* 58). Different

views (such as the idea of Trinity, or God's grace), are reasonable people even though these ideas might seem illogical or irrational to nonbelievers. As Nussbaum says: "I think that this demand for a particular sort of grounding for a view is disrespectful to religious citizens who think that faith is a very good basis for their views, in fact the best basis" (ibid., 32). As long as people are reasonable, Nussbaum asks, why exclude their comprehensive doctrines as being deemed unreasonable, according to Rawls' requirements?

conceptions of the world can reasonably be elaborated from different points of view. It is a mistake to suppose that "our differences are rooted solely in ignorance and perversity, or rivalries for power, status, or economic gain" (ibid.). Hence, it is not in general unreasonable to subscribe to any reasonable comprehensive doctrine (*PL* 60). Reasonable people see that the burdens of judgment set limit to what can reasonably be affirmed. Therefore, they see that it is unreasonable for the state to use political power to repress comprehensive doctrines that are not unreasonable.[28]

Another objection raised by Farrell and Tham is grounded in their claim that "[t]he precepts of natural law are meant to be universally binding principles of reason recognized by anyone with use of reason. In this case, there should be virtually universal agreement on fundamental moral issues" (Chap. 2 of this volume). They follow Aquinas in holding that every rational person can make the right normative judgment on a moral or social issue. They point out that "if NLT is correct, it should not be possible for there to be foundational disagreements that are both reasonable and irresolvable." However, Rawls holds that because of the burdens of judgment, it is false that reasonable people with the two moral powers can always arrive at the same conclusion.[29] The very "fact of reasonable pluralism" in our society flies in the face of Farrell and Tham's claim.

The third objection made by Farrell and Tham is that Rawls wrongly subscribes to the idea of "epistemic abstinence." Joseph Raz attacks Thomas Nagel's view expressed in "Moral Conflict and Political Legitimacy" for having committed "epistemic abstinence."[30] Is Rawls committed to it? Rawls says that "[s]ome conflicting reasonable judgments (especially important are those belonging under peoples' comprehensive doctrines) may be true, others false; conceivably, all may be false" (*PL* 58). Because of the burdens of judgment, we have to be tolerant of different comprehensive doctrines, as long as they are reasonable.[31] Contrary to Farrell and Tham, Rawls says that "being reasonable

[28] John Rawls, *Political Liberalism*, 61.

[29] John Rawls, *Political Liberalism*, 58.

[30] Thomas Nagel, "Moral Conflict and Political Legitimacy," *Philosophy & Public Affairs*, Vol. 16, No. 3 (Summer, 1987), 215–240. I have discussed Nagel's argument in Chap. 1 of this volume.

[31] John Rawls, *Political Liberalism*, 58.

is not an epistemological idea (though it has epistemological elements)." (*PL* 62). It is rather part of "a political ideal of democratic citizenship" (*PL* 62).

Finally, let us consider Farrell and Tham's view that "[w]e deem a belief reasonable not only because of its content but also because of its grounds." Their point is that a comprehensive doctrine supplies the ground for our holding a certain view, and that self-standing public reasons—precisely because they are self-standing and not backed by any doctrine—are therefore not good enough. Should we agree with them? The answer is no. The reason is that if we hold proposition *p* regardless of whatever reasonable comprehensive doctrine we might hold, *p* is more certain. Consider the biblical tale of the Good Samaritan. You happen to encounter someone in dire need (e.g., he is seriously injured), and you know you can help him at the cost merely of some inconvenience to yourself. Should you help him? On any plausible view—viz. whether you are a utilitarian, a Kantian deontologist, a Scanlonian contractualist, a Confucian, a Natural Lawyer, or a Marxist—you should help him. Is the proposition that you should help this person in dire need any less valid or any less credible because this is what any plausible or reasonable moral view would prescribe? Clearly, the answer is no. On the contrary, if you should help him on any plausible moral view, this conclusion is more credible.

A General Objection to NLT

According to Farrell and Tham, NLT disagrees with Rawlsian political liberalism. First, NLT holds that it is possible to tell whether an action is in accordance with the natural law because we can discern whether the action clashes with the objective order of human goods. It is, on this view, also possible to tell whether a (positive) law should be enacted to prescribe actions that protect the common good and political society, and to proscribe those which undermine them. In other words, NLT holds that there is but one comprehensive doctrine with its "conception of the good to be recognized by all citizens who are fully reasonable and rational" (*PL* 134). Therefore, NLT maintains that we should investigate moral issues without bracketing what political liberalism holds to be "comprehensive doctrines." Instead, insofar as political liberalism favors the kind of "self-standing" common language of public reason, NLT rejects this idea. NLT holds that its own use of reason is public reason, albeit the NLT conception of public reason. (But it is not clear to me how its reason can justifiably be called *public* reason.) NLT maintains that practical reason can lead us to determine whether a kind of action is detrimental to the common good. Therefore, it claims that we can arrive at the right decision through rational deliberation.

Second, as Farrell and Tham claim, NLT holds that whether an act is morally right or wrong is to be judged in accordance with whether the act is compatible with the objective order of human goods. The problem with this view is that, according to some comprehensive doctrine, such as Catholicism, the ultimate good is eternal salvation. This means that the objective order of human goods depends on the truth or falsity of some religious comprehensive doctrines (such as Catholicism, Protestantism, Islam, Buddhism), which cannot be shown to be true or false by rational argument. For instance, according to the Catholic Church, *extra ecclesiam nulla salus* (there is no salvation outside the Church), but this is not a view that would be held by an atheist.[32] Whether there is salvation depends on whether Catholicism is true. If God does not exist, the doctrine of eternal salvation is presumably false. Aquinas may hold that we can discover through reason that there is a God, and that He is good, and so on. Yet, it would be difficult to persuade non-Catholics by rational argument that Catholicism is true. Conceivably, many (perhaps most) people come to believe in Christianity not through rational arguments, but because of revelation or personal experience.

Farrell and Tham's claim that according to NLT we could reach the same ethical conclusion about the moral rightness (or wrongness) of an act by means of rational assessment of it against the objective order of goods, and that we would reach the same conclusion as the Church teaches us, is doomed. For a Catholic assessment of an act may depend on the Catholic conception of the objective order of goods, which

[32] My view here is inspired by Rawls. As he anticipates, some religious people might argue that "outside the church there is no salvation, and therefore a constitutional regime cannot be accepted unless it is unavoidable.… When there is a plurality of reasonable doctrines, it is unreasonable or worse to want to use the sanctions of state power to correct, or to punish, those who disagree with us" (*PL* 138).

conception includes eternal salvation, and consequently the assessment cannot be carried out by rational argumentation alone.

In the history of moral philosophy, we know that there are various incompatible views about the goods and various normative ethical views. We are certain that no parties espousing any religious, moral, and philosophical doctrine could be shown to be mistaken in short compass, or at all. If history is anything to go by, no consensus on such doctrines will be achieved within our lifetime. Certainly, no resolution of social and political issues can be achieved in a timely fashion by resolving incompatible comprehensive doctrines (and their accompanying conceptions of the good). This is not to deny that there are any moral truths, however.

Missing Links

There seem to be two missing links in the argument presented by Farrell and Tham. One has to do with the assertation—by Farrell and Tham and also by MacIntyre—that for Aquinas fundamental precepts, such as "Do not lie" and "Do not kill innocent people," apply to all killings of innocent people and also all lies, regardless of the circumstances. Why should we accept this view held by Aquinas to be correct? Surely, Farrell and Tham need to show us the steps by which Aquinas comes to this conclusion (For my argument that this absolutist view is mistaken, see the section "moral contextualism" later in this chapter).

Another missing link is that even if, as held by Aquinas and Farrell and Tham, all killings of innocent people are wrong, how does this lead us to the conclusion that medical assistance in dying is wrong? First, let me point out that whereas euthanasia is killing, MAID is assisted death (where a patient who receives a lethal pill may not take it).

Second, even if killing an innocent person is wrong, why should we think that MAID is wrong? Let us remember that we are talking about a terminally ill patient, who is in unbearable pain, and has persistently and desperately wanted to die. When we say that it is wrong to kill an innocent person, despite the good consequences that might follow from it (e.g., imagine killing a tyrant's only child might prevent some bad consequences), we are talking about an innocent person who has no reason to die, does not want to die, and it is not in her interest to die. In other words, it may not be permissible to kill someone who does not want to die. But surely there are exceptions.

One exception is self-defense. In a usual case, a villain is zeroing in to kill his victim, who kills him first on grounds of self-defense. But it has been argued, persuasively to my mind, that the standard argument in self-defense can be extended, *mutatis mutandis*, to two other cases. The first case is an "innocent aggressor," someone who (although zeroing in to kill his victim) is innocent because he has been involuntarily intoxicated. Is it permissible for the victim to kill him first?[33] The second case is an "innocent threat," because the person on his trajectory to cause you to die is not even doing anything other than, say, falling onto you. He is innocent because he was blown off by an unusually strong gust of wind, or alternatively pushed from behind by a villain. He is merely a threat and not an aggressor because he is not an agent in his fall. Is it permissible for you to kill him?[34] This case is analogous to a fetus who is endangering its mother's life. Judith Thomson argues that it is permissible for you to kill an innocent aggressor or even an innocent threat. In an ectopic pregnancy (where a fetus is implanted in the fallopian tube), the Roman Catholic Church permits an abortion so that a section of the fallopian tube is removed with the result that the fetus will die. Here, the fetus which is endangering its mother's life is analogous to the innocent threat.[35] Whereas the Roman Catholic Church relies on the dubious

---

[33] Judith Thomson, "Self-Defense," *Philosophy and Public Affairs*, Vol. 20, No. 4 (Autumn, 1991), pp. 283–310. In Thomson's example, a villain has injected a drug into an innocent person by force. This innocent person, as a result, has become crazy temporarily. This "innocent aggressor" is trying to kill you by running his truck over you.

[34] Ibid. Suppose you are enjoying a sunbath on the beach, below a cliff. A fat man on top of the cliff is blown off by an unusually strong gust of wind, or alternatively pushed from behind by a villain. You realize that he will fall on you and crush you to death, unless you blow him up (thereby killing him) with an anti-tank gun. (Suppose you are locked into your sundeck and cannot move.) Although the fat man would kill you, if you do nothing, he does so only in the same sense that a falling rock will kill you. Neither the rock nor the fat man is an agent. Therefore, unlike the "innocent aggressor," he is only an "innocent *threat*." (I modify this example slightly, which is from Thomson.)

[35] Ibid.

doctrine of double effect,[36] Thomson bases her argument on self-defense (against the innocent threat).

Most important, when we think that killing an innocent person is wrong, we have in mind the scenario where killing an innocent person for the sake of either our self-interest or promoting the collective utility cannot be right. We assume in the background that she has a good life, or does not want to die, or it is not in her interest to die. However, we should come to a different conclusion, or *at least* suspend judgment, when we are facing a terminally ill patient, who is in unbearable pain and who desperately wants to die. In this different scenario, it can be in her interest to die. But we cannot kill her, or assist her to die, if she is not terminally ill, or if she does not want to die. Our personal autonomy should be respected. However, if the three aforementioned conditions are satisfied, then the precept that "killing an innocent person is wrong" should not apply.

Finally, contrary to Farrell and Tham's claim, no reasonable thinker would say that, if abortion or MAID becomes legal, a physician is obligated to perform the procedure to abort a fetus or to assist a patient to die. Only willing physicians are needed to help because no physician should be pressed into service against their religious or moral belief that abortion or MAID is wrong. If they prefer not to help for whatever reason, their reluctance must be respected.

The Hippocratic Oath

Another argument advanced by Farrell and Tham is that physicians have sworn the Hippocratic Oath which forbids them from deliberately killing any patient. To participate in MAID means that a physician will kill, or will be involved in the killing of, the patient, according to this argument. And this is disallowed.

Note that there are different versions of the Oath. One version might say that a physician must not kill his patient, whereas another might say

[36] For a critique of the doctrine of double effect, see part (B) of Chap. 1 of this volume, where I articulate and defend the Thomson/Scanlon thesis that permissibility is independent of the agent's intention.

that a doctor must not harm or act against the interest of his patient. If it could be in the interest of a patient (appropriately qualified) to die, then MAID does not harm her. Consequently, a physician who participates in his patient's MAID is not acting against the Oath. Moreover, in some cases (such as the British case of Diane Pretty, who was diagnosed with motor neurone disease) the person who could assist in a patient's suicide is the spouse who, not being a physician, has not sworn the Oath and is not bound by it.

The main problem with the position maintained by Farrell and Tham is, however, that it puts the cart before the horse. Consider the following two possibilities: (1) Should we follow a set of rules and be bound by them because the rules are good rules? Or (2) should we keep the rules because we have promised that we would follow them? I believe (1) is correct, but they hold that (2) is right. There is a genuine difference between (1) and (2). For if one day we discover that the rules are not perfect, and that there exist previously unforeseeable circumstances which the rules should not cover, we should change the rules. This is compatible with (1). However, (2) implies that we must abide by the rules even if there are previously unforeseeable cases which the rules, if we had been able to foresee and had been asked, should not cover. I submit that (2) is implausible.

Consider this scenario. You have always been a most honest person because that was what you learned at school, church, and home. One day, an angry villain comes looking for Mary, your neighbor. He is holding a rifle and asks you where she is. You realize that if you tell him what you know, Mary will likely be killed. Should you tell the villain what you know? Or is it permissible for you not to tell him the truth? Clearly, it is permissible, perhaps even obligatory, for you to not tell the villain the truth, because when we learned about the importance of truth-telling, our teachers or parents did not have this sort of case in mind.

In law, there are "implied terms" in a contract. If you purchase some fruits from a grocery and they turn out to be rotten, can you return them? Yes, of course. This is because had you asked the grocery-keeper whether you could return the fruits if they turned out to be not edible, he would have said "Of course." It is an implied term that the fruits are edible (or merchantable).

Similarly, suppose John promised to take Jane out to dinner on her thirtieth birthday on July 30, 1914, in Sarajevo. Two days before Jane's birthday, however, WWI broke out. Was John still obligated to take Jane out on her thirtieth birthday? The answer must be no because the obligation to fulfill a promise is not absolute. Another way of saying the same thing is that there is an implied term, or a *force majeure* clause, that releases John from his obligation.

2 Reply to Fan

The main steps of Ruiping Fan's argument for his Confucian conception of public reason can be summarized as follows:

1. Whether a normative view or a social practice is right depends on its ethical tradition—that is, past and current moral beliefs.
2. The dominant moral tradition in various East Asian countries or regions (such as China, Japan, South Korea, Taiwan, Hong Kong, and Singapore), notwithstanding some pluralist values existing therein, has been Confucianism.
3. Therefore, various East Asian countries should hold and practice Confucianism.
4. Rawls' liberal conception of public reason is unsuitable to China because (a) China has always been a Confucian country, (b) religious toleration—for which political liberalism was devised—has always existed in China, and (c) individualist values—which Rawls' theory supports—are not suitable for China.
5. Chan's Confucian theory of public reason is problematic because it admits the liberal notion of "civility."
6. Kim's Confucian theory is imperfect because he has smuggled the Western "democratic value of public equality" into his theory.
7. A Confucian theory of public reason must contain Confucian virtues, such as humanity (*ren*), appropriateness (*yi*), propriety (*li*), filial piety (*xiao*), and harmony (*he*).
8. It should also contain a list of basic liberties and rights (namely, the rights to life, to liberty, and to formal equality)—which are universal

and therefore not liberal or Western (because these rights can be derived from Confucian virtues of humanity and appropriateness).

9. Confucian rights are virtue-based, whereas liberal rights are grounded in the idea of individual autonomy.
10. An elderly person should have a right (based on the virtue of filial piety) to care from her adult children.
11. Reasons can be given as to why MAID (or MAS, medically assisted suicide) should not be legalized, including that:

    (a) A legal right to MAID would "contradict the central requirement of basic Confucian virtues that an innocent human life must be honored and respected" because MAID "cannot consistently be understood as manifesting the honoring or loving of an innocent human life, because this is … to intentionally allow, affirm and even will the providing of medical professional assistance with patients to destroy their lives while medicine is, and ought to be, set up in such society for saving their lives and relieving their suffering."
    (b) Often, a request for euthanasia or MAID is "a call for help."
    (c) When an ill patient is suffering in the hospital, filial piety dictates that his adult children not obey his request for MAID, but instead seek better medical and nonmedical means to take care of him.
    (d) Although further medical treatment, which only prolongs the patient's suffering, can morally be withheld or withdrawn, MAID is disallowed.
    (e) In a Confucian society, parents are selfless and aim to lessen the burden of their children. If MAID were legalized, patients (who are kind to their children) would choose MAID as something they should accept.

In Fan's argument, three things call for comment. First, he says that because China has been practicing Confucianism for centuries, Chinese should continue to be Confucians. Second, various options (e.g., theories of Rawls, Chan, and Kim) are rejected because they are, or are mixed with, Western or liberal theory. Third, filial piety is a special virtue in East Asia, and a filial adult child should not allow his/her parent to opt for MAID under any circumstances.

I shall comment on these in turn, and devote more space on the first point, which is grounded in moral relativism.

Moral Relativism

Fan assumes that a country, nation, or culture that practices Confucianism should continue to practice it. He then asserts that, as a matter of fact, citizens of East Asian nations practice Confucianism.[37] Therefore, these nations should continue to practice Confucianism. The assumption that a country that has practiced a certain moral code should continue to practice it is an idea of moral relativism. Although Fan accepts a certain minimal list of rights and liberties, which he believes are universal and nonliberal, his argument that China (and other East Asian countries) should practice Confucianism because they have always been Confucian countries invokes the idea of moral relativism. It is therefore fitting to address this idea here.

Moral relativism is the view that what is morally right or wrong can be evaluated with respect to the *beliefs* of a social group. It is often invoked to explain variations of moral codes between different cultures: for instance, compared with Westerners, East Asians (appear to) value filial piety more. As a sociological phenomenon, variations in such moral outlooks are often benign: different cultures just do things in different ways. Yet, how do we justify such variations of moral outlook among different cultures, if these outlooks are in disagreement?

Perhaps one natural explanation is moral relativism. According to this view, if people in a social group *think* that doing X is right, then it *is* right. It follows that two cultures could disagree and yet both be right.[38] When

[37] It is an open question whether, or to what extent, people in Hong Kong (for instance) subscribe to, or practice, Confucianism. I leave aside Fan's sociological thesis that East Asians generally practice Confucianism in this Chapter. I shall take this up again in Chap. 7.

[38] Gilbert Harman is perhaps the most well-known contemporary proponent of moral relativism. See his "Moral Relativism Defended," *Philosophical Review* 84, no. 1 (1975): 3–22.

a social group is a culture, moral relativism says that what is morally right or wrong depends on the *beliefs* in that culture. Thus, on this view, it is right for contemporary East Asians to value filial piety highly, and also right for contemporary Westerners not to value filial piety so highly.

The problem with moral relativism is, I think, that it is a false view. Moral relativism entails that we cannot criticize any practice in any culture other than our own. But this is absurd, because this would entail that just because people of another culture hold certain different ethical beliefs, they are therefore justified to do anything that we consider morally horrendous and are immune to any criticism from without (as long as they consider it right). The most obvious examples are societies where racism or sexism is widely believed and practiced, not to mention some societies that have allowed slavery. The most extreme example is, of course, Nazi Germany, where many German people participated or acquiesced in the persecution of Jews. This persecution, which began in 1933, culminated in the Holocaust (1939–1945).[39] Not only should we protest against such evil acts in the strongest terms, but we should also resent racist and sexist attitudes even if and when they do not cause the taking of any lives.

If moral relativism has unacceptable consequences, we need to ask: how can we reconcile the fact that there are benign moral variations across cultures, and the fact that some practices (such as genocide) are totally unacceptable even if those engaged in the practice believe that it is permissible? How should cultural differences bear on our ethics or morality?

A Distinction: Critical Morality and Positive Morality

To tackle this issue, it is important to distinguish two quite different senses of the term "morality"—namely "positive morality" and "critical morality."[40] Positive morality is the morality actually accepted and shared

[39] This was a systematic process of ethnic cleansing carried out on Hitler's order, in which the Nazis murdered around six million Jews and would have wiped out the entire Jewish population of Europe were it not for Germany's defeat in WWII.

[40] H. L. A. Hart, *Law, Liberty and Morality*, 20.

by a given social group.[41] Thus, it is part of East Asians' positive morality that children should be filial to parents, whereas part of Nazi Germans' positive morality was that it is permissible to kill and dispose of people who (they thought) were of an inferior race.[42]

Another sense of the term "morality" is critical morality, which is the objectively right morality that can be backed up with reasons, principles, and arguments.[43] When practitioners of practical ethics and bioethics write their papers for journals, they are typically theorizing in critical morality, and seldom regard their own works as merely as part of positive morality. An assumption that ethicists and bioethicists make, even if not consciously, is that there is objectivity in ethics, and that they are striving to establish an objectively true moral view, even if that is an ideal which might never be attained.[44]

To establish what is positive ethics and positive bioethics, all we need to do is to carry out a (statistical) survey of people's beliefs and attitudes. The positive morality of Nazi Germany was that it was permissible to kill people who they believed were of an inferior race. The conclusion we can draw is that *bare* positive morality—or positive morality *per se*—does not have any justificatory force. In other words, simply because a particular social group believe that certain practices are right (or wrong), it does not follow that such a practice is indeed right (or wrong).

When ethicists and bioethicists argue that a practice is impermissible (in the critical sense), they provide reasons as to why this practice is wrong. Thus, when we argue that ethnic annihilation is extremely evil,

---

[41] Ibid.

[42] The Nazis believed that the Jewish race was inferior, even though this belief was false and unfounded.

[43] In 1963, H. L. A. Hart defined "critical morality" as consisting in "the general moral principles used in the criticism of actual social institutions including positive morality" (*Law, Liberty and Morality*, 20). With the benefit of meta-ethics since 1963, we should accept a more robust conception of critical morality which includes not only moral principles but also moral reasons and arguments.

[44] Two Point. First, I do not believe that a moral principle must be universal in the sense that it applies to everyone regardless of their circumstances. See the discussion on moral contextualism in Reply to Farrell and Tham, and also below. Second, the idea that applied ethicists presuppose moral objectivity is inspired by Thomas Nagel, who holds that our moral experience reflects that there are objective reasons. See his *The Last Word* (Oxford: Oxford University Press, 1997), Chap. 6, *The View from Nowhere* (New York: Oxford University Press, 1986), Chap. 8, and *The Possibility of Altruism* (Oxford: Clarendon Press, 1970).

we put forward moral reasons, principles, and arguments to show that this is so, even if most of those complicit in an act of genocide (e.g., in Nazi Germany) believed that the practice was permissible. This shows that what most people believe about a practice *by itself* does not translate into any reason for us to accept it, or to condemn it.[45]

Critical (Bio-)ethics

Part of our conception of critical ethics is that moral reasons can be more or less objectively right or wrong. Of course, the way in which moral reasons can be objective is different from the way in which science is objective. And mathematical objectivity differs from scientific objectivity and moral objectivity.[46]

A moral view is judged plausible or implausible not only by reference to how plausible its premises and conclusion are but also by examining whether there are any unacceptable consequences flowing from this view. This is especially true in the case of utilitarianism or, for that matter, moral relativism.[47]

---

[45] Obviously, it does not mean that by offering *any* (critical) reason we will have arrived at the correct conclusion. This is because the reason we have provided may well turn out to be mistaken, either because we may have been mistaken about the facts, or about our moral reason. Even if our reasons and conclusion seem plausible, they may eventually turn out to be mistaken, perhaps because the moral theory on which we have been relying might turn out to be false.

[46] This is a huge topic. For a particularly illuminating account, see T. M. Scanlon, *Being Realistic about Reasons*, 2013, esp. Chap. 2.

[47] According to the utilitarian theory, we should do what maximizes utility. This moral theory is a plausible starter. Yet on a closer examination, this theory seems mistaken. Utilitarianism is not mistaken because it emphasizes the importance of consequences—since any plausible theory must take account of consequences. It is mistaken because it singles out consequences as the *only* relevant consideration. A counter-example to utilitarianism is this. If we had to choose between saving a life, or curing many people's minor headaches, but not both, we should save the life, even if the people suffering the headaches are sufficiently numerous that curing them would maximize utility. On this point, see T. M. Scanlon, *What We Owe to Each Other* (Cambridge, MA: Harvard University Press, 1998), Chap. 5, Sect. 9, 229–241.

Cultural Differences

Although positive morality *per se* does not provide any moral reason or justification as to whether, or why, a practice is morally permissible or not, it does not follow that people of different cultures cannot disagree with each other and yet all be justified or right. This is because the circumstances in which they are situated may be different. For instance, it might be entirely justifiable for East Asians to stick to certain tenets of filial piety, whereas it is entirely permissible for Westerners to not follow some of these tenets.

The strictest tenets of filial piety were put forward in Ancient China, and they covered various aspects of life. But I shall focus on merely five of the most important tenets. Filial piety consisted particularly in:

(1) Showing love and respect toward one's parents
(2) Supporting them financially in their old age
(3) Caring for them whenever they need help, and especially in their old age
(4) Producing a male heir
(5) Mourning parents for three years after their deaths[48]

Nowadays, East Asians do not adhere to (5). Most of them—certainly those who live in cities—no longer accept (4), as evidenced by the fact that most people are just as happy to have a daughter instead of a son.[49] In fact, many contemporary East Asian married couples opt not to have any children at all.

Moreover, East Asians today subscribe to a modified version of (5), namely:

[48] Children should stay in a hut next to their parent's tomb for three years. See: https://quizlet.com/77808650/china-chapter-14-section-3-flash-cards/.

[49] As I recall, there was a survey (reported in *Sing Tao Evening News* more than 24 years ago) that Hongkongers who aim to have one child have a slight preference for a daughter to a son. (*Sing Tao Evening News* closed in 1996.).

(5a) A filial person should show concern for sick parents and sorrow at their deaths.[50]

So East Asians today accept (1), (2), (3), and (5a). However, note that Westerners also accept (1),[51] (3), and (5a) as they also would show concern if their parents are sick and sorrow at their deaths. Insofar as support includes financial support, East Asians seem more ready than Westerners to support their parents.

To recap, our observations are: first, Ancient Chinese held (1), (2), (3), (4), and (5). Second, contemporary East Asians hold (1), (2), (3), and (5a). Finally, Westerners subscribe to (1), (3), and (5a). Why does this not show that moral relativism is correct, or that positive morality *per se* has justificatory force, or both?

Moral Contextualism

How can we account for our intuition that the beliefs and practices of Ancient Chinese, contemporary East Asians, and contemporary Westerners with regard to filial piety are all justifiable, even if their positive moralities (regarding filial piety) are different (and conflicting)?

Recall that moral contextualism is the highly plausible view that what is right or wrong depends on the circumstances—or the context. For instance, we believe that in general lying is wrong. Do we believe that lying is wrong in all circumstances? Those who believe this are moral absolutists.[52] Their view is opposed to moral contextualism.

The problem with moral absolutism is that it is a false theory. If it were a correct theory, then we would have to accept absurd consequences. Suppose a Nazi German soldier is searching for innocent Jewish people to take away. Without success, he comes to you and asks: "Have you seen

---

[50] See "Filial Piety," *Wikipedia the Free Encyclopedia*: https://en.wikipedia.org/wiki/Filial_piety#Adaptation.

[51] Western philosophers have written many articles on filial piety. These articles have appeared in various learned journals, including *Journal of Philosophy, Australasian Journal of Philosophy, Philosophical Quarterly, Journal of Applied Philosophy*, and *Southern Journal of Philosophy*.

[52] Recall that moral absolutism is the view that if an act of Type A is wrong, it is wrong regardless of the circumstances.

any Jews around here?" You know his purpose, and you believe that you can save the Jews without incurring any danger on yourself by lying: "No, I have not seen any Jews around here." If moral absolutism were correct, you should not lie even in this case. According to moral absolutism, you must tell the soldier the truth regardless of the circumstances. But this is very hard to believe. For we believe that you should lie to the Nazi German solider under such circumstances. Such a belief can be accounted for by moral contextualism.[53]

Moral Contexts and Cultural Differences

I have argued that positive morality *per se* has no justificatory force. But do cultural differences matter to ethics, or bioethics? Before answering this question, we need to ask the following: what exactly is a culture? And what lies behind cultural differences?

I take it that a culture consists of certain traditions and customs, and also a set of common beliefs and practices, among other things. Some of these beliefs might be religious or secular, but they likely involve some positive morality. So, if two different cultures have different positive moralities, such differences *by themselves* cannot justify their different stances toward filial piety (because positive morality *per se* has no justificatory force).

But it is important to note that *different circumstances or contexts* might underlie different positive moralities held by two cultures. These differences might include the fact that the two cultures are at different stages of technological development, or have different types of economic structure (e.g., feudalism and capitalism), for instance. Thus, we can explain why Ancient Chinese subscribed to the strictest tenets of filial piety, whereas contemporary East Asians endorse a more relaxed version. For one thing, contemporary East Asians no longer hold the kind of sexism characteristic of ancient agrarian, feudal modes of production, where a man's labor power represented a greater productive force than a woman's

[53] In 1964, when Hong Kong was facing the worst drought in its recorded history, water was supplied for only 4 hours every 4 days. In such circumstances, watering the lawn was certainly wrong (when there was not enough water given for human consumption). Once the drought was over, watering the lawn again became both permissible and even right.

because what counts in a feudal agrarian society (where serfs work in the field) is physical strength.

For another, contemporary Westerners do not hold that children have to support parents financially because most Western societies have developed social security or welfare system. Such institutions do exist in Mainland China[54] and in Hong Kong, but they are considered to be "inadequate," "incomplete," and "limited."[55] (To my mind, it is an exaggeration to say that *only* East Asians value filial piety.)

If I am correct, such an explanation enables us to understand—and also to justify in this case (at least)—as to why contemporary East Asians and contemporary Westerners hold slightly different views regarding filial piety. But it does not follow that whenever there is a social *explanation* of an event (e.g., the Holocaust), then this event is justified. For instance, after WWI Germany was faced with humiliation and challenges. Even though these circumstances might (partially) *explain* why the Holocaust occurred, they do not *justify* it. Whether a social group is justified to uphold its culture, tradition, and positive morality ultimately depends on critical morality, that is, whether there are sufficient moral reasons to justify them.

To summarize, I have argued the following. First, moral relativism is false. Second, positive morality *per se* does not justify anything. Third, even though positive morality *per se* does not justify anything, the fact that different cultures hold different *positive* moral outlooks might reflect the underlying fact that people in these cultures were facing different circumstances (or contexts). Fourth, different circumstances might justify different moral outlooks.[56] Fifth, different circumstances do not justify just about anything.[57] In particular, different circumstances might—but do not necessarily—justify different moral beliefs and practices. Whether this is so eventually depends on whether the different circumstances can provide any reasons for a different moral outlook.

---

[54] I thank Kai-Yee Wong for pointing this out to me.

[55] See Li Jiange et al., "Social Security Reform in China: Issues and Options," January 27, 2005 (https://economics.mit.edu/files/691).

[56] That is why East Asians and Westerners have somewhat different views concerning their financial obligations toward their parents.

[57] Thus, Germany's humiliating defeat in WWI and its severe economic difficulties since then did not justify the Nazis' racist ideology and their criminal and evil behavior in carrying out the genocide during the Holocaust.

Liberalism or Western Theory

Fan holds that a Confucian theory should not include anything liberal or Western. He has not explained why Confucianism should not be contaminated with liberalism or Western theory. *But we can guess.*

Before we make the guess, let me say that one would have thought that the important thing about a theory is whether it is right, not whether its pedigree is right. For one thing, the relative equality of the genders (which Fan acknowledges to be a good thing) in contemporary China has been due to Communist rule, not Confucianism. If anything, Confucianism stresses the supremacy of men, not the equality of the genders.[58] In traditional Confucianism, the following are truistic: "Filial piety is the most important of all virtues" (百行孝為先). "There are three ways to be unfilial; having no sons is the worst" (不孝有三, 無後為大). In a well-known passage, Confucius says: "Only women and commoners are difficult to get along with" (唯女子與小人難養也).[59] In traditional or Classical Confucianism, a woman would follow "her father before her marriage, her husband after her marriage, and her son during her widowhood" (在家從父, 出嫁從夫, 夫死從子).[60] It was widely regarded that a virtuous woman is one without talent (女子無才便是德; an alternative translation: a woman should stay at home). In the Song Dynasty, it was widely believed that "to starve to death is a small matter, but to lose one's chastity is a big matter" (餓死事小, 失節事大).[61] Fung Yu-lan writes of Dong Zhongshu's (or Tung Chung-Shu's) Confucian view, which has been the ruling ideology in Han Dynasty (202 BC–220 AD) and subsequent dynasties: "Tung selects three [major human relationships] and call them three *kang*. The literal meaning of *kang* is a major

[58] Furthermore, prior to 1994 in Hong Kong, women were not permitted by law (which followed Chinese customary law of Qing Dynasty) to inherit properties in certain villages of the New Territories. See K. Chan, "Women's Property Rights in a Chinese Lineage Village," *Modern China*, Vol. 39, No. 1 (January 2013), pp. 101–128. This tradition or positive morality had existed for hundreds of years.

[59] Confucius, *The Analects*. An alternative translation is: "Women and Commoners are difficult to cultivate and educate." But what Confucius says immediately after this sentence makes this translation unlikely.

[60] See: https://en.wikipedia.org/wiki/Confucianism#Women_in_Confucian_thought.

[61] Patricia Buckley Ebrey, *Women and the Family in Chinese History* (Routledge, 2002), 10–12.

cord in a net, to which all the other strings are attached. Thus the sovereign is the *kang* of his subjects, that is, he is their master. Likewise the husband is the *kang* of the wife, and the father the *kang* of the son."[62] Therefore, the kind of Confucianism that allows for relative equality of the genders is not the traditional form of Confucianism, but rather a version that has been influenced by Marxian and/or Western thoughts. If so, why is it important to have a form of Confucianism that contains no element of liberalism? (I suppose that by "Western theory," which Fan thinks is bad, he means liberalism and not Marxism.)

The most likely explanation as to why Fan wants to have a version of Confucianism that is purged of any remnants of liberalism is that, for him, liberalism stresses individual autonomy and is hence conceptually inseparable from individualism. And for him, individualism is morally bad. Therefore, liberalism is also bad.

For what reason does Fan think that liberalism is inseparable from individualism, and why is individualism bad? I think there are two reasons. First, Fan believes that a liberal would approve of illicit practices, such as prostitution and gambling, which (according to him) a Confucian society may also tolerate. Second, Fan says that consensual parent–child incest (where the child is an adult) is morally evil, even if it does not do any damage to the psyche. Fan seems to think that prostitution, gambling, and consensual incest are bad or evil, and yet liberalism seems to approve of them because liberalism maintains that there should be a right to gamble and to visit a prostitute. J. S. Mill argues in favor of a *legal* decriminalization of—but (*contra* Fan) not a moral right to—prostitution and gambling in much the same way that, according to Fan, a Confucian society tolerates these practices. According to Mill, that gambling and prostitution should not be criminalized does not mean that the criminal law should allow a pimp or a gambling house to advertise their trades because gambling and prostitution are borderline cases. This being so, Mill's attitude toward gambling and prostitution does not seem to be very different from its Confucian counterpart. (See J. S. Mill, *On Liberty*, Chap. 5.)

[62] Fung Yu-Lan, *A Short History of Chinese Philosophy* (New York: Macmillan, 1948), 197. See also Fung Yu-Lan, *History of Chinese Philosophy* (Princeton: Princeton University Press, 1953), Vol. II, Chap. 2.

Furthermore, the fact that X is not criminalized does not mean that it is virtuous or moral to have X. A liberal who believes that we should learn from Aristotle and Kant (as regards virtues and other things) would hold that the conditions under which gambling and perhaps prostitution be allowed by law should be heavily circumscribed. A liberal would say that children should have the moral and legal rights to food and education, and that citizens should have *basic* liberties namely, "political liberty (the right to vote and to be eligible for public office) together with freedom of speech and assembly; liberty of conscience and freedom of thought; freedom of the person along with the right to hold (personal) property; and freedom from arbitrary arrest and seizure as defined by the concept of the rule of law" (*Theory* 61). Rawls does not recognize a *basic* right to acquire means of production, or to accumulate money as much as possible, or a right to earn money without paying tax. It is, moreover, doubtful that liberals would recognize a *moral* right to gambling or prostitution.

Has any liberal argued in favor of consensual incest? I do not know of any. But perhaps Fan's reasoning is that a liberal accepts Mill's harm principle, and that the harm principle has the unpalatable consequence that a voluntary act between consenting adults should be lawful. In other words, the example of incest between consenting adult siblings—let alone parent–child incest—serves as a critique of liberalism because incest of any form is objectionable.

Fan's challenge forces a liberal to face a dilemma. Either a liberal would say that incestual practice should be decriminalized, or not. On the one hand, it seems counter-intuitive to say that incestual practice should be decriminalized. On the other hand, how can a liberal plausibly hold that such a practice should be criminalized? Peter Railton has argued that consenting adult siblings (having used double contraceptives) have harmed their own psyche by engaging even in a one-off act of incest.[63] If Railton's argument is accepted, a twentieth-century liberal—one who is in favor of some degree of paternalism (such as H. L. A. Hart)—would be against decriminalizing consensual incest.[64]

[63] Peter Railton, "The Affective Dog and Its Rational Tale: Intuition and Attunement," *Ethics* 124 (July 2014), 813–859.

[64] A twentieth-century liberal (such as H. L. A. Hart) is different from a nineteenth-century liberal (such as J. S. Mill) in that the former (i.e., Hart), but not the latter, accepts paternalism. See H. L. A. Hart, *Law, Liberty and Morality*, Stanford University Press, 1963.

My own view is that the wrongfulness of incest must be suitably linked to the risk of genetic deformity to the offspring of incest. It is true that if one of the consenting adult siblings is sterile, incest between them would not cause genetic deformity because no conception would follow. Yet it does not follow that the law should allow this kind of "safe" incestual practice because after a community has *internalized* certain emotions (e.g., indignation) against incestual conduct, it would be difficult for the community to revise its internalized emotions toward "safe" incestual practices. To explain this point with an analogy: suppose someone has recently espoused vegetarianism on moral grounds (as she subscribes to the view that killing animals, in particular nonhuman mammals, for food is morally wrong.) Yet she would still find meat tempting. Is it a good idea for her to look for roadkill to gratify this taste? I think the answer is no and, I hope, the reason is not difficult to see. A vegetarian practice is not only an idea in thought but requires discipline, habituation, and internalization. Eating roadkill might do violence to our vegetarian discipline and internalization and is therefore not a good idea. A similar case can be made against "safe" incestual practice. Furthermore, there are other worries. For one thing, if this kind of "safe" incestual practice is allowed, siblings might manipulate each other, not to mention father–daughter incest. For another, there is a big difference between homosexual practice and "safe" incestual practice. It is believed that about 5% of all men have inborn homosexual inclination. However, to the best of my knowledge, no one has claimed that there is inborn incestual inclination. I realize that my view here needs elaboration and defense. But that will have to be the subject for another occasion.

Another reason why Fan argues against liberalism and individualism is that, whereas a liberal person would act with only her own self-interest in mind, a Confucian would consider his family as a whole. The idea that a Confucian should act on the consensus of the family does not strike me as a traditional Confucian view. (I shall return to this below.) But I should entertain the question whether a liberal is too concerned with her own individual self-interest in an objectionable way.

Suppose that Xiao Bing, a lone child of a family in China, has done extremely well at the public examination prior to university. Any department of the best universities would admit him. His family would like him to study medicine because that is a respectable profession and pays well.

However, he himself would like to study philosophy because he thinks that philosophical issues are deep, profound, and extremely important and also because he enjoys thinking about them. On Fan's view, the whole family should decide whether Xiao Bing should study which subject. (In traditional Confucianism, the father would decide whether Xiao Bing will study which subject, or even whether he should go to university at all.) Should Xiao Bing's father think that it is best for Xiao Bing to study medicine, then in a traditional Confucian family that is the end of the matter.

The film *Dead Poets Society* explores the issues of parental authoritarianism in its story of the tribulations of a white boy in contemporary America. He was doing well in Grade 12 in a private preparatory school before heading to university. His father wants him to give up everything in order to study and get admitted to the best university and eventually become a medical doctor. His father tells him that after becoming a doctor, he can do what he likes. But his father's authoritarian rule was so oppressive and suffocating that the son commits suicide in protest.

Nowadays we would think that a child's wish or desire to major in her preferred subject should not only be taken into account, but her strong interest should be the first factor to consider. This is so for the following reasons. First, if she is interested in a subject, she will tend to do well—at least compared with those subjects in which she is not interested. Second, if she is interested in a subject, then (everything else equal) she will *enjoy* studying it. Such enjoyment *per se* is a good reason to let her decide what subject to major in. Third, the subject in which she wants to major will have an impact on her life. If she wants to become a musician instead of a physician, it is her life and she should have a big say as to what she can major in. After all, she will have no one else to blame should it turn out that she subsequently dislikes her major.

These reasons are in favor of personal autonomy that one should have a say in what one would become, for instance. Although personal autonomy should be respected and a factor to be taken into account in family decisions or social policy, I do not think that it has overriding or even any importance in every issue. Suppose someone wants to try out heroin or cocaine; it does not mean that his personal autonomy should be accorded any weight at all. On the other hand, a terminally ill patient suffering unbearable pain is going through the final lap of her life that might not be worth living, her autonomy to want to die is necessary for MAID to be allowed. If she does *not* want to die, no one should mention MAID to her, let alone force her to die. On the other hand, if a 17-year old wants to receive MAID because of unrequited love, his autonomy should be accorded no weight at all because his unhappiness should be deemed temporary.

Is Rawls' Theory Individualistic?

It is sometimes said that Rawls' liberalism is an individualist theory because he views a society as "a cooperative venture for mutual advantage," the basic rules of which "specify a system of cooperation designed to advance the good of those taking part in it."[65] This might suggest that Rawls' view is "individualistic" in an objectionable sense, insofar as it involves seeing social institutions simply as instrumental to individuals' private interests.[66] As T. M. Scanlon points out, this is not an accurate description of Rawls' view:

> When Rawls says that social institutions are to be justified on the ground that they are a fair system for advancing "the good" of the individuals taking part in them, the relevant idea of an individual's "good" is a broad one, including all of the aims that an individual has reason to want to promote, whether or not these aims are "self-interested" ones that involve private benefits. An individual's aims may include, for example, realizing some ideal of social life, or living up to the tenets of his or her religion and promoting its wider acceptance.[67]

So understood, Rawls' theory cannot be viewed as individualist in any objectionable sense. (I shall have more to say about whether Rawls' theory is individualist in Chap. 7.)

---

[65] *A Theory of Justice*, rev. ed., p. 4.

[66] See T. M. Scanlon, "Some Main Points in Rawls' Theory of Justice." https://www.academia.edu/44023083/Some_Main_Points_in_Rawls_Theory_of_Justice_1.

[67] Ibid.

One of Fan's main objections to Rawls as an individualist philosopher has to do with Rawls' illustration of "balancing of political values" in the case of abortion. Rawls says that "as an illustration" of "a reasonable balance of political values" in the issue of abortion, we should consider three important political values: "the due respect for human life, the ordered reproduction of political society over time, including the family in some form, and finally the equality of women as equal citizens."[68] Rawls believes that "any reasonable balance of these three values will give a woman a duly qualified right to decide whether or not to end her pregnancy during the first trimester."[69]

Fan objects to Rawls as follows:

> It is evident, from my view, that both the three values as he formulates as well as the way of balancing them as he conducts are characteristic of an individualistic moral feature that is embedded (and perhaps shared) in comprehensive Western doctrines, either religious or secular, that generally assume and accentuate individual independence, authority or autonomy rather than communitarian or relationalistic propriety as emphasized in certain Eastern doctrines (such as Confucianism) in making a decision such as abortion. While it emphasizes the equality of women in the issue, it overlooks the value of a united father-mother relation, the importance of the role of the father in the decision of abortion as well as the enormous concern of the mother's parents as implied in certain Eastern non-individualistic moral views, such as [the] Confucian view. (Chap. 3 of this volume)

Fan does not object to the equality of the genders. Rather, according to Fan, Rawls should not have allowed the woman the sole right to decide whether to abort during the first trimester. What is preferable, Fan says, is "a united father-mother relation, the importance of the role of the father in the decision of abortion as well as the enormous concern of the mother's parents as implied in certain Eastern non-individualistic moral views, such as Confucian view." I have no idea why the woman's parents should weigh in on the abortion decision (according to Confucianism or

[68] Rawls, *Political Liberalism*, 243, note 32.
[69] Ibid.

not). It might be that the woman's mother might need to help out in caring for the children if the woman needs to work months after her maternity leave. If this is the objection, however, it is purely practical and can hardly be said to be Confucian. In traditional Confucian families, which were found in only patriarchal societies, the father has a say—and indeed the overriding say—on practically any issue of importance within a family. Because of this uneven balance of power, a joint father–mother decision on abortion will translate into the father's decision. After all, in traditional Confucian societies, girls usually did not receive any education and hence were illiterate; only the boys had any real chance of attending school (depending, of course, on the social class to which their families belonged).

Fan seems to accept the equality of women as equal citizens. In Chap. 3, he says or implies that we should respect human lives.[70] It is not clear whether he accepts "the ordered reproduction of political society over time, including the family in some form" since he has not expressed opinion on this issue. His main objection appears to be that Rawls allows the woman to have the sole decision on whether to have an abortion during the first trimester, whereas Fan claims that the father also should have a say. I have already commented on this part of his argument.

As I pointed out in Chap.1, it is not clear how Rawls arrives at the conclusion that "any reasonable balance of these three values will give a woman a duly qualified right to decide whether or not to end her pregnancy during the first trimester." Rawls reiterated that he had not presented an "argument," but only an "illustration." [71] Notwithstanding this, it is not clear—even as an illustration—how Rawls arrives at his conclusion. One would think that various factors are also relevant, such as whether a fetus is a human being, at what point a fetus acquires moral status, whether the fetus is the result of a rape, whether the couple have used contraceptives, and whether the fetus is endangering the woman's life. Admittedly, some of these (such as when a fetus becomes a human

[70] Fan says in Chap. 3: "The first is that human lives are most noble (*gui*) or valuable among all types of lives in the world so that innocent human lives must be respected and protected."

[71] John Rawls, Political Liberalism, 243, n. 32. Rawls emphasizes that he only intended his point as an "illustration," and not as an "argument." See his "The Idea of Public Reason Revisited," in *The Law of Peoples*, 169 n. 80.

being or when it acquires moral status) might involve controversial metaphysical claims. I myself have argued elsewhere that the central moral problem of abortion (except in exceptional cases, such as when the mother's life is endangered) is irresolvable,[72] and I shall not contradict myself here by saying that this issue can be readily resolved. Notwithstanding Rawls' illustration, we should accept his general view that politically divisive issues bordering on constitutional essentials (such as abortion) should be adequately discoursed in terms of public reasons in the public forum in a community, where the majority view in an elected legislature should be enacted as legitimate law.

Let us return to Fan's view that in allowing the woman the sole right to decide whether abortion should be performed—instead of allowing also the woman's husband and her parents to have a say—Rawls is an individualist because the woman is an individual, whereas the woman's husband and her parents constitute a family, and that is not individualist. I find this reasoning problematic because whether to abort or not will concern the woman the most and therefore, if and when abortion is morally permissible, she should have the right to decide whether to abort. Allowing her husband and her parents to weigh in would only undermine her say as well as complicate the matter in an objectionable way. (I am not, of course, denying that they can express their opinions.) Similar things, *mutatis mutandis*, can be said about the case of Xiao Bing.

Should a Confucian Object to MAID?

Fan puts forward three reasons for his conclusion that in a Confucian society, MAID should not be legalized. First, Fan says that a legal right to MAID would "contradict the central requirement of basic Confucian

[72] "Abortion and Degrees of Personhood: Understanding Why the Abortion Problem (and the Animal Rights Problem) are Irresolvable," *Public Affairs Quarterly* 11, no. 1 (January 1997), 1–19. I have also argued that the problems of animal experimentation and nonvegetarianism cannot be resolved in "Animal research, Non-vegetarianism, and the Moral Status of Animals—Understanding the Impasse of the Animal Rights Problem," *Journal of Medicine and Philosophy*, vol. 27, No. 5 (2002), 589–615. However, I argue that despite these problems, one should be a quasi-vegetarian if one is indifferent between eating meat and, say, seafood. See my "Toward Quasi-vegetarianism," in Hon-Lam Li and Anthony Yeung, eds., *New Essays in Applied Ethics* (UK: Palgrave Macmillan, 2007), 64–90.

virtues that an innocent human life must be honored and respected." Second, Fan says that a Confucian should treasure his parents' life. Therefore, a filial adult child should not want his parent to have MAID because he would want his parent to live longer. Third, Fan argues that if MAID were legal, Confucian parents would feel obligated to ask for MAID to avoid becoming a burden on their children[73]—even if, had MAID not been legalized, they would not have wanted to die. Regarding the first reason, I have replied to a similar contention raised by Farrell and Tham, and I have nothing to add here. So I shall turn to the other two reasons.

Assuming (for the sake of discussion) that Confucianism is a true moral doctrine, I do not think that a Confucian should object to MAID. This is because those patients qualified to receive MAID must (1) be terminally ill, (2) be going through unbearable suffering, and (3) be competent and have insistently asked to receive MAID. Under such circumstances, it could and even would be in the parent's interest to die. This does not mean that a hospital should mention MAID to her (let alone asking her to consider MAID carefully), unless she first expresses her desperate desire to die and persists in making such requests. A filial son or daughter should respect their parent's wish to die, especially if dying is in her interest.

I accept that there are cases where someone who says she wants to die but is actually seeking care and attention. I am not considering these cases because there should be a way to screen off these cases from those where terminally ill patients suffer unbearably and earnestly want to die. Moreover, I accept that there are cases where the right thing to do is to upgrade the palliative care, and not to offer MAID straightaway, to a terminally ill patient who wants to die. I am focusing on those cases where even the best palliative care is unable to give a terminally ill patient a decent survival with a manageable amount of pain. In such a scenario, a terminally ill patient can rightfully ask for MAID, and should be offered it.

[73] It is conceivable that some patients would seek MAID because they do not want to be a burden on society, especially if they rely on social provision of medical care. But this would not be a justified reason to seek MAID.

Finally, according to Fan's Confucianism, if medical "treatments may not help the patient recover but simply prolong the patient's suffering, they may legitimately be withheld or withdrawn."[74] But no plausible reason is given for allowing this kind of "passive euthanasia," which is often much more cruel than MAID or active (voluntary) euthanasia. For instance, Jean Davis, 86, starved herself to death and found her experience "intolerable" but the only way she could die legally in Britain.[75]

3 Concluding Remarks

When Rawls talks about public reason, his aim is to provide a road map such that reasonable people of different religious faiths and conflicting moral and political allegiance can discourse about constitutional essentials and matters of basic justice—as well as controversial issues that border on them (e.g., abortion and MAID)—in a way that is normatively satisfactory. His idea is to bracket our conflicting comprehensive doctrines since, being a political liberal, he believes that there is more than one reasonable comprehensive doctrine. Public reason refers to "ideals and principles, standards and values" (*The Law of Peoples* 144) that are self-standing, and do not need to be backed up by comprehensive doctrines. This requirement of discoursing in terms of public reasons is especially important for public officials, judges, legislators, and candidates for public office. This does not guarantee unanimity, of course. But it does ensure that after discourse and deliberation of the relevant issue in terms of public reason, a legal enactment expressing the majority's view will be

[74] See Ruiping Fan, Chap. 3 of this volume.

[75] According to Guardian, "The debate around voluntary euthanasia has gained new urgency after it emerged that a 86-year-old woman starved herself to death because she believed it was the only way she could legally exercise her right to die. Before she died, Jean Davies said she was going through the 'intolerable' experience because the government had failed to reform the law on assisted suicide, leaving her no legal alternative. Davies died on October 1, five weeks after she stopped eating and a fortnight after she decided to stop drinking water. She said she feared taking a drug overdose in case it did not work. Four weeks into her fast she told the Sunday Times about her experience. 'It is hell. I can't tell you how hard it is. You wouldn't decide this unless you thought your life was going to be so bad. It is intolerable,' she said" (Alexandra Topping, "Assisted Dying," *The Guardian*, 19 October 2014). https://www.theguardian.com/society/2014/oct/19/right-to-die-campaigner-starved-herself-jean-davies.

legitimate law. This is the way to resolve issues pertaining to the basic structure of society, according to political liberalism.

By comparison, NLT as expounded by Farrell and Tham lacks a strategy to resolve different opinions regarding basic political issues in society. NLT insists that there is but one correct answer to every issue, and that a rational person should be able to arrive at the correct answer. Even if we agree that there is a correct answer to every moral issue, *it does not mean that we know the answer*. Consequently, those who disagree might still be reasonable.[76]

Fan's Confucianism is grounded in a version of moral relativism, despite the fact that he acknowledges a list of basic nonliberal rights and liberties which are universal. He pushes for a kind of family decision in bioethical issues that he thinks is especially suitable for East Asian countries. And he argues that MAID should not be legalized. Fan does not envisage that there might be disagreements within a Confucian society. As far as I know, his view on MAID would clash with most Confucians' view, according to which a filial son or daughter should respect their parent's earnest decision to choose MAID if she is terminally ill, going through unbearable suffering, and desperately wants to die. Acting against a parent's strong desire to die in dignity—to be rid of unbearable pain during her last lap in life—is unfilial, according to many Confucian scholars.[77] My point is that there are different versions of Confucian ethics, and Fan's version—or at least the way he puts the theory into practice on the topic of filial piety—is by no means the standard view, nor is it a defensible view.

One way in which NLT and Fan's Confucianism both try to resolve the problem of MAID is to limit internal disagreement by limiting the size of the community, or the values of citizens. According to Farrell and Tham's NLT, deliberation should be carried out in a *polis* or at any rate within a tradition, as MacIntyre recommends. Yet, isn't this an acknowledgment that NLT is not suitable for a cosmopolitan metropolis, or the modern condition of "reasonable pluralism"?

---

[76] In a criminal trial, either a Defendant has committed the offense, or he has not. In the context of uncertainty, members of the jury might disagree and yet could all be reasonable, even though we know that one side must be mistaken.

[77] I have personally talked to various Confucian scholars in Hong Kong and abroad. But I shall not name them here without their prior permission.

According to Fan, Confucianism is only for societies which are already Confucian. So the problem of internal disagreement is preempted from the outset. Yet, I doubt that the issue arising from "the fact of reasonable pluralism" (Rawls) can be so easily resolved. First, even within China, Korea, or Japan—not to mention Hong Kong, Singapore, and Taiwan—there are people of different religious faiths, and with different moral and political allegiances. In Japan, Korea, Taiwan, and Singapore, many issues are deliberated and debated before votes are cast. Their political systems are genuinely democratic.[78] China is very different because it is not a constitutional democracy (in Rawls' sense).[79] Confucianism has been severely criticized by Chinese intellectuals since the May Fourth Movement in 1919, and was severely criticized again by the Chinese Communist Party (CCP) during the Cultural Revolution (1966–1976). To the extent that Confucianism has been revived since the 1980s, it is a watered-down version. The fact is that traditional Confucianism defends and stabilizes feudalism, an economic structure which has underpinned—and has been functional for—the development of productive forces in the agrarian mode of production in China for two millennia. The lack of genuine democracy in China today is the result of the joint influence of Confucianism and Marxism. (In this respect, Marx's idea of the dictatorship of the proletariat coheres with Confucianism, which supported and *stabilized* absolute monarchy or feudalism for thousands of years).[80]

Whether Hong Kong and Singapore are Confucian enough, they are the products of the Common Law of Great Britain because the rule of law exists in these two cities.[81] The democracies of Japan, South Korea, and more recently, Taiwan, were very much the results of American influence. China is the only major country in Asia that has not been influenced by

[78] I wish to qualify my claim about Singapore because I am not certain about the extent to which Singapore's democracy is genuine. See Chap. 7.

[79] Burton Dreben clarifies that by "constitutional democracy," Rawls means constitutional liberal democracy. See Burton Dreben, "On Rawls and Political Liberalism," in *Cambridge Companion to Rawls*, ed. Samuel Freeman (Cambridge University Press, 2006).

[80] See Chap. 7 of this volume. For a particularly illuminating account of historical materialism, see G. A. Cohen, Karl Marx's Theory of History: A Defense, (Princeton: Princeton University Press, 1978), chaps. 2-4, 6-8, and 10.)

[81] There is a question whether the rule of law still exists in Hong Kong, since July 1, 2020. I shall not consider this question here.

the Common Law or Western democracy. Instead, the main influences on China in the twentieth century have been the Russian legal system (which was influenced by the Continental law of Germany), Marxism, Confucianism, and, to a lesser extent, Taoism, Buddhism, and atheism.

I do not deny that these East Asian countries are Confucian to some extent. But they are also the product of other multifarious influences. Therefore, reasonable pluralism obtains in these places. (I shall elaborate on this in Chap. 7.)

As Marx observed in 1848, as capitalism develops, the civilization or culture in different peoples are becoming more and more alike.[82] The project of keeping China—let alone other East Asian countries—purged of liberal ideas is doomed. In these places, we find pluralism of values.

In any modern society, there are two kinds of conflict: conflicts of interest and conflict of values. Nonliberal views of either the left or the right may find these conflicts to be evidence of social pathology.[83] However, Rawls is liberal in two senses. First, he finds such conflicts unavoidable. Second, he holds that even if conflicts could be avoided, the state would have to use oppressive force to secure unanimity of values—Rawls calls this "the fact of oppression" (*PL* 37)—and this is too much a sacrifice of individual liberty. As Rawls puts it, even if everyone in a nation has the same set of values—the same comprehensive doctrine—force is necessary to maintain stability of the basic structure. He thinks that the Inquisition was not an accident (*PL* 37). Similarly, in Imperial China, anyone who said anything that the emperor found disagreeable was liable

[82] Marx says: "The bourgeoisie, by the rapid improvement of all instruments of production, by the immensely facilitated means of communication, draws all, even the most barbarian, nations into civilisation. The cheap prices of commodities are the heavy artillery with which it batters down all Chinese walls, with which it forces the barbarians' intensely obstinate hatred of foreigners to capitulate. *It compels all nations, on pain of extinction, to adopt the bourgeois mode of production; it compels them to introduce what it calls civilisation into their midst, i.e., to become bourgeois themselves. In one word, it creates a world after its own image*" (emphasis added).

"National differences and antagonism between peoples are daily more and more vanishing, owing to the development of the bourgeoisie, to freedom of commerce, to the world market, to uniformity in the mode of production and in the conditions of life corresponding thereto" (Marx and Engels, *Manifesto of the Communist Party*, Parts I and II).

[83] I owe this point to T. M. Scanlon, "Some Main Points in Rawls' Theory of Justice."

to be executed.[84] The reason why oppressive force by the state is necessary to maintain a comprehensive doctrine might have a number of different causes, some of which are trivial ones. But Rawls points out that reasonable people disagree because of the "burdens of judgment."[85] As Rawls believes, political liberalism rests its case on the premise that reasonable people can disagree both on theoretical and practical matters, and that toleration is the best way to try to resolve the disagreement by debating the issues via public reasons. But the ideas of public reason propounded by Farrell and Tham, and Fan are unsatisfactory in this respect, because they do not recognize that there could be *reasonable* disagreement among reasonable and rational citizens, and consequently they do not attempt to resolve such disagreement. Insofar as Rawls' concept of "public reason" has to do with providing a common language to those espousing different reasonable comprehensive doctrines to discourse and debate politically divisive issues, it is not clear why the theories put forward by Farrell and Tham, as well as Fan, would qualify as *public* reason. Their disagreement with political liberalism is not merely a difference in *conception* of public reason, but also the very *concept* of public reason. Otherwise, all ethical views, for example, deontology, utilitarianism, virtue ethics, Buddhist ethics, and Taoist ethics, would qualify as "public reason,"[86] and if so, the term "public reason" would lose its point and meaning.

---

[84] Rawls says: "[A] continuing shared understanding on one comprehensive religious, philosophical, or moral doctrine can be maintained only by the oppressive use of state power. If we think of political society as a community united in affirming one and the same comprehensive doctrine then the oppressive use of state power is necessary for political community. In the society of the Middle Ages, more or less united in affirming the Catholic faith, the Inquisition was not an accident; its suppression of heresy was needed to preserve that shared religious belief. The same holds, I believe, for any reasonable comprehensive philosophical or moral doctrine, whether religious or nonreligious. A society united on a reasonable form of utilitarianism, or on the reasonable liberalisms of Kant or Mill, would likewise require the sanction of state power to remain so. Call this 'the fact of oppression'" (*PL* 37).

[85] This might have to do with difficulties in weighing empirical and scientific evidence, the weight of considerations, the vagueness of moral and political concepts, how we assess evidence and weigh moral and political values, and the difficulties of setting priorities by social institutions (*PL* 56–57).

[86] I thank Bonnie Steinbock for supplying the examples of deontology and utilitarianism.

5

Replies to Li and Fan

Dominic Farrell LC and Joseph Tham LC

In Chap. 2, we defended the idea that public reason is essentially truth-directed political inquiry and decision making, and that this is the conception of public reason with which the natural law tradition works. Here we shall reflect on the preceding defenses of rival conceptions of public reason, namely, political liberalism (Chap. 1) and Confucianism (Chap. 3), particularly when it comes to bioethical issues in general and, more specifically, medically assisted suicide (MAS).

1 Political Liberalism, Public Reason, and Bioethics

In Chap. 1, Hon-Lam Li sets out a qualified defense of the Rawlsian conception of public reason. Furthermore, Li considers and defends its applicability to bioethics and the specific issue of MAS.

D. Farrell LC (✉) • J. Tham LC
Pontifical Athenaeum Regina Apostolorum, Rome, Italy
e-mail: Dfarrell@upra.org; jtham@legionaries.org

H.-L. Li, M. Campbell (eds.), *Public Reason and Bioethics*,
https://doi.org/10.1007/978-3-030-61170-5_5

The first part of Hon-Lam Li's chapter aims at articulating Rawls's notion of public reason and, drawing on T.M. Scanlon, addresses some of the concerns it faces. Li thereby lays useful groundwork for the issues discussed in the other chapters.

We do not have any substantial objections to Li's presentation of Rawls's account of public reason. In our view, however, it does not add any new perspectives that make it necessary for us to revise or further specify our earlier objections to political liberalism and its conception of reasonableness.

It may be helpful though to flesh out some of the concerns that proponents of the sort of political perfectionism, embodied in the natural law tradition, have with political liberalism.

We can better identify where the natural law tradition differs from political liberalism by considering where they stand regarding the following two sets of opposing theses.

(I) *Perfectionist Thesis.* The principles of justice can only be justified with reference to a certain conception of the good life.
(II) *Anti-Perfectionist Thesis.* The principles of justice can be justified without any reference to a conception of the good life.
(III) *Consequentialist Thesis.* Persons do not have any rights that cannot be legitimately trumped or overridden for the sake of the general welfare.
(IV) *Anti-Consequentialist Thesis.* Persons have certain rights that cannot be legitimately trumped or overridden for the sake of the general welfare.

Political liberalism defends the priority of the right over the good. Specifically, this means that it holds (II) and (IV).[1] Political perfectionism, on the other hand, defends the priority of the good over the right. It is committed to (I) instead of (II). However, it can be either consequentialist or anti-consequentialist, depending on whether it is then

[1] Michael J. Sandel, *Liberalism and the Limits of Justice*, 2nd ed. (Cambridge: Cambridge University Press, 1998), x, henceforth *LLJ*.

committed to (III) or to (IV) instead. Consequentialism defends (III), the natural law tradition (IV).

Now political liberalism purports to be self-standing in that it does not ground the priority of the right over the good [(II) and (IV)] in a comprehensive doctrine but in 'the fact of reasonable pluralism' that characterizes modern democracies. By this it means that modern liberal democracies work despite their pluralist make-up, with people holding different conceptions of the good. They work because people are reasonable and give priority to the right over the good. Reasonable people do so to ensure their cooperation with one another in society.

Michael Sandel objects, however, that the existence of reasonable pluralism does not constitute sufficient grounds for giving priority to the right over the good. To hold that the right has priority over the good in a society of free persons, the political liberal must make two assumptions. The first assumption is that those people who are free to think for themselves will disagree on matters of religion and morality, namely, the good life. This explains why there is reasonable pluralism. The second assumption is that people who are free to think for themselves will not disagree, on due reflection, over the principles of justice. It is only warranted to claim that the right has priority over the good if people, on due reflection, simultaneously disagree about the good and agree over the principles of justice.[2]

However, as Sandel points out, even in the sort of society which political liberalism purports to describe, there are many disagreements about justice. For example, there is the debate between egalitarians and libertarians on distributive justice. Many of the former defend Rawls's difference principle, whereas the latter, such as Robert Nozick and Milton Friedman, reject it. This is not a disagreement over how a principle applies to a situation or problem. It is a disagreement over the principles of justice themselves. However, the political liberal cannot rule out the libertarian as unreasonable in advance. This means that the 'fact of reasonable pluralism', as Rawls describes it, does not exist. People who are committed to liberal democracy do disagree over which principles of justice are constitutive. In that case, it is not warranted to rule out considerations

[2] Sandel, *LLJ*, 203.

concerning the good when it comes to public reasoning. To rule those considerations out, reasonable people would have to disagree about the good while agreeing over the principles of justice. However, they do not agree over the principles of justice.[3]

Furthermore, as we pointed out in our chapter, it is unreasonable to bracket comprehensive doctrines for the sake of social cooperation and keep them out of the domain of public reason.[4] While Rawls does not contest the truth of any comprehensive doctrine that can fit in with the overlapping consensus of a modern liberal democracy, he does claim that, as a rule, political values trump the non-political considerations of a comprehensive doctrine.[5] The problem is that there is not always a clear boundary between the two. Indeed, political liberals sometimes unconsciously draw on their own comprehensive doctrine to determine what counts as a political value and a non-political one.

Consider the example with which Li opens his chapter. There are two Catholic politicians—let us abstract from party politics and call them A and B—each of whom accepts their church's teaching on the wrongness of abortion. However, B, unlike A, believes that his personal views on the matter do not entitle him to vote against the existing pro-abortion legislation or promote its repeal. The political liberal can take B to be reasonable and respectful of public reason if and only if both A and B believe abortion is wrong on grounds that are wholly internal to their shared comprehensive doctrine (their Catholic faith, in this case). Now, A and B believe that abortion is wrong because they both accept the moral teaching of their religion. Nevertheless, not all religious moral beliefs are founded exclusively on reasons that are internal to its creed. Jews and Christians, for example, accept the Decalogue and believe that murder is wrong. True, one of their reasons for believing this is internal to their religion. They believe that murder is wrong because God has taught them that it is. They do not necessarily believe, though, that murder is wrong only in virtue of a divine command (*malum quia prohibitum*). Rather, they believe that the divine commandment against murder is truth-apt

[3] Sandel, *LLJ*, 202–210.

[4] Sandel, *LLJ*, 196–202.

[5] John Rawls, *PL*, 146.

and designates an actual moral fact. It proscribes an action that we have reason to reject regardless of the divine commandment. In so doing, it indicates, albeit implicitly, what those reasons are (*prohibitum quia malum*). At least, a Jew or Christian will generally understand the commandment in this way: in the light of the biblical teachings on the dignity of humans and their capacity for moral agency (Genesis 1:26–31; Psalm 8; Sirach 15:14). They can assume, therefore, that they have the same reasons to deem murder wrong as a non-believer. These reasons are not wholly internal to their comprehensive doctrine but, insofar as they are based on some supposedly incontrovertible facts, are also external to it to some extent. Consequently, these reasons can be assumed to be normative for any right-thinking person who gives the underlying facts due consideration. Furthermore, should the reason in question regard an action that bears decisively upon the cohesion and welfare of society, it is relevant to public policy.

Hence, A and B believe that abortion is wrong both on grounds that are internal to their faith and on others that are external to it. They believe that human life begins at conception and that, as a result, abortion constitutes the direct killing of an innocent human being and a violation of the fundamental human right to life. Moreover, they believe that any right-thinking person should come to the same conclusion when considering the relevant facts. Their belief in the wrongness of abortion cannot be characterized as a non-political value. It is not reducible to an esoteric religious belief. It regards the fundamental issue of the right to life. Consequently, it is politically relevant and an object of public reason. Both A and B can legitimately argue for pro-life policies. In this case, it is B rather than A who is inconsistent. The political liberal may object, though, that A is wrong to pursue pro-life policies and B right to abstain from doing so because there is no overlapping consensus in society on the wrongness of abortion. Consequently, a pro-life policy is one which could be reasonably rejected and so does not pass the test of public reason. However, the same is true of pro-choice policies. At this point, the political liberal in favor of pro-choice policies could argue that they are a necessary requirement of a commitment to freedom, equality, and fairness. They safeguard a pregnant woman's equality and autonomy. Even so, the pro-choice liberal cannot deem A unreasonable. This is because A is arguing from the same principles in defense of values to which the pro-choicer does not attach the same importance. For A, pro-choice policies are unfair and inegalitarian, whereas pro-life policies are necessary to safeguard a human embryo's equal claim to the right to life. When it comes to substantive moral issues, therefore, the boundaries between comprehensive doctrines, especially religious ones, and public reason are not as clear-cut as the political liberal may assume. Paradoxically, should the political liberal deem B to pass the test of public reason by supporting pro-choice legislation, and A to fail that test by opposing such legislation, then it is the political liberal who may be unwittingly using a comprehensive doctrine to determine what counts as political and non-political values.

Indeed, the case of A and B confirms one of the main contentions of our chapter: that public reason cannot bracket comprehensive doctrines but must engage in truth-directed inquiry into second-order questions.

The second part of Li's chapter, on the other hand, makes a political liberal case for the legalization of medically assisted death. The overall strategy is to show the inconsistency of the arguments of those who oppose its introduction. Two observations are in order, though.

First, it is imprecise and misleading to describe the bioethical issue as 'medically assisted death' (henceforth MAD). Without any further stipulations, this term does not designate a definite intentional action but the whole range of ethically appropriate ways, such as palliative care and a good bedside manner, in which physicians assist patients to die. It can only designate ethically appropriate forms of assistance because it refers to a certain kind of professional service. By definition, a profession consists in the duly licensed provision of a highly specialized service in a disinterested and morally upright way. Li, on the other hand, takes MAD in a more specific sense: a physician's deployment of medical expertise to directly bring about, in the least distressing way, the death of a terminally ill patient whose existence is unbearable and who, after repeatedly giving due reflection to the matter, has requested this intervention each time. Furthermore, let us assume that such an act constitutes, along with palliative care and a good bedside manner, a variety of MAD, in the broader sense of the term. Now, within an argument in support of said act, this assumption constitutes the conclusion, not one of the premises. In this case, MAD will designate both a class of intentional actions, which includes palliative care and a good bedside manner, and a specific kind of intentional action: that by which a physician directly kills the terminally ill patient who repeatedly requests this. However, the same term cannot designate both the subject and the predicate of the conclusion of a categorical syllogism. To disambiguate this conclusion and the underlying syllogism, it is necessary to adopt a term that refers primarily and exclusively to a physician's direct killing of the terminally ill patient who requests this. In other words, it is necessary to employ a term that does not allow for the undue conflation of this practice with the kinds of action, whereby medical practitioners provide care for dying patients without directly and voluntarily killing them. The term 'medically assisted suicide' (MAS) satisfies these requisites, whereas 'medically assisted death' does not. For this reason, it is preferable to the latter.[6]

Our second observation regards Li's survey of the arguments against the legalization of MAS. This survey is incomplete in at least one significant regard. It does not address those arguments against MAS that are both non-theist and non-consequentialist. As a result, it does not fully succeed in accomplishing what it sets out to do: prove that there are no compelling arguments, suitable for the forum of public reason, against the legalization of MAS.

The upshot of Li's survey is that the opponent of MAS is committed to at least one of the following three arguments: (1) the religious argument, (2) the slippery slope argument, and (3) the argument from abuse. The main premise of the first argument is that God does not authorize the taking of one's own life. The second argument, in its theoretical variant, holds that a state can only legalize clear-cut, morally licit acts but since there is no clear criterion for distinguishing between justified and unjustified cases of MAS, by legalizing MAS, will effectively give license to engage in abusive instances. Similarly, the third argument allows for cases in which MAS is legitimate but rules out its legalization on consequentialist grounds: such a law would open the door to abuses that outweigh and trump the possible benefits.

---

[6] 'Medically-assisted suicide' (MAS) is still a problematic term. Strictly, the term 'medical' designates the activities proper to that profession, namely, ethically appropriate ones. MAS, on the other hand, is a contested category. Only derivatively, namely, in the light of an appropriate argument, will it designate an action that is proper to the medical profession.

The latter two arguments are consequentialist. They rule out the legalization of MAS on the consequentialist grounds that such a law will foreseeably cause more harm than good. Furthermore, these two arguments are non-theistic. They do not base their normative conclusions on claims about what God demands or deserves. The first argument, on the other hand, is theistic and non-consequentialist.

Li raises some valid objections to each of these three arguments. He does not consider, though, a further class of argument: that which is both non-consequentialist and non-theistic. This is significant as far as the natural law tradition is concerned. The non-consequentialist arguments that said tradition proposes are mainly non-theistic rather than theistic. This is apparent even in a document as markedly religious as John Paul II's encyclical letter *Evangelium vitae*.

In that document, John Paul II argues that MAS is wrong. The first reasons adduced are that it is contrary to love of self and constitutes a renunciation of one's responsibility toward others. Only after, does John Paul II consider God's sovereignty (nn. 65–66). He adduces arguments that are non-consequentialist and non-theistic before he gets round to the 'religious argument'. Nor is the former kind of argument exclusive to the natural law tradition. Kant, for example, deems suicide to be a wrongful abdication of one's moral responsibility (*MM* VI:422–424).

In our chapter, we outlined two non-consequentialist, non-theistic arguments against MAS. In our view, these are the natural law tradition's main arguments against MAS. The one holds that MAS is an improper use of medical expertise; the other holds that it is a kind of wrongful killing. Neither argument is theistic. Instead, each analyzes whether MAS falls under a certain kind of intentional action: 'medical treatment' and 'murder', respectively. However, Li does not take the non-theistic and non-consequentialist arguments against MAS into account. As a result, he does not establish that there are no compelling, publicly debatable arguments against the legalization of MAS, but only that some are either unsatisfactory or based on reasons peculiar to a specific comprehensive doctrine.

It may still be possible though for there to be some agreement between political liberalism and the natural law tradition. For example, in Chap. 13, Michael Campbell draws on Rawls's conception of public reason and reaches conclusions very similar to ours on its limited potential, if not downright unsuitability of political liberalism's when it comes to legislating on bioethical issues. As he points out, many key bioethical problems involve substantive issues that fall outside the limited scope of the political conception of justice. Our own chapter attempts to illustrate this point in greater detail. Furthermore, Campbell appears to suggest that the state should refrain from engaging in extensive legislation on these matters but allow for greater freedom of action. We agree with this, though maybe on somewhat different grounds. In reaching these conclusions, Campbell illustrates that a political liberal has reasons, very similar to the ones we outline, to be skeptical about the applicability of Rawlsian public reason to legislation on controversial bioethical issues.

Moreover, Rawls's concept of public reason is Neo-Kantian. On the one hand, it is rooted in Kant's thought, as Rawls repeatedly indicates. On the other hand, it is not concerned with providing an exact account of Kant's own thought and its internal coherence, nor with remaining faithful to it when applying it to contemporary debates. Instead, it selects and develops those aspects of Kant's thought that it deems useful for resolving a set of theoretical problems. It disregards, therefore, positions and arguments of Kant, such as those on sexual morality (*MM* VI:277–280, 424–425) and suicide (*GMS* IV:421–422; *MM* VI:422–424), that it deems outdated, just as Neo-Aristotelian virtue ethicists do not take on board the Stagirite's views on slavery and women. At any rate, just as a Neo-Kantian political theory, such as political liberalism, needs to be distinguished from the great philosopher's actual thought, so too is it necessary to distinguish between the Neo-Kantian conception of public reason and Kant's. This is important because the objections we have raised against the former may not necessarily apply to the latter.

For example, if we abstract from the historical contingencies of Kant's Enlightenment project, there is, we believe, a broad convergence between his conception of public reason, as described by Terence Tai (Chap. 10), and that of the natural law tradition. For Kant, autonomy is the quality of the will of one who acts on principles taken as true and rational from a first-person perspective. However, as Tai points out, the positive laws of society are justified if and only if they are fit for acceptance by an autonomous agent. Assessing their compatibility with autonomy and duly revising them is an ongoing process. It is accomplished if informed persons express their critical thoughts publicly and thereby stimulate a deliberation within society at large. In this case, Kant is proposing much the same kind of truth-directed inquiry that we defended. Furthermore, like the natural law tradition, he believes that law is narrower in scope than morality and is restricted to right (*Recht*). In these two regards, at least, his conception of public reason converges with that of the natural law tradition.

2 Confucianism, Public Reason, and Bioethics

Kant's thought is rooted in the Western tradition, while Rawls's takes political liberalism to be rooted in the background culture of the modern liberal democracies of the West, such as that of the United States. Their conceptions of public reason are not necessarily suited to the countries of East Asia. Perhaps there is a conception of public reason that is both rooted in their culture and suited to their political life.

This question is at the center of Ruiping Fan's chapter. He outlines a reconstructionist Confucian conception of public reason, argues that it is the one best suited for policy debates in China and East Asia, and applies it to the test-case of MAS.

In his account of Confucian virtues, Fan underscores the rights and reciprocal duties of families and their members, and how these bear on MAS. We find this perspective convincing, complementary to our analysis of MAS, and a refreshing alternative to liberalism's generally unilateral focus on individuals. Furthermore, as Catholics, this perspective is familiar and not foreign to us. It is virtually identical to the Catholic Church's teaching on the family's status as the fundamental unit of society and on the duty of adult children to care for their aged or infirm parents. However, we find his analysis of MAS broadly compelling not because it is Confucian, but because it strikes us as being right in many regards. This bring us to the central difference between our conception of public reason and Fan's. We agree with Fan that public reasoning in East Asia should be based on and informed by Confucian values rather than the liberalism typical of Western democracies. Confucianism encapsulates the public reasoning that has been built up and tested over time within East Asian countries. As such, not only should it be taken seriously rather than dismissed lightly in favor of Western liberal values. Rather, it constitutes the proper starting point for public reasoning in East Asian countries. However, public reasoning is, as we have argued, truth-directed inquiry of the political variety. While it is inevitably informed by a tradition or comprehensive doctrine, it cannot be constrained definitively by it. As a truth-directed inquiry, it must be open to following the evidence or better argument, even if this means parting company at certain junctures from the tradition that informs it. So, while we agree with Fan on the need for public reasoning to be informed by Confucianism in certain parts of East Asia, we believe that he prioritizes its Confucian substance over truth-directed inquiry. In our view, public reasoning is only consistent if its character as a truth-directed inquiry takes precedence over the tradition that legitimately informs it.

For Fan, the Confucian conception of public reason is more suitable than political liberalism when it comes to China or any other country whose moral culture is predominantly Confucian. Furthermore, he proposes that his account of Confucian public reason is more adequate than that of Sungmoon Kim or the moderate Confucian perfectionism of Joseph Chan. Unlike us, he does not dispute the validity of the liberal conception of public reason outright. He grants that it may be suitable for Western liberal democracies, just as the Confucian conception of public reason is the right one for many East Asian countries. In his view, the framework by which public reasoning is conducted legitimately within a political community (nation states in the modern world) is not universally valid and stance-independent, but derives from and coheres with that society's cultural history and predominant ethical system. Let us call this a communitarian conception of public reason. Furthermore, Fan distinguishes this communitarian conception of public reason from a pragmatist one. This is because the predominant ethical framework of a nation, Confucianism in the case of China, gives that nation's conception of public reason a specific substantive content that a purely pragmatist argument would not.

Significantly, Fan finds support for this communitarian conception of public reason in Rawls. On the one hand, Rawls holds that political liberalism is proper to the modern liberal democracies of the West. On the other hand, he distinguishes well-ordered peoples that are liberal from those are decent. Of course, Rawls might assume that decent peoples have reason to become a modern liberal democracy. He is, after all, a liberal rather than a communitarian. Fan argues, though, that China has does not have a reason to become a modern liberal democracy. He is not a liberal but a communitarian in the aforementioned sense. Political liberalism, he argues, is rooted in historical circumstances that are peculiar to the West but extraneous to China and other East Asian countries. Consequently, the West has developed an individualist conception of society. Liberalism is an expression of this individualism. Under the influence of Confucianism, on the other hand, China has retained a communitarian conception of society (to be distinguished from a communitarian conception of public reason). In other words, Fan contests the purported neutrality of political liberalism. In a variant of one of the main 'communitarian' objections to Rawls and like-minded liberals, he treats political liberalism as a loaded rather than a neutral concept, and, more specifically, as an outgrowth of an atomistic concept of the person. However, he does not follow up on this charge with an analytical explanation of why an atomistic conception of society is theoretically inconsistent in ways that the communitarian view is not.[7] For his purposes, it suffices to notice that a liberal democracy and the political liberal's conception of public reason are incompatible with the predominantly Confucian moral culture of China.

Nevertheless, Fan also recognizes certain standards that are independent of a society's cultural history and dominant ethical framework. He accepts Rawls's claim that a society is decent if and only if it is committed to peace in foreign policy and, in second place, has a legal system which secures and safeguards a minimal set of human rights.[8] Furthermore, these rights may be implicit rather than explicit. They are not explicit, for

[7] Charles Taylor provides a good example of an analytical argument against the individualist conception of the person on which liberalism is based. Charles Taylor, "Atomism," in *Philosophy and the Human Sciences*, ibid., (Cambridge: Cambridge University Press, 1985), 187–210.

[8] John Rawls, *The Law of Peoples* (Cambridge, MA: Harvard University Press, 1999), 65.

example, in a Confucian society. Rights-discourse is extraneous to its tradition. However, they are implicit in a Confucian society's virtue theory. At any rate, Fan takes the decency requirement to be a universally binding constraint on legitimate cultural pluralism when it comes to the basic structure of political society. In this case, he appears to hold a qualified communitarian conception of public reason. He holds that public reasoning can only be carried out within a decent society (decency condition) and according to a framework that derives from and coheres with that society's cultural history and predominant ethical system (communitarian conception of public reason).

By making the decency condition a constrain on the communitarian conception of public reason, Fan appeals to a standard that transcends and constrains the Confucian moral culture of China. But on what grounds does decency constitute a universally valid constraint on the communitarian conception of public reason? The reasons which Rawls gives, and which Fan presumably shares, have to do with the dignity and nature of persons. This means that certain features of human nature enjoy normative priority over Confucianism and are universally valid. If so, one must also specify at what point and on what grounds these universally binding features of human nature cede normative priority to the communitarian conception of public reason and, in this case, to specifically Confucian values. In other words, it is necessary to specify at what point and on what grounds public reason shifts gear from truth-directed inquiry to a communitarian framework, or shifts gear from one kind of truth-directed inquiry, based on a correspondence theory of truth, to another kind, based on a coherence theory of truth. In our view, Fan does not provide such an explanation and so does not show convincingly why public reasoning should be conducted according to a communitarian framework rather than the sort of truth-directed inquiry that the decency requirement entails.

Conducting public reasoning according to demands of a truth-directed inquiry does not mean that China or other East Asian countries should disregard their own moral culture. We accept as a matter of fact that such countries will, and even should, take their own moral culture as the parting point for policymaking. This is what has happened in the West. However, we believe that, since public reason is truth-directed inquiry of the political variety, the dominant moral tradition will function as its necessary parting point and source but not as its definitive constraint. Hence, public reasoning in a country of a Confucian tradition, whether on bioethical issues or others, should begin by drawing on Confucian ethics, taking seriously the positions and claims that it embodies, while examining them critically. The aim of such a critical examination is not to disprove them but to see whether the claims of Confucian ethics are true. Indeed, taking the cautionary measure of presuming them to be true until proven otherwise is consistent with truth-directed inquiry. No doubt many will withstand the test.

We have outlined several points of agreement and disagreement between our proposed conception of public reason and the alternative ones that have been defended in this volume. Moreover, we have mentioned some in favor of treating public reason as truth-directed inquiry of the political variety. We are also aware that more work is needed to show that the correspondence theory of truth that we suppose is more consistent than a coherence of pragmatist one. Still, engaging in these further inquiries to get at the truth of these matters confirms our claims about the character of confronting and debating conflicting views.

6

Replies to Li and Farrell–Tham

Ruiping Fan

My chapter provides a reconstructionist Confucian conception of public reason. In contrast, Hon-Lam Li argues for a liberal political conception, and Dominic Farrell LC–Joseph Tham LC defend a natural-law-tradition conception. Whereas my conception is constructed only for Confucian-influenced East Asian societies to adopt, both Li and Farrell–Tham appear to be confident enough to intend to guide all societies alike by their respective favored conceptions. In recommending their conceptions, they do not bother to bear "the burdens of judgment" by considering the particular cultural and historical conditions of a specific society with which any proposed account of public reason is bound to be intertwined. In my view, however, differing from the notion of pure reason for theoretical inquiry that may be independent of any particular society, a conception of public reason is primarily meant for a society to appeal to in order to justify particular policies and laws (bioethical policies and laws included)

R. Fan (✉)
Department of Public Policy, City University of Hong Kong,
Kowloon, Hong Kong
e-mail: safan@cityu.edu.hk

H.-L. Li, M. Campbell (eds.), *Public Reason and Bioethics*,
https://doi.org/10.1007/978-3-030-61170-5_6

made for the society. This kind of justificatory task is inevitably culturally and historically conditioned, for it requires that the political concepts and principles to be used for the task be reasonably acceptable to all living persons in the society.

In the first part of his chapter, Li purports to offer "a qualified defense of John Rawls' political liberalism and his idea of public reason." By "a qualified defense" he seems mainly to mean that the Rawlsian idea of public reason should be supplemented by Thomas Scanlon's contractualism. Although he is aware that the Rawlsian idea of public reason, as Rawls acknowledges, is constructed from the cultural and historical conditions of modern liberal democratic societies in the West, Li does not find it necessary to offer any "qualification" for the idea to be applicable to other societies, such as mainland China or Hong Kong in East Asia, while these societies are culturally and historically different from the West. Instead, he seems simply to take Rawlsian liberal public reason as global public reason for all regions, East Asian societies included, without recognizing that establishing global public reason would require a different ground than the overlapping consensus of modern Western liberal democratic political culture which Rawls and his followers have offered. In contrast, global public reason would require a shared political culture across the globe. Is there such shared global political culture available? My chapter answers no, indicating that East Asian societies affirm a set of legitimate Confucian moral and political concerns that substantially differ from modern Western liberal democratic values. While I hold that the Confucian tradition must accept a minimal conception of human rights and basic liberties, this conception should be developed out of the fundamental rationale of the Confucian virtues and should not contradict any central requirement of the virtues. My chapter accentuates that a proper conception of public reason for East Asian societies cannot, and should not, accept the full-fledged liberal conception of individual rights and liberties. Instead, it must integrate the central requirements of the virtues that are still vibrantly practiced in contemporary East Asian societies. No matter whether I am right regarding this matter, nobody should simply assume that there exists a shared global political culture to ground a universal conception of public reason for all societies in the world. Li seems to take it for granted that the Rawlsian liberal conception of public reason serves as a global standard for public reason, applicable not only to Western societies but also to non-Western societies, such as Hong Kong, in which he lives, and he takes it to constitute an adequate ground on which to formulate policies regarding bioethical issues such as abortion and medically assisted suicide (MAS) in Hong Kong.

Li stresses that on the Rawlsian liberal conception of public reason, political values and public reasons are self-standing, in the sense that they do not presuppose any comprehensive doctrine. However, he fails to distinguish two different meanings of "self-standing" at stake: that "they do not presuppose any comprehensive doctrine" does not mean that "they are not supported by any comprehensive doctrine." Rawls has clearly conceded (e.g., at the conclusion of his paper 'The Idea of Public Reason Revisited') that in order for such values and reasons to stand as public reasons for modern Western societies, they have to be supported by the reasonable comprehensive doctrines present in modern Western societies. It seems that Li intends to downplay this fact. He fails to recognize that the separation between the Rawlsian conception of public reason and the comprehensive doctrines operating in the modern West is far less radical than he believes. For example, by analyzing Rawls' comment on the issue of abortion in terms of his conception of public reason, my chapter illustrates that Rawls' intended neutrality of aim does not stand separable from the shared individualistic moral feature of Western comprehensive doctrines, such as liberal comprehensive doctrines. This is the case even if Rawls is not disingenuous regarding the neutrality of aim toward different Western comprehensive doctrines in constructing his conception of public reason. The problem is that the shared individualistic moral feature among Western comprehensive doctrines is not possessed by the comprehensive doctrines present in non-Western societies, such as Confucian comprehensive doctrines operating in East Asian societies. As a result, Rawls' conception of public reason may not have been designed to be neutral between Western comprehensive doctrines (such as comprehensive liberalism) and Eastern comprehensive doctrines (such as comprehensive Confucianism). Accordingly, Li is not justified in applying it to East Asian societies without offering qualifying arguments.

In the second part of his chapter, Li addresses the issue of medically assisted suicide (MAS) by using the Rawlsian liberal conception with reference to Scanlon's contractualism. Li argues that MAS should be legalized if certain conditions are met. I think his argument is gravely deficient not so much because he has drawn this conclusion, but because he has failed to consider certain important moral and political values. In particular, he lists "autonomy," "the badness of suffering," and "the futility of continued suffering" as the political values in favor of MAS, and enumerates "the religious view that only God can decide who should die," "the slippery slope argument," and "the argument from abuse" as political considerations against MAS. It should be interesting for readers to compare these values he raises for his argument with the following concerns that I propose in my chapter for considering whether MAS should be legalized in East Asian societies: (1) the due respect for the patient's request for assistance for suicide; (2) the cherishing of the patient's life as demanded by the central requirements of basic virtues such as *ren* (humanity), *yi* (righteousness), *li* (ritual propriety), *xiao* (filial piety), and *he* (harmony); (3) the proper professional role of the physician as implied by the virtues of *ren*, *yi*, *li*, *xiao*, and *he*; and (4) the likely effects of such legalization on elderly people in these societies. Li may want to charge that I have not given sufficient weight to the values of "the badness of suffering" and "the futility of continued suffering," but I have at least considered them in my argument and attempted to balance them against other values (see Section VI of my chapter). In his case, however, he simply refuses to take into account the values of the virtues, as shown in my (2), (3), and (4) above, which are operative in East Asian societies such as Hong Kong.

Following Scanlon's contractualism, Li holds that "an act, policy or law is permissible if and only if it can be justified to everyone affected by it." If this liberal contractual aspiration is practicable, I think it has to start from a particular place in the world. For example, to determine an act, policy, or law about MAS, "everyone in every place" is too abstract; it is much more sensible and feasible for both Li and myself to consider "everyone in Hong Kong," as we are living in Hong Kong and ruled by the law of Hong Kong. In a recent online bioethics course I taught at the City University of Hong Kong in March 2020, I followed the politically correct lecture style of "value-clarification" and introduced the pros and cons of MAS to my 82 undergraduate students from Hong Kong. The class ended up, unsurprisingly, being divisive regarding whether MAS should be legalized in Hong Kong. The main reason offered by the students against legalization was that it is morally evil to intentionally help kill an innocent human patient in the medical context, even if this is requested by the patient in the terminally ill and suffering situation. This reason seems to resemble the first central moral and political requirement of the Confucian virtues (that human lives are most noble (*gui*) or valuable among all types of lives in the world so that innocent human lives must be respected and protected, regardless of individual autonomy) that I lay out in my chapter, although I did not introduce it to the students in my class. Moreover, two-thirds of the students contended that adult children should have a right to veto their parent's request for MAS. The reasoning for this right offered by the students was as follows: given the intimate parent–child relationship naturally formed in the society of Hong Kong, the adult child should undertake filial obligations (*xiao*) to the parent, and this obligation should grant the child the veto right in order to safeguard the parent's wellbeing in the medical context. Although the survey result of my class may not be reliably representative of the entire society of Hong Kong, it does suggest that the way of life of Hong Kong's younger generation remains significantly affected by the Confucian virtue of filial piety, notwithstanding the liberal contract-based ethical understanding of this relation. No matter whether Li agrees with such reasons offered by my students, he should have taken them into account in order to render his MAS legislation proposal "justified to everyone affected by it" in Hong Kong. He should not have simply taken the Rawlsian liberal idea of public reason to be the proper conception for Hong Kong.

Central to Farrell–Tham's natural-law-tradition (NLT) idea of public reason is not liberal political reasoning, but rather, as they explicate in their chapter, essentially truth-directed political enquiry. From this NLT idea, public reason belongs to the category of natural law reasons, whose truth is of divine origin, but is not divinely revealed. Rather, as Farrell–Tham emphasize, natural law reasons are accessible to anyone who gives due consideration to the relevant facts for accepting them, and public reasoning starts from a few fundamental NLT theses, such as "a voluntary action is (morally) good if and only if it is not contrary to the objective order of human goods." Hence, the normative requirements of objective human goods constitute pre-political moral constraints on legislation, and public reasoning for legislation should be conducted according to the criteria of rational enquiry. Accordingly, Farrell–Tham contend, on the NLT conception, public reason must be a form of truth-directed enquiry that inevitably pursues the cognitive value of truth, by contrast with the Rawlsian liberal political conception which suspends commitment to the idea of a single right political order.

In particular, Farrell–Tham draw on the distinction made by Alasdair MacIntyre between philosophical and practical problems of intractable moral disagreement. Public reasoning on such disagreement, as they see it, has to involve a resort to theoretical enquiry which must, as required by the precepts of rational deliberation (i.e., the precepts of NLT), not be restricted to a debate between rival political philosophies, but must extend to their deeper, underlying philosophical and cultural differences. Although they note that MacIntyre does not think such shared rational deliberation can take place within the politics of the modern state (for reasons such as that there is no longer a shared conception of the common good and accepted modes of rational enquiry in the modern pluralist state), they appear to be more optimistic than MacIntyre. They keep insisting that the NLT conception of public reason be adduced from the fundamental theses of NLT along with MacIntyre's account, and it be used to direct all pluralist societies in the contemporary world.

I think Farrell–Tham have made a legitimate point that public reasoning should not abstain from pursuing moral and political truths. Political liberals may have gone too far in claiming the necessity of epistemic abstinence in political deliberation in contemporary society, although we are indeed facing moral pluralism. As Farrell–Tham indicate, what we are facing is not only conflicting views on first-order matters, such as abortion and MAS, but also a complex circumstance in which legislation on such first-order matters is fraught with contentious second-order questions, such as questions of personal identity and the grounds for a rights-claim. Such deeper philosophical and cultural issues, as Farrell–Tham emphasize, should not be bracketed when individuals are interested in pursuing truth by engaging in sincere and serious public discussion and reasoning with each other. For example, as Farrell–Tham illustrate, legislation on abortion pertains to the right to life, which is in turn connected with the metaphysical question of what constitutes a human person. To this last question there is the animalist conception of personhood in contrast with the psychological conception. It would be both disingenuous and unfair to exclude these profound questions from entering public discussion and deliberation for legislation on abortion. Indeed, I have learned this significant point ("public reasoning as truth-directed enquiry") from Farrell–Tham's chapter. Although my chapter does not touch on this point, I don't think my reconstructionist Confucian conception of public reason would have a problem in accepting it as a necessary goal of the conception.

That said, I don't think pursuing the truth should be the only goal for a suitable conception of public reason. Believing in the cognitive value of moral truth is one thing, but believing that one has rationally proved a particular view or theory as the moral truth (so that anyone who does not accept it is irrational) is quite another. Confucianism would accept the former but not the latter. Even if a framework of all normative reasons (or all natural law reasons in Farrell–Tham's case) can be employed to pursue truth as the only important goal, public reasons are only part of normative reasons and must set additional social goals to pursue as well. In particular, a conception of public reason must be used to seek such non-cognitive values as societal stability, cohesion, and peace, along with the cognitive value of moral truths. This is to say, the overall goals of a suitable conception of public reason for a society must include both cognitive and non-cognitive values, which need to be balanced against each other in order to contribute to policy discussion, formulation, and justification in the society.

In my view, an appropriate balancing method is not to set down an absolute order of priority among these values but to adopt different ranking strategies for two different tasks involved in using public reasons: the task of policy discussion and the task of policy justification. For the task of policy discussion, more weight should be given to cognitive than non-cognitive values. That is, individuals should be free to appeal to any comprehensive doctrines they believe to be true in constructing their arguments and attempt to convince others to formulate a policy or law as they see proper. Indeed, in my chapter I suggest dropping Rawls' theoretical and practical requirements (in his guidelines of inquiry for evaluating whether a comprehensive doctrine is "reasonable," as well as for conducting public discussion) so as to facilitate sincere and honest public discussion and deliberation. For example, under my Confucian conception of public reason, those beliefs of the virtues that are still vitally accepted and practiced in East Asian societies (such as the two central requirements of the virtues I lay out in the chapter) are certainly legitimate to appeal to in public discussion. However, the beliefs of the virtues that are only embraced by a small group of individuals should also be legitimate to use in public discussion, as long as the users hold them to be true. Of course, every individual is free to choose or not to choose a particular idea or doctrine to be articulated in public discussion, depending on one's practical judgment regarding whether it is strategically helpful to be used for convincing others to accept one's alleged truth and support one's favorite policy decision.

For the task of policy justification, however, more weight should be given to non-cognitive than cognitive values in the use of public reason. That is, when government attempts to provide justification for a policy or law that is already made by a government branch through a procedure consistent with public reason, controversial comprehensive doctrines should be avoided as far as possible. For example, while both the animalist and psychological notions of personhood may legitimately be adopted and articulated by individuals in public discussion regarding legislation on abortion, neither of them should be cited by the government in justifying an already enacted piece of legislation. Given such legislation, to cite incompatible notions in justification cannot help but entail that one of the two notions is held by the government to be superior to the other. But the government does not have the capacity to judge which notion is superior, and neither would it be fair to do so. Instead, this task of justification should preserve non-cognitive values such as societal stability, cohesion, and peace, rather than seeking moral truth. Hence, its arguments should be constructed in terms of much less controversial public reasons, such as those beliefs of rights and virtues that are generally accepted in the society. Of course, the cognitive value of moral truth can continue to be pursued through ever on-going public discussion.

In sum, the NLT conception of public reason upholds a truth-directed, epistemological ideal, the Rawlsian liberal political conception upholds a justice-directed, non-epistemological ideal, and my reconstructionist Confucian conception upholds a comprehensive ideal, in pursuit of both cognitive and non-cognitive values. Deep cultural and comprehensive philosophical doctrines should not be excluded from the scope of public reason, so that citizens can engage in sincere and un-rhetorical exchanges in public discussion and deliberation. In fact, Farrell–Tham never claim that all moral and political truths established through NLT rational arguments be maintained in the policy and law of a society at any cost. They stand ready to compromise the value of such truths for the sake of other values, such as social stability, cohesion or peace. They admit that public reason is more limited in scope than natural law reasoning. Moreover, they also admit that even if a law is made against fundamental human goods, one may still have reason to respect the political authority and the legal system legitimate in virtue of their contribution to social cohesion, peace, and justice. What they fail to recognize, however, is that there are different understandings of substantive truths just as there are distinct religious and moral traditions in society. In my view, both the Confucian and natural law traditions hold a cognitivist and realist position toward moral statements (which are apt for truth and falsity and many of which are in fact true), and the Confucian tradition should learn from the natural law tradition (that has developed a well-articulated system of rational beliefs and arguments) and conduct a cross-tradition dialogue.

However, this does not mean that Confucian tradition must accept all fundamental theses set forth by the natural law tradition. Even if both traditions engage in truth-directed enquiry, they may not arrive at the same moral conclusions or put forward similar discursive arguments regarding moral and political issues. Moreover, in building and recommending a conception of public reason for a society, one is essentially concerned with certain non-cognitive values (such as social cohesion, peace, and justice), as well as the cognitive value of moral truths. One will have to engage a society's particular cultural and philosophical doctrines that are affecting and even shaping the cohesion, peace, and justice of the people in the society. This is to say, complete public reasoning cannot, and should not, be truth-directed only. It must also be directed by non-cognitive values. That is why I have prosposed my version of a reconstructionist Confucian conception of reason for East Asian societies.

7

Further Reflections

Hon-Lam Li

In this chapter, I shall respond to criticisms raised by Farrell and Tham in Chap. 5 and those by Fan in Chap. 6. I shall also reflect on Rawls' political liberalism and issues arising therefrom.

(A) Political Liberalism: A Synopsis

1. Before we address the charges leveled at Rawls and my articulation of his political liberalism in Chap. 1, let us remind ourselves of the reasons for which Rawls wrote his book, *Political Liberalism*.[1] There were two

I thank Michael Campbell, Win-chiat Lee, and Alastair Campbell for comments. I am grateful to Sin-Yee Chan and especially Kwong-loi Shun and Chung-yi Cheng for discussions on the relation between Confucianism and absolute monarchy. I am indebted to Allen Wood for an idea on the discussion of Velleman's article (see Note 31). I thank Kai-Yee Wong, Leo Cheung, and Alexandre Erler for comments on the final section of this chapter, "Is Political Liberalism Simply the Product of Western Hegemony?" I thank Lei Zhong for his contribution to Note 58.

[1] This will go toward answering the criticisms raised by Farrell and Tham in Chap. 5.

H.-L. Li (✉)
Department of Philosophy and CUHK Centre for Bioethics, The Chinese University of Hong Kong, Shatin, Hong Kong

H.-L. Li, M. Campbell (eds.), *Public Reason and Bioethics*,
https://doi.org/10.1007/978-3-030-61170-5_7

reasons. First, Rawls wants to see whether, and how, a constitutional liberal democratic society could be stable over time *for the right reasons.* Although Rawls continues to be convinced of the truths of his positions regarding full autonomy, the good of the community, his philosophical accounts of the nature of agency and practical reason, of moral objectivity, moral justification, and moral truth proposed in *A Theory of Justice* (henceforth "*Theory*"), he realizes that these positions are not *publicly* justifiable to citizens in a pluralistic society.[2] This is because citizens hold different comprehensive doctrines—about "what is of value in human life, and ideals of personal character, as well as ideals of friendship and of familial and associational relationships, and much else that is to inform our conduct, and in the limit to our life as a whole" (*Political Liberalism*, henceforth "*PL*," 13).

When persons have realized their capacity for a sense of justice and for a conception of the good[3] ("the two moral powers") and have the powers of reason, they are free. Their having these powers to the requisite minimum degree to be fully cooperating members of society makes them equal (*PL* 19). Free and equal citizens in a constitutional liberal democracy have an enduring desire to honor fair terms of cooperation.

The sources of disagreement between reasonable persons, or "the burdens of judgment," are due to (a) the difficulty in assessing scientific and empirical evidence relating to an issue; (b) our different views in according the proper weight to some consideration; (c) the vagueness of concepts, and not only moral and political ones; (d) our disparate total experiences, which shape our assessment of evidence and our weighing of moral and political values; (e) the difficulty in agreeing on a conclusion in the face of different kinds of normative considerations of different forces on both sides of an issue; and (f) "there is no social world without loss" and we are forced to set priorities (*PL* 56–68, 197). Because of these various factors, many of our most important judgments are made under circumstances in which no conscientious persons, even with full powers of reason, can be expected to agree on the same conclusions.

[2] See Samuel Freeman, *Rawls*, (Abingdon and New York: Routledge, 2007), 324–6.

[3] A conception of the good is to be understood broadly to include a conception of what is valuable in human life. See *PL* 19.

This gives rise to the epistemic ideas of reasonable disagreement and reasonable comprehensive doctrines. A reasonable comprehensive doctrine (a) involves theoretical reason and cover major religious, philosophical, and moral aspects of human life in a more or less consistent and coherent way; (b) organizes recognized values in a consistent way and gives certain values a particular primacy and weight; and (c) normally belongs to a tradition of thought and tends to evolve slowly in light of what it sees as good and sufficient reasons (*PL* 59). A reasonable comprehensive doctrine must also recognize the burdens of judgment, and consequently other political values, such as liberty of conscience.[4] Thus, Rawls suggests that Catholicism has shown itself to be a reasonable comprehensive doctrine: since the Second Vatican Council (1962–65), the Church has modified its doctrine to accommodate many scientific and moral views in the modern world (*LP* 154, n52).[5] But the literal interpretation of the Bible by the fundamentalists—namely, that God has created the fossils of the dinosaurs to test our faith—would not be reasonable.[6] Similarly, the doctrine of astrology, even if held by otherwise reasonable persons, would not be a reasonable doctrine.[7] In the same vein, libertarianism—the view that extreme inequality is compatible with distributive justice (e.g., held by Robert Nozick)—is also unreasonable. (I shall return to this point below.)

A disagreement is reasonable if it is a disagreement between reasonable doctrines. Although Rawls' theories are normative and hence not meant to be descriptive of the real world, he holds that as a result of the "burdens of judgment" and the long-term outcome of the work of human reason under free conditions, there exists a plurality of reasonable

[4] The reason for this is that free and equal citizens would propose and accept a fair system of cooperation among themselves. This system would rule out anyone's trying to use the state's power to favor a comprehensive doctrine, or to impose its implications on the rest. See *Restatement*, 191–2.

[5] Samuel Freeman, *Rawls*, op cit., 350. See also Antonella Piccinin, "Rawls and Catholicism: Towards Reconciliation?" in *Cultural and Religious Studies* 7, no. 1, (January 2019): 50–6. D doi: 10.17265/2328-2177/2019.01.004

[6] Samuel Freeman, op. cit., 350–1.

[7] Cf. Martha Nussbaum, "Perfectionist Liberalism and Political Liberalism," *Philosophy & Public Affairs* 39, no. 1 (WINTER 2011), 3–45, esp. 5, 26, 29.

comprehensive doctrines in liberal well-ordered societies. As a result, even fully rational and reasonable persons often cannot agree on religious, philosophical, and moral matters. This is known as the "fact of reasonable pluralism" (*PL* 153).

As I described in Chap. 1 of this volume, Rawls proposes that reasonable citizens holding reasonable comprehensive doctrines have different reasons of their own to agree on a political conception of justice. In *Political Liberalism*, Rawls takes this conception to be justice-as-fairness and its two principles of justice. But in his later work, "The Idea of Public Reason Revisited," he allows the political conception to be any member of "a family of liberal conceptions" because they satisfy the criterion of reciprocity. So a liberal conception should (a) have a list of basic rights, liberties, and opportunities; (b) have an assignment of special priority to those basic rights, liberties, and opportunities, especially with respect to claims of the general good and perfectionist values; and (c) contain measures that enable all citizens an adequate all-purpose means to make effective use of their freedoms (*LP* 141). The reason why Rawls rejects libertarianism as an unreasonable doctrine is that it fails to fulfill condition (c), and hence it violates the criterion of reciprocity.[8] Rawls holds that any member of a family of liberalism must meet condition (c)—as well as conditions (a) and (b)—so that the basic structure of society can be stable *for the right reasons*. Yet, a libertarian regime would not have stability for the right reasons.

Apart from his concern to stabilize society for the right reasons, Rawls also wrote *Political Liberalism* in order to establish the conditions of legitimacy for the exercise of political power in a constitutional liberal democracy.[9] On Rawls' view, legislators, executive officials, and judges—especially appellant and supreme court judges—as well as those running campaigns for high office (such as Joe Biden and Paul Ryan in 2012), should articulate their political positions in terms of public reasons, which are self-standing and are not derived from any comprehensive doctrines. Public justification is therefore "not simply valid reasoning, but argument addressed to others: it proceeds correctly from premises we accept to con-

[8] See my detailed discussion in Chap. 1.

[9] See Samuel Freeman, Rawls, op. cit., 324–5.

clusions we think they could also reasonably accept" ("The Idea of Public Reason Revisited," in *The Law of Peoples*, henceforth "*LP*," 155).[10] Citizens and officials who can discharge their moral responsibility of public justification are said to have fulfilled their "duty of civility" (*LP* 155).

However, Rawls initially limited occasions on which public justification is necessary to those where "constitutional essentials and matters of basic justice" are concerned (*PL* 214). This seems too restrictive for the following reason. Given Rawls' narrow definition of "constitutional essentials," important constitutional issues, such as abortion and medical assistance in dying, may fall outside the ambit of "constitutional essentials."[11] This is despite the fact that Rawls and five other preeminent philosophers submitted an *amicus* brief to the Federal Supreme Court of the United States in which they argued that US citizens should have a constitutional right to medical assistance in dying.[12] In *Justice as Fairness: A Restatement*, Rawls duly addressed this problem. He said that if an issue "borders" on a constitutional essential, and "can be the cause of deep conflict" (*Restatement* 117), it should be discoursed in the public forum in terms of public reasons only.

According to Rawls, if officials and citizens fulfill the duty of civility by deliberating on an issue of importance in terms of public reasons, "the legal enactment expressing the opinion of the majority is legitimate law" even if each citizen thinks that the law enacted is not the most reasonable (*LP* 137). Apart from claiming that citizens of different reasonable comprehensive doctrines would converge on a political conception of justice, Rawls does not claim that people who discourse in public reasons would agree on any controversial public issue, such as abortion or medically assisted death.

[10] References to pp. 129-180 of *LP* refer to "The Idea of Public Reason Revisited." References to pp. 1–128 are to "The Law of Peoples," contained in *LP*.

[11] Rawls' concept of "constitutional essentials" includes "(a) fundamental principles that specify the general structure of government and the political process" and "(b) equal basic rights and liberties of citizenship that legislative majorities are to respect: such as the right to vote and to participate in politics, liberty of conscience, freedom of thought and of association, as well as the protections of the rule of law" (*PL* 227).

[12] See Ronald Dworkin, et al., "Assisted Suicide: The Philosophers' Brief," *New York Review of Books*, March 27, 1997, Issue; https://www.nybooks.com/articles/1997/03/27/assisted-suicide-the-philosophers-brief/?printpage=true

Now that we have reminded ourselves of Rawls' theories, we are in a better position to answer the objections raised by Farrell and Tham.

(B) Response to Farrell and Tham

2. With respect to what I wrote in Chap. 1, Farrell and Tham make the following charges. First, they say that "the idea of public reason is essentially truth-directed political inquiry and decision making," and that Rawls' political liberalism is not concerned with truths. On their view, "it is unreasonable to bracket for the sake of social cooperation comprehensive doctrines and keep them out of the domain of public reason." Certainly, Rawls holds that his own comprehensive doctrines put forward in *Theory* are all true. To insist that the "idea of public reason is essentially truth-directed," as Farrell and Tham do, is simply to deny the "fact of reasonable pluralism."

Moreover, to insist that the role of public reason is truth-seeking, as Farrell and Tham do, does not mean that the speaker knows which belief is true, and which is false, given the fact of reasonable pluralism (because of the burdens of judgment). As Rawls explains:

> Since many doctrines are seen to be reasonable, those who insist, when fundamental political questions are at stake, on what they take as true but others do not, seem to others simply to insist on their own beliefs when they have the political power to do so. Of course, those who do insist on their beliefs also insist that their beliefs alone are true: they impose their beliefs because, they say, their beliefs are true and not because they are their beliefs. But this is a claim that all equally could make; it is also a claim that cannot be made good by anyone to citizens generally. So, when we make such claims others, who are themselves reasonable, must count us unreasonable. And indeed we are, as we want to use state power, the collective power of equal citizens, to prevent the rest from affirming their not unreasonable views." (*PL* 61)

Thus, Rawls holds that reasonable persons see that the burdens of judgment set limits on what can reasonably be justified to others, and so they uphold a certain form of liberty of conscience and freedom of thought. For this reason, it is "unreasonable for us to use political power, should we possess it, or share it with others, to repress comprehensive views that are not unreasonable" (*PL* 61).[13]

3. Furthermore, quoting Michael Sandel, Farrell and Tham argue that "[p]olitical liberalism must assume not only that the exercise of human reason under conditions of freedom will produce disagreements about the good life but also that the exercise of human freedom under conditions of freedom will *not* produce disagreements about justice," and that "[t]he political liberal cannot rule out the libertarian as unreasonable in advance." As I mentioned above, Rawls holds that libertarianism is an unreasonable doctrine (because it does not seek to provide all citizens an adequate all-purpose means to make effective use of their freedoms), and that a libertarian regime cannot be stable for the right reasons.

4. In his review of *Political Liberalism*,[14] Sandel argues that Stephen Douglas' position is similar to that of political liberals' in their common desire to bracket controversial normative issues. Thus, Sandel might think that the debate between Stephen Douglas and Abraham Lincoln parallels the recent debate between Joe Biden and Paul Ryan. In other words, he might think that Stephen Douglas proposes a policy of state neutrality in favor of keeping slavery, and that this is similar to Biden's stance, just as Abraham Lincoln's abolitionist position aligns with Paul Ryan's. I believe the two issues are not analogous.[15]

First, Abraham Lincoln said that "Douglas ignored the basic humanity of blacks" and that "slaves did have an equal right to liberty."[16] Trying to

[13] Samuel Freeman says that the reason for which Rawls does not want to use the term "true" in *Political Liberalism* to describe not only comprehensive doctrines but also the political conception of justice (such as liberal principles of justice) because the concept of truth is a contested one in philosophy, and Rawls wants a political conception to be neutral to comprehensive doctrines, including various philosophical doctrines, such as moral skepticism, moral emotivism, and other forms of moral noncognitivism. See Freeman, op. cit., 355–60, esp. 356.

[14] Michael Sandel, "Review of *Political Liberalism*," *Harvard Law Review* 107, no. 7, (May, 1994), 1765–94, esp. 1777–83.

[15] Farrell and Tham made this criticism in their penultimate draft, but have since dropped it. I am responding to this criticism because Sandel's argument is an important one directed against Rawls and it is instructive to see where it fails.

[16] Debate at Ottawa, Illinois, Lincoln quote, August 21, 1858.

prove that Lincoln was an abolitionist, Douglas charged Lincoln with his (Lincoln's) previous statements (prior to the debate in Charleston) "that the declaration that all men are created equal applies to blacks as well as whites."[17] Clearly, Lincoln's view represents a basic principle of liberalism, and is consistent with public reason and the liberal principle that all people have equal right to liberty.

Second, Douglas proposed the doctrine of "popular sovereignty," which was a controversial political doctrine according to which the people of federal territories—and not Congress—should decide for themselves whether their territories would enter the Union as free or slave states.[18] Douglas' doctrine would be an unreasonable doctrine because it violates the principle that everyone should have basic rights, liberties, and opportunities, and because it violates the principle that society must provide an adequate all-purpose means to everyone to make effective use of their freedoms. Because of its unreasonableness, Douglas' motion would never make it to the stage for voting, according to political liberalism. Furthermore, the voting itself was problematic because blacks had no right to vote.[19]

I therefore believe that Biden's position is actually close to Lincoln's, in that they both hold that violation of people's basic right is wrong, and that the slaves are people. Sandel's argument might have greater force against someone who wants to bracket any controversial issues whatsoever. But this view has no force even against Nagel's view (discussed in Chap. 1) because Nagel holds that such bracketing only applies to controversial moral issues that are private or self-regarding. Nagel would argue that slavery is not a self-regarding issue.

5. Farrell and Tham raise another criticism of liberalism. Citing Charles Taylor's paper, "Atomism," they write: "Charles Taylor provides a good example of an analytical argument against the individualist conception of the person on which liberalism is based."[20] Three points are in order.

---

[17] https://en.wikipedia.org/wiki/Lincoln–Douglas_debates

[18] Encyclopedia Britannica: https://www.britannica.com/topic/popular-sovereignty

[19] Rawls holds that Lincoln's view was reasonable (even if not the most reasonable), but Douglas' was not (*LP* 174).

[20] Chapter 5 of this volume, Note 7.

First, Taylor's charge of atomism was directed neither at Rawls, nor at liberalism. Taylor's target was Robert Nozick, and Nozick's libertarian theory as expressed in *Anarchy, State, and Utopia*.[21] In fact, neither Rawls' name, nor the term "liberalism," appears in Taylor's paper.

Second, I have argued in Chap. 4 that Rawlsian liberalism is not individualist in any objectionable sense. For one thing, Rawls' preferred political conception of justice asks us to care for the least advantaged members of society, so that any social inequality must be to their advantage. For another, he stresses the importance of "social cooperation among equals for mutual advantage" (*Theory* 14), or of "the common good"[22]—understood as "certain general conditions that are in an appropriate sense equally to everyone's advantage" (*Theory*, 246). Thus, a government should "aim at the common good, that is, at maintaining conditions and achieving objectives that are similarly to everyone's advantage" (*Theory*, 233). That stance is clearly not individualistic.

Is it individualist to have a fully adequate scheme of basic liberties which is compatible with a similar scheme of liberties for all (*PL* 291), and to assign special priority to them over claims of the general good? It does not seem so. In rejecting the view that basic rights and liberties could be traded off for greater economic and social benefits, Rawls' "appeal has been to the common good in the form of the basic equal liberties of the representative citizen" (*Theory*, 243).

Yet Rawls bends over backward to concede that there is a minimal sense in which these basic liberties might be individualistic. As alluded to above, Rawls endorses Isiah Berlin's view that "there is no social world without loss" (*PL* 197), which means (in Berlin's words): "Some among the Great Goods cannot live together. That is a conceptual truth. We are doomed to choose, and every choice may entail an irreparable loss."[23] No society can include all forms of life. Rawls elaborates: "That there is no social world without loss is rooted in the nature of values and the world, and much human tragedy reflects that. A just liberal society may have far more space than other social worlds but it can never be without loss" (*PL*

[21] Robert Nozick, *Anarchy, State, and Utopia*, (Basic Books, 1974).

[22] See John Rawls, *Theory*, 101, 123, 178, 232, 233, 243, 246, 425, 472, 478, 489.

[23] Isiah Berlin, "The Pursuit of the Ideal," in *The Crooked Timber of Humanity* (New York: Knopf, 1991), 13, cited by John Rawls, *Political Liberalism*, 197.

197*n*). Thus, Rawls points out, where some religious sects oppose the culture of the modern world and want to lead their lives away from its unwanted influences, a question regarding their children's education arises. Should the education of these children include knowledge of their constitutional and civic rights so that "they know the liberty of conscience exists in their society and that apostasy is not a legal crime" (*PL* 199)? Rawls answers in the affirmative, and leaves us with an open question whether the passing of a certain historical way of life is to be regretted (*PL* 199).

That said, is political liberalism unjustly biased against certain comprehensive conceptions, or biased in favor of individualistic conceptions? That would be so, Rawls says, "only if, say, individualistic ones alone can endure in a liberal society, or they so predominate that associations affirming values of religion or community cannot flourish, and moreover that the conditions leading to this outcome are themselves unjust" (*PL* 199). This is not a society compatible with political liberalism.

6. We come to Farrell and Tham's final objection: Kant, who inspires Rawls, objects to suicide (and hence would also object to MAS). Farrell and Tham do not say why Kant objects to suicide. Kant's objection to suicide is stated in the *Groundwork of the Metaphysics of Morals*. To illustrate the first formulation of the Categorical Imperative, Kant says that no maxim that allows suicide can be universalized, and hence no such maxim is permissible. In illustrating the second formulation of the Categorical Imperative, Kant simply says that one who commits suicide to avoid the pain of illness would be to use oneself *merely* as a means, and consequently that suicide is impermissible. But neither of these illustrations amounts to an argument because Kant simply assumes that suicide is impermissible.[24]

[24] One would draw very different conclusion about these formulations of the Categorical Imperative if one holds that suicide is permissible. For instance, one might think that the first formulation is problematic because there are so many false negatives. The maxim that I want to be a better-than-average student, and the maxim that I want to donate more money than an average citizen would do to help the poor, cannot be universalized. See Allen Wood, *Kantian Ethics* (New York: Cambridge University Press, 2008), Chap. 4.

In what follows, I shall examine J. David Velleman's well-known Kantian view against assisted suicide.[25] It is grounded in the sanctity of life. Kant, who holds that whereas nonhuman animals have "price" and are replaceable, human beings have "worth" and are irreplaceable because of the equal dignity that all have in virtue of their rational nature. He also says that a human being who commits suicide in order to avoid a painful death is using herself as a mere means.[26] However, Kant does not tell us why one who commits suicide must be using oneself as a mere means (which is impermissible), as opposed to using oneself as a means (which is permissible).

Picking up Kant's thread, J. David Velleman weaves an argument against suicide and assisted suicide, as well as euthanasia. Velleman argues that the case for suicide and assisted suicide is grounded in the acceptance of two "principles." One principle—which Velleman accepts—is that "there is a presumption in favor of deferring to a person's judgment on the subject of his own good" ("A Right of Self-Termination?" henceforth "Self-Termination").[27] The other principle—which Velleman rejects—is "the principle that a person has the right to end his life solely on the grounds of the benefits he will thereby obtain or the harms that he will avoid" ("Self-Termination" 608). Thus, while Velleman grants that we should defer to the patient regarding how bad her pain is, and whether (given the pain) her life is worth living, as well as whether it is in her interest to commit suicide, he rejects the view that because it is in the interest of a patient to commit suicide it is permissible for her to do so.

Velleman bases his view on the premise that the interest of any person counts only because she is a person—only because she "counts"—and consequently whether a person counts is not an issue on which we need to defer to her, since it is not an interest or a value "for" her, but rather an interest-independent value or a value "in" her. This value—"dignity"—that all people have equally is by virtue of their rational nature, or their "personhood." Our dignity, or the value of being a person, "is therefore

[25] I draw on materials used in "What We Owe to Terminally Ill Patients: The Option of Physician-Assisted Suicide," *Asian Bioethics Review*, vol. 8, no. 3 (September 2016), pp. 231–34.

[26] Kant, *Groundwork of the Metaphysics of Morals*, ed. Allen Wood.

[27] J. D.Velleman, "A Right of Self-Termination?" *Ethics* 109, no. 3 (April 1999): 606–28, at 607.

something larger than any particular person who embodies it" ("Self-Termination" 612). He further explains:

> [V]alue *for* a person stands to value *in* the person roughly as the value of means stands to that of the end: in each case, the former merits concern only on the basis of concern for the latter. And conditional values cannot be weighed against the unconditional values on which they depend. The value of means to an end cannot overshadow or be overshadowed by the value of the end, because it already is only a shadow of that value, in the sense of being dependent upon it. Similarly, the value of what's good for a person is only a shadow of the value inhering in the person, and cannot overshadow or be overshadowed by it. ("Self-Termination" 613)

Unlike the question of how bad the pain is, or whether one's life is worth living, the question on the value of one's dignity is a question which one is not privileged to answer. Therefore, "respect for a person's autonomy does not require deference to him on questions of his dignity, as it does on questions of his good" ("Self-Termination" 612-3).

Velleman further argues that someone's interest counts only because she (the person) counts, which means that we must presuppose that she has dignity or the value of her personhood. Thus, an important value that has been overlooked in the liberal view is dignity, which transcends interests but is also presupposed by interests that count. Therefore, a view that takes account of only a person's harm and benefit in deciding whether to commit suicide—or whether to smoke cigarettes—has left out an important value (namely, dignity) in the debate. Yet, dignity is a value that is incommensurable with interests since it is an "absolute value" that interests (or "relative values") must presuppose.

Velleman thinks that his account can explain why someone who sells herself as a slave for money is violating her own personhood. In addition, just as "the anti-Semitism of a self-hating Jew" is self-denigrating, so someone who wants to terminate her life on the grounds that life is not worth living is denigrating the value of personhood, according to him.

Moreover, someone who encourages impermissible conduct (such as suicide[28]) is engaging in something that is impermissible.

Velleman concedes that if a patient's rational faculties are decaying, at some point she will lose her dignity, and his argument would not apply to her. At another point in his paper, Velleman comes close to conceding that if the pain a patient is suffering in "is truly unbearable pain," then "[t]o find pain unbearable is to find it thus destructive not just of one's well-being but of oneself. … If his pain is truly unbearable, then he isn't his rational self any longer: he is falling apart in pain" ("Self-Termination" 618).[29]

Velleman's argument is problematic. First, on his view, one's interest counts only because one—in virtue of one's dignity (or rational nature)—also counts. This would entail that the interests of a nonhuman animal (e.g., a dog) would not count at all. This is very counterintuitive. Living in a pre-Darwinian era, Kant did subscribe to such a conclusion. We now realize that we are not that different from other animals because nonhuman mammals are rational to some degree and have psychological properties that are not totally different from human beings.

Velleman's argument also goes astray in another way. What is the role of dignity in ethics? Dignity could be understood to play the same role as the more familiar terms of "inherent value" (Tom Regan) and "moral status" (Mary Ann Warren). When we weigh claims of ours against each other, or alternatively weigh human interests against interests of nonhuman animals, we need a way to compare these claims or interests. What moral weight should we give to human and nonhuman interests?[30]

---

[28] Velleman thinks that suicide is morally impermissible, even though it is legally permissible in most jurisdictions.

[29] Velleman further explains: "I don't think that we serve the patient well, in these circumstances, by claiming broad rights of self-determination in his name. He may indeed be entitled to help in dying, and he will certainly have to participate in the relevant decisions. But let us keep in mind that these decisions would be premature if the patient were not already in the twilight of his autonomy, where self-determination is more of a shadowy presumption than a clear fact" ("Self-Termination," 618–9).

[30] Most nonhuman animals have moral status, albeit to a lesser extent than the moral status of human beings. The more plausible view is that human beings have the highest moral status, followed by other apes, then other mammals, then birds, and reptiles, and so on.

Moreover, if we understand "value in a person" as moral status—something more akin to *qualification* than an intrinsic value in itself which we must protect at all costs—it does not follow that it is impermissible for one to shorten one's life in order to live a better one.

The Kantian principle that a person is an end in itself may imply that the lives of persons have value, even great value (at least most of the time), but the value of a person is not to be identified with the value of their continued existence. Moreover, even though the dignity of persons implies that a person's life is of great value and not to be sacrificed lightly, then if what has dignity is rational nature (the capacity of a person to direct their own life), then respecting a person's dignity entails respecting their choices over their own body and life. A person who chooses to die when continued life is degrading to them is making a decision that should be respected.[31]

Contrary to Velleman's contention, it seems that we deny people's dignity when we deny persons the option to end their own lives *if the only alternative is extreme pain leading to certain death*. Therefore, Velleman's argument that even when we are suffering excruciating pain and desperately want to die, we must continue to suffer in pain just to "respect" our dignity, or value of personhood, is totally unconvincing.

Telos Revisited

7. Let me respond to some criticisms of my view raised by Farrell and Tham.[32] They argue that I have misinterpreted John Finnis' view. I do not agree. Although I am not an expert in natural law theory, I believe that Finnis writes clearly.[33] As he says:

> [Andrew Koppelman] is right in thinking that Grisez, George, Bradley and I reject as fallacious (and never argue on the basis of) any proposition like

[31] I am indebted to Allen Wood for the ideas in this paragraph.

[32] See Chap. 8 of this volume.

[33] Finnis was once my tutor in Jurisprudence at University College, Oxford. Although he and I hold different views on sexual ethics, he is a very clear writer, as acknowledged by Rawls (who thinks that Finnis is an "especially clear" writer) (*LP* 142), a superb teacher (based on my own experience), and is widely acknowledged to be perhaps the most important scholar on natural law theory.

> "natural functions or tendencies are moral standards and ought to guide deliberation and choice."[34]

Finnis does not rely on the argument that human beings have some natural functions (*ergon*), or some ultimate or final end (*telos*). He even thinks that this kind of argument is fallacious. As I pointed out in Chap. 4, the reason for which Grisez, Boyle, and Finnis reject this argument is that "the reality of free choice is incompatible with the supposition—for instance, of Aristotle—that there is a single natural end of human life."[35]

Moreover, Farrell and Tham say that "Li cites an article whose title signals that the authors intend to show, in typical natural law fashion, that humans have ultimate ends." Unfortunately, they do not say which article they have in mind. Furthermore, even if what they say is correct, we cannot assume that there is anything wrong with endorsing part of an article without endorsing its main thesis.

I take it that (contrary to Farrell and Tham) saying that humans have a *telos* is more than saying that "there are various incommensurable basic human goods (life, friendship, marriage, play, knowledge of the truth, religion)." Darwin can say that humans have no preordained *telos*, but that there exist various goods for human beings (e.g., health). Finnis holds that there are seven "basic human values," namely, life, knowledge, play, aesthetic experience, sociability (friendship), practical reasonableness, and religion (*Natural Law & Natural Rights*, Oxford: Oxford University Press, 1982, 85–90). But he rejects the view that "the conception of human good"—even that entertained by Aristotle and Aquinas—is "dependent upon" "a teleological conception of nature" (*Natural Law & Natural Rights*, 52). In other words, Finnis maintains that there are seven "basic forms of good" for humans without endorsing a teleological conception of nature or that humans have a natural function or a *telos*. (For discussion on *telos* and why the idea that a human being has a *telos*

[34] John Finnis, "Law, Morality, and 'Sexual Orientation'," in his *Human Rights and Common Good: Collected Essays*, Volume III, (Oxford University Press, 2011).

[35] German Grisez, Joseph Boyle, and John Finnis, "Practical Principles, Moral Truth, and Ultimate Ends," *American Journal of Jurisprudence*, 32:1, 99–151 at 101. For a contrary view, see Edward Feser, "Natural Law Ethics and the Revival of Aristotelian Metaphysics," in *Cambridge Companion to Natural Law Ethics*, ed. Tom Angier, 2019.

is implausible in a post-Darwinian era, see my discussion in Chap. 4 of this volume.)

Moral Contextualism and Abortion Revisited

8. It is unfortunate that in writing "By moral contextualism, we mean, and Li appears to have in mind, the view that that φ-ing, a specific kind of intentional action, is neither morally good nor bad in or of itself but only good or bad under a specific set of conditions or circumstances," Farrell and Tham have mischaracterized my view on moral contextualism. Moral contextualism is the view that whether an act is right or wrong depends on context. But it does not follow that "a specific kind of intentional action … is neither morally good or bad and in or of itself but only good or bad under a specific set of conditions or circumstances." In Chap. 4, I wrote that "we believe that in general lying is wrong," followed by the rhetorical question, "Do we believe that lying is wrong in all circumstances?" I also wrote that "if Aquinas is a moral contextualist, then he could have accepted that abortion might be wrong in most cases, but not wrong when the mother's life is endangered by the fetus' continued growth in her fallopian tube." There might be a certain act that is neither right nor wrong, but only right or wrong in certain contexts. Lying and killing do not belong to this class of acts because these acts are generally wrong, but might be justified or excused under certain circumstances.[36]

Farrell and Tham say that the Roman Catholic Church does not permit abortion of any kind, and that I am mistaken about the Church's view. Yet, they admit that some theologians hold that abortion is justified where continued pregnancy would endanger the mother's life, for example, in the cases of ectopic pregnancy (i.e., where the fetus is implanted in the mother's Fallopian tube) or where the pregnant mother has cervical cancer. Yet, they argue that termination of the pregnancy in such cases, by removal of the mother's Fallopian tube or her uterus, does not constitute abortion as long as the death of the fetus is merely foreseen but not intended as an end or as a means to an end. In Chap. 1, I have argued in

[36] For an illuminating discussion on the ideas of justification and excuse in the law, see H. L. A. Hart, *Punishment and Responsibility*, Chapters 1–3 (Oxford: Oxford University Press, 1968).

favor of the Thomson/Scanlon thesis that intention does not alter the permissibility (or impermissibility) of an act, and will have nothing more to add here, except that if the Thomson/Scanlon thesis is correct, the doctrine of double effect (DDE) is false (because DDE claims that intention could make a difference to *permissibility*, which is denied by the Thomson/Scanlon thesis).[37]

Another reason why DDE does not work is that, if a fetus is a person (as acknowledged by Farrell and Tham, and indeed by the Roman Catholic Church), while the mother may resort to her right of self-defense in order to justify her abortion (even though a fetus is an innocent threat[38]), the fetus (being a person on their view) must also have the right of self-defense. It is true that a fetus cannot do anything to stop its mother from going to the abortion clinic, but its father (especially if he believes that the fetus is a person) can exercise his right of self-defense (which covers his child)[39] and can seek an injunction from the court to stop his wife from having an abortion. Indeed, it is arguable that because the state should protect its citizens, it should protect even fetuses (if they are persons).

The true justification as to why abortion in these two cases is permissible is, I believe, that the mother has a greater degree of personhood and hence a higher moral status, or alternatively, that the mother is certainly a person but whether a fetus is a person is only a matter of probability.[40] The only claim with which my view conflicts is the dogma that a fetus is

[37] I hold that intention can alter the culpability of a wrongful act, but I agree with the Thomson/Scanlon thesis that intention does not alter the permissibility (or impermissibility) of an act. See Chap. 1 of this volume.

[38] I discussed the issue of self-defense against an innocent threat in Chap. 4. In "Self-defense," *Philosophy & Public Affairs* 20, no. 4 (Autumn, 1991), 283–310, Judith Thomson argues that a person endangered by an innocent threat has the right of self-defense. On the other hand, Michael Otsuka and Jeff McMahan argue that an innocent threat has the same status as an innocent bystander and cannot justifiably be killed. See Michael Otsuka, "Killing the Innocent in Self-Defense," *Philosophy & Public Affairs* 23, no. 1 (Winter 1994), 74–94; Jeff McMahan, "Self-Defense and the Problem of the Innocent Attacker," *Ethics* 104 (1994): 252–90. My own view is that either can kill the other based on an agent-relative interpretation of self-defense, but it does not follow that the fetus' father (who is a devout Catholic) or the state cannot apply an injunction from the court to stop the mother from having an abortion.

[39] In the Common Law, the right of self-defense covers one's children under an imminent attack.

[40] This assumes that moral status is proportional to personhood in this case. See my "Abortion and Degrees of Personhood," *Public Affairs Quarterly* 11, no. 1 (Jan 1997), 1–19.

a (full) person upon conception. But while I believe that Catholics generally accept this view, there is no reason why non-Christians should accept it.

9. In any case, Farrell and Tham's—or the Catholic Church's—strategy appears to be this: what most people consider as justified abortion (where the mother's life is endangered) is redefined as not abortion at all, just as killing in self-defense is not killing (and certainly not murder[41]) because the fact that these acts are justified immediately place them in a different category than killing or abortion.

No one would say that killing in justified self-defense is murder because murder is by definition unjustified killing as defined by the law. But is killing in self-defense not killing? They have to say that killing in self-defense is not killing in order to maintain the moral absolutist view that all killings are wrong. To say that killing in self-defense is not killing is, however, absurd.

More important, Farrell and Tham's strategy of argument by redefinition reaches its limits. They had to gloss over the case where one should lie to a Nazi German soldier looking for the Jews to send to the gas chambers. (In this case, DDE would not help them because you are lying as a means to the end of saving innocent people's lives, and DDE would disallow this.) To maintain moral absolutism, should they say that lying in this case is wrong (which would be extremely implausible)? Or should they say that this is a white lie and therefore is not lying? But why is this a white lie? *The answer is precisely that the context or circumstances make it so*. In other words, for most people, the context or circumstances in which an act is performed would contribute to the act being permissible or not. For Farrell and Tham (and according to them, the Catholic Church), the context or circumstances in which an act is performed would contribute to its reclassification (e.g., whether telling untruths is lie or white lie). The force or persuasiveness of moral contextualism is inescapable because, whether one says that killing in self-defense is justified killing, or whether it is not killing at all (so as to defend moral absolutism), one has to say

---

[41] Farrell and Tham say that killing in self-defense is not murder (which is, of course, true). But they have to claim that killing in self-defense is not killing in order to say that killing of all kinds is wrong.

that "killing" in self-defense is not wrong. (I put killing in scare quotes to indicate that Farrell and Tham might not regard this as killing at all.)

But their move by reclassifying a type of action as something else (e.g., by classifying killing in self-defense as not killing, or presumably classifying lying to the German Nazi soldier in order to save innocent lives as not lying) does not solve their problem. Unless they hold that lying to save innocent people's lives is impermissible (because it is lying), they must hold that this is a white lie and therefore not really a lie. But if they hold this view, why do they not also take the view that medical assistance in dying (where the patient is terminally ill, suffering unbearably, desperately wants to die in dignity) is not killing or not suicide at all?

From all these cases, it is clear that if we wish to defend moral absolutism, we would have to redefine or reclassify justifiable cases—of killing or lying, for instance—as falling outside of its original category. The advantage of moral contextualism is clear: even though some acts (e.g., killing, lying) are usually wrong, the context can make a difference. Therefore, MAID in a case where the three conditions hold (e.g., terminally ill) should be, or at least could be, permissible.

Moral Disagreement and Scientific Disagreement

10. One further minor point in Chap. 8: Farrell and Tham write:

> Li deems such a [MacIntyrean] project [of focusing on principles] infeasible because, unlike science, it cannot test rival theories through common and well-defined method of experimentation. This, we suggest, is not a compelling objection.

Farrell and Tham have misunderstood my view. My point was that in clashes between scientific paradigms, there are "critical experiments" that can settle the dispute. One well-known critical experiment is the Michelson-Morley experiment, which attempted to detect the existence of ether in which light waves were supposed to travel. The failure to detect ether in this experiment later paved the way for Einstein's special theory of relativity. My point was that there is probably no single "moral thought-experiment" or single argument that can knock down any moral theory. Notwithstanding this, I hold that there is moral objectivity and moral progress, and that these rely on reasons and arguments (which are more or less persuasive). In Chap. 4, I said that moral objectivity is different from scientific objectivity or mathematical objectivity.[42]

11. Farrell and Tham purport that virtually all my criticisms of their views are based on misinterpretations or misunderstanding of what they say. Apart from noting that this is extremely unlikely, I shall not try to disentangle myself from such claims, except to note the following:

In Chap. 8, Farrell and Tham write: "Nor have we claimed that Rawls is committed to epistemic abstinence." However, they wrote the following in Chap. 2:

> There are problems, though, with adopting *epistemic abstinence* toward FEF and treating these as principles that are not to be asserted as true.... (emphasis added).
>
> [Political liberalism] cannot excuse itself by appealing to *epistemic abstinence* (emphasis added).
>
> The problem is that the political liberal's *epistemic abstinence* is incompatible with the abovementioned requirements of rational enquiry (R1 and R2) (emphasis added).

Although it is possible that I might have misunderstood some of their claims, it is extremely implausible to say that I have misunderstood virtually all of their claims with which I disagree. Consequently, I believe we should take their claim with a pinch of salt.

(C) Response to Fan

12. In Chap. 6, Fan imputes several claims to what I wrote in Chap. 1. Fan says: "[B]oth Li and Farrell-Tham appear to be confident enough to intend to guide *all societies* alike by their respective favored conceptions"

---

[42] I also applauded T. M. Scanlon's effort to work out his position on these matters in his *Being Realistic about Reasons*.

(emphasis added). This attribution is simply false. For one thing, I have not made the claim that Fan attributes to me. For another, I have never taken the view that a political idea might be applicable to all kinds of society.

Based on this false imputation, Fan says that "Li does not find it necessary to offer any 'qualification' for the idea to be suitable for other societies, such as [M]ainland China or Hong Kong." He adds that "Li fails to recognize that the separation between the Rawlsian conception of public reason and the comprehensive doctrines operating in the modern West is far less radical than he believes."[43] Consequently, Li "is not justified in applying [Rawls' conception of public reason] to East Asian societies without offering qualifying arguments."

I believe that any normative idea—or any moral reason or political principle—that is applicable to a case, A, must also be applicable to other cases (e.g., B, C, D, …) that are *relevantly similar* to A. This might be called "the universality of reason."[44] This does not mean that a moral reason, or a political principle, if valid, must be applicable to all societies. This is because various societies might not be in relevantly similar circumstances. Thus, while constitutional democracy may be the right political structure for many countries in the twenty-first century, it was hardly the appropriate basic structure for feudalism in China or Europe, say, 1000 years ago. Why? The short answer is given by "moral contextualism."[45] What is right in circumstances A must be right for circumstances B, if A and B are relevantly similar. But the same kind of act or practice might not be right for circumstance C, if C is *not* relevantly similar to A. Even though constitutional liberal democracy is suitable for

[43] As Fan argues, "Rawls has clearly conceded (e.g., at the conclusion of his paper 'The Idea of Public Reason Revisited') that in order for such values and reasons to stand as public reasons for modern Western societies, they have to be supported by the reasonable comprehensive doctrines in modern Western societies." Earlier in this chapter, and also in Chap. 1, I have explained that citizens (in a constitutional liberal democracy) holding different and irreconcilable reasonable comprehensive doctrines have their different reasons, grounded at least in part in the idea of toleration, to support the liberal political conception of justice. The reasonable comprehensive doctrines in a liberal democracy might still be very different from each other. Without citing Rawls, Fan claims that Rawls has made *new* concessions. I believe, however, that this is what Rawls has been arguing all along, and does not represent any new concession.

[44] T. M. Scanlon, *What We Owe to Each Other*, 73–4, 367, 371–2.

[45] For a full discussion of moral contextualism, see Chap. 4 of this volume.

advanced, industrialized nations, including the United States, the United Kingdom, and Germany, it might not be suitable for what Rawls calls "burdened societies."[46]

The long answer is that, according to Karl Marx's theory of history (or historical materialism), which for me has some plausibility, the technological level of a society determines its type of economic structure, which in turn determines its ideological superstructure.[47] Thus, an agrarian economy will have feudalism as its type of economic structure, together with a feudal ideology—one that legitimates absolute monarchy. Confucianism emphasizes the five cardinal relationships (五倫), viz., the hierarchical relationship between a ruler and his subjects, between father and son, between husband and wife, between brothers, and between friends. According to historical materialism, a society that is capable of mechanical machinery will have a capitalist economic structure, together with an ideology suitable for capitalism. Thus, Marx says: "The hand-mill gives you society with the feudal lord; the steam-mill society with the industrial capitalist" (Marx, *The Poverty of Philosophy*).

If historical materialism is correct, it would not make any sense to condemn feudal societies that existed 1,000 years ago for not having practiced capitalism or for not having embraced a capitalist ideology, famously encapsulated in the slogan "Freedom, Equality, Property and Bentham."[48] For the same reason, it would not make any sense to insist that constitutional liberal democracy should apply to any feudal society in the Middle Ages. In other words, given a certain level of technological development in a society, only certain type of economic structure and ideological superstructure can be realized.[49] In this sense, Marx says:

[46] Two points here. First, "burdened societies" are those "whose historical, social, and economic circumstances make their achieving a well-ordered regime, whether liberal or decent, difficult if not impossible" (*LP* 90). "[W]hile they are not expansive or aggressive, [they] lack the political and cultural traditions, the human capital and know-how, and, often, the material and technological resources needed to be well-ordered" (ibid., 106). Second, I am not saying that Rawls would agree with me here. He agrees with Sen that a just society would go quite a long way to resolving problems, such as famine (ibid., 107).

[47] The most analytical, clear, and cogent account of historical materialism is found in G. A. Cohen, *Karl Marx's Theory of History: A Defence* (Princeton: Princeton University Press, 1978).

[48] Karl Marx, *Capital* (New York: International Publishers, 1967), vol. 1, chap. VI, 176.

[49] It is possible to claim that liberal democracy is better than absolute monarchy, but that liberal democracy was not realizable in feudalism. Consider this analogy: traveling by plane is faster than traveling by boat, but traveling by plane was not realizable in feudalism.

"Right can never rise above the economic structure of a society and its contingent cultural development."[50]

Again, if historical materialism is correct, questions arise as to whether Confucianism, which served so well as the dominant or ruling ideology in feudal China, can still serve its function in industrial China today. I think that the cardinal relationships between the ruler and the subjects, as well as between husband and wife, will need to be drastically revised. First, the kind of relationship between an absolute monarch and his subjects—in which the monarch could legitimately demand his subjects to die for whatever reasons—no longer exists. Second, agrarian feudalism provided a background against which the ideology of male superiority (男尊女卑) flourished; wives were required to submit to their husbands.[51] The feudal preference for boys over girls is much less important in major Chinese cities now than it used to be, but still has a considerable hold in rural China today. Again, this relic of feudalism will gradually vanish as its relevance in an industrializing China diminishes.

13. Fan's overall argument combines a kind of relativism with a kind of universalism—certain basic opportunities and rights that are "universal human rights."[52] First, he argues that the kind of political institution suitable for a nation depends on its prevailing culture. Thus, given that various countries or regions he listed (including China, Japan, Korea, Taiwan, Singapore, Vietnam, and Hong Kong) are "Confucian-influenced," they should practice Confucianism. As I argued in Chap. 4, I find this argument curious, as this struck me as an unargued form of relativism. Fan assumes without argument that if a culture practices a certain form of

[50] Marx, *Critique of the Gotha Program*, in *Marx Selections*, ed. Allen W. Wood, (New York: Macmillan, 1988), 190.

[51] Quoting from, "孟子言「夫婦有別」, 此固指男女內外有別, 相互對待態度亦有別。《禮記. 禮運》言「夫義、婦聽」即是女子于歸之後, 事夫之道以順為德, 此與《禮記. 郊特牲》「男帥女、女從男, 夫婦之義」的意義相近, 《左傳》昭公二十六年言「夫和而義, 妻柔而正」義亦雷同。, 但在古代男尊女卑的社會, 「夫義」不是「婦聽」的必要條件, 夫妻關係, 實際建立在「婦聽」, 妻子以順為道, 以順為德的單方退讓基礎" (https://zh.wikipedia.org/wiki/五倫#夫婦有別).

[52] See Chap. 3 of this volume. Fan says that he takes this idea from John Rawls, *The Law of Peoples*, 65. The universal basic human rights comprise "the right to life (to the means of subsistence and security); to liberty (to freedom from slavery, serfdom, and forced occupation, to a sufficient measure of liberty of conscience to ensure freedom of religion and thought); to property (personal property); and to formal equality as expressed by the rules of natural justice (that is, that similar cases be treated similarly)" (*The Law of Peoples*, 65).

thought, tradition, or way of life, then it should continue to practice it, and any extraneous (moral or political) ideas outside of this culture should be blocked.[53] But this unargued premise is without any merit. If it were plausible, the Nazi regime in Germany could have argued on the same basis that their "culture," which involved the extermination of six million Jews, was morally permissible. Such an argument for Nazism would be absurd because its *form* is unacceptable. The *form* of this argument is no less absurd when the subject matter is Confucian culture. Consider this argument: "Because the people in Confucian-influenced nations believe in and practice Confucian virtues, they should continue to practice Confucianism." This argument is unacceptable because it is an instantiation of the form: "because people believe in and practice a certain culture, they should continue to practice it," which is unacceptable. Fan makes no attempt to argue that Confucian virtues are valuable, but instead relies on an abstract argument according to which Confucian virtues are valuable for Confucian-influenced nations, whereas liberalism may be good for Western countries.

Second, despite his relativistic inclination, Fan curiously allows a strand of universalism in his theory, viz., "universal human rights" mentioned above. If people should universally have the rights to life, to liberty, to property, and formal equality, why should Confucian virtues be valuable to only people of Confucian-influenced countries? Why shouldn't the virtue of filial piety be valued in Western countries as well?[54] I believe that just as the unspeakable evil of genocide is universally condemned as horrendous, so the virtue to take care of one's parents when they need help is treasured as such both in Asia and in Western countries.[55] As President Obama said: "[T]hrough the struggles of slaves and immigrants, women and ethnic minorities, former colonies and persecuted religions, we have learned better than most that *the longing of*

[53] I assume this is one of the reasons for which Fan says that Confucian-influenced nations should be free from liberalism.

[54] According to Fan, a liberal would characterize the parent–child relation in the West as "contract-based." But this characterization is nothing but a caricature of such a relation.

[55] Although filial piety is a virtue in the East and West, I qualified this statement in Chap. 4 with the observation that in the West older people are often supported by welfare and social security.

freedom and human dignity is not English or American or Western—it is universal and it beats in every heart."[56]

This is not to deny that what is relevant in one nation might not apply in another nation, if and when their circumstances are *relevantly different*. As I discussed in Chap. 4, the same kind of act (e.g., lying), though generally impermissible, may be permissible under some special circumstances (e.g., if a German Nazi soldier asks you about the Jews). Fan should, but does not, provide an account as to when, or how, some considerations (e.g., Confucian virtues) apply only to Confucian-influenced nations, but not to others. Therefore, I conclude that his argument is unsatisfactory.

14. Suppose for the sake of argument that Fan is correct in contending that Confucian-influenced nations should practice Confucian virtues. Clearly most countries are influenced by more than one philosophy or religion. Japan, for instance, is influenced not only by Confucianism but also by Shintoism and Buddhism. In fact, Shintoism and Buddhism are the main religions that influence the Japanese, more than Confucianism does.[57] Further, Japan has been a constitutional liberal democracy since 1947. So even if it were the case that the people of a nation should continue to practice its prevailing moral and political beliefs, it would follow from Fan's view that Japanese citizens should practice the beliefs of Shintoism, Buddhism, Confucianism, and constitutional democracy. But such a conclusion would not provide a way out when there is more than one form of belief in society. If these constellations of beliefs conflict, which body of belief should prevail, and why? Which belief should a citizen follow?

In Mainland China, Marxism-Leninism is the official ideology. There are numerous Institutes of Marxism-Leninism in the country, including those in many universities. Students from primary school onward are encouraged to learn from the "good thinking, good way of life, and good

[56] Emphasis added. On May 25, 2011, Barack Obama addressed the British Parliament in Westminster as to how the special relationship between the United States and Great Britain can continue to help the two nations serve as catalysts for global action as the world faces a new series of threats and challenges.

[57] If we google "what are the main beliefs in Japan," it pops up Shintoism and Buddhism, not Confucianism.

character" of Comrade Lei Feng, who died at the age of 22 after a life which exemplified Communist virtues. At the same time, many Chinese believe in Buddhism and Taoism, either as philosophies or as religions, or both. Finally, many college students have found liberal ideas attractive.[58]

Furthermore, while 82% of the Chinese population lived in rural areas and worked in agriculture in 1978, this figure has fallen to 39% in 2020. In the past 42 years, China has transformed its agrarian economy into an industrial one. If historical materialism is approximately correct, feudalism is functional for the development of technology (and other productive forces) found in an agrarian economy. Thus, remnants of feudalism still remained in China: apart from male superiority and the preference of boys over girls, *guanxi* (關係), or personal connection, helped to get things done much more easily. When obstacles appeared to be insurmountable, they might be overcome by *guanxi*, which might involve personal favors or occasionally even bribes.[59] Are *guanxi* and the accompanying corruption the "culture" of feudal remnants that have legitimized themselves? (This is one implication of Fan's argument: that a culture legitimizes itself and continues to do so.) My question is: even if Fan's main argument were correct, how could the Chinese people—torn between Maxism-Leninism, Confucianism, Buddhism, Taoism, liberalism, and remnants of feudal ideology and practices—choose which "subculture," or form of belief, to follow?

The same is true of Hong Kong. Hong Kong has been ruled by the British for well over a century from 1842 to 1997. As a British colony, Hong Kong had no truly democratic institutions until several years before the handover in 1997.[60] But it has an independent and highly regarded judiciary. Most important, the laws and legal precedents that have been followed in Hong Kong are predominantly the Common Law of Great Britain. Further, with the establishment of Independent Commission Against Corruption (ICAC) in 1974, blatant corruption has disappeared in Hong Kong. What set Hong Kong apart from Mainland China is the rule of law practiced in Hong Kong and legal

[58] I thank Lei Zhong for this information about the influence of liberalism in universities in Mainland China.

[59] See Desmond Shum, *Red Roulette* (New York: Scribner, 2021), especially Chaps. 12 and 13.

[60] Martin Lee, Q. C., has argued that although Hong Kong did not have democracy, democracy was practiced in the United Kingdom, and hence that British democracy has an indirect effect for Hong Kong.

consciousness among the Hong Kong people. Further, although Hong Kong has a population of 7.5 million, one million people are Buddhists and another one million people are Taoists. Christians (Protestants or Catholics) amount to 900,000.[61] About 34% of the people in Hong Kong do not have a religion. We do not have any statistics about how many people in Hong Kong are Confucians. Surely, many Chinese people are Confucian-influenced, but many are also influenced by Buddhism, Taoism, and Christianity. Many secondary schools in Hong Kong, and certainly most of the better ones, are run by the Jesuits or the Protestants. The language of instruction in most of the secondary schools is English; they follow the British curriculum. Most secondary students in Hong Kong are considerably more familiar with the Bible than with the Confucian classics. My question to Fan is the same: to which constellation of beliefs should people in Hong Kong defer if and when their different beliefs conflict? For instance, Confucius asks his disciplines to respect ghosts and gods and stay away from them (敬鬼神而遠之), and asks them rhetorically how we could know the afterlife if we do not know even about life itself (未知生、焉知死?). On the other hand, Christianity asks us to believe in Heaven and Hell and in the Trinity, whereas Buddhism says that we should believe in karma and reincarnation. Even if Fan's theoretical presumption that a people should believe in and practice its culture were correct, which set of beliefs should people in Hong Kong espouse or follow? I believe that "the fact of reasonable pluralism" exists in Japan and Hong Kong, and probably also in China itself.

15. Let us revisit Fan's argument that a Confucian should reject medically assisted suicide (MAS), or (which I call) medical assistance in dying (MAID). Fan argues in Chaps. 3 and 6 that anyone who accepts the Confucian virtues of "*ren* (humanity), *yi* (righteousness), *li* (ritual propriety), *xiao* (filial piety), and *he* (harmony)," and also accepts the Confucian view that life is noble or valuable (*gui*), must conclude that MAID should be rejected. But Fan does not provide any persuasive argument for reaching this conclusion. Of course, we all accept that people's lives are noble and valuable, and that killing them would normally be very wrong. However, there are various kinds of circumstances under which killing is justified or

[61] December 24 to 26 are public holidays in Hong Kong, as is the birthday of Buddha in May.

not wrong. One can justifiably respond with deadly force against an aggressor in self-defense. The Roman Catholic Church allows abortion when implantation occurs in a Fallopian tube, or when a pregnant woman suffers from cervical cancer.[62] Some might even say that capital punishment is justified for those who have committed especially heinous crimes.[63] My point is that those who concede the right to kill in self-defense, and permit abortion when the mother's life is at stake, also recognize that life is noble or valuable. They hold that killing under these circumstances is permissible, *despite this recognition*. Such a view coheres with moral contextualism. Unless one subscribes to moral absolutism, which is highly implausible, there seems no plausible way to reach the conclusion that MAID is impermissible, given *only* the premise that life is noble or valuable.[64]

16. Furthermore, I would like to note that Chinese patients in Hong Kong with "acute illness" generally believe that patients have the right to decide to end their lives, and that (voluntary) active and passive euthanasia are morally acceptable.[65] Since Fan mentioned his students' response in his class in Chap. 6, let me point out that my students are generally very much in favor of the legalization of MAID.[66] Perhaps the conclusion to

---

[62] I recognize that Farrell and Tham deny that the kind of medical procedure to save the mother with the foreseen consequence that the fetus will die is not abortion. For me, this is just a play on words. Most English speakers would call this abortion.

[63] For the record, let me say that I myself reject capital punishment. See Hon-Lam Li, "Contractualism and the Death Penalty," *Criminal Justice Ethics* 36, no. 2 (2017): 36:2,152–82, DOI: https://doi.org/10.1080/0731129X.2017.1358912

[64] I discuss moral absolutism in Chap. 4. Fan also says that Chinese cherish their parents' lives. Even with this additional premise, we do not get the conclusion that MAS is impermissible.

[65] See R. C. S. Lam and Wai-Tong Chien, "Attitudes of Acutely Ill Patients towards Euthanasia in Hong Kong," *The Open Nursing Journal,* 2007, vol. 1, 1–5.

[66] First, at the invitation of the Medical Students' Association at the Chinese University, I gave a talk on the permissibility of MAID. Before I started talking, I asked for students' views on the permissibility of MAID when a patient is terminally ill, suffering excruciating pain, and desperately wants to die (henceforth "special condition"). All students raised their hands to indicate that they believed that MAID would be morally permissible under the circumstances. On another occasion, when I was invited by a lecturer at the Faculty of Medicine of the Chinese University of Hong Kong to express my view about MAID in her class, joined by two physicians who oppose MAID. At the beginning of the class, virtually all students (who were studying medicine) were in favor of the view that PAS should be permissible when the patients meet the "special condition." By the end, after they heard my argument in favor of MAID, and two physicians' arguments strongly opposed to MAID, the number of students in favor of MAID was reduced to about 60 or 65% in favor of MAID. Finally, in my class ("Applied Ethics") taught at my university in the spring of 2020, I asked the students to indicate their views before and after my lecture on MAID. Virtually every student was in favor of MAID when patients are in the "special condition." What do I want to show? I want

draw is that students are subject to the sway of their instructors, and therefore that we should take Fan's anecdote and mine with a pinch of salt.

I would like to reiterate that several Confucian scholars I have spoken with on separate occasions in Hong Kong or North America are of the view that Confucian filial piety does not entail that MAID for one's parent should be rejected, if it is in her (the parent's) interest—or is at least not against her interest—to pass away (e.g., if she is terminally ill and is suffering unbearably), *and* if she desperately wants to die. We recognize that human lives are noble, and we cherish our parents and wish them a long life, but that does not mean that we can permissibly deny them the moral right to have MAID. These scholars rightly hold that refusing consent to MAID in such circumstances would actually be *unfilial.* This reasoning also accords with commonsense.

To avoid misunderstanding, let me enter two caveats. First, I do not claim that MAID should be legalized in every society. Instead, as I argued in Chap. 1, because of the Practical Slippery Slope argument and concerns over abuse, whether MAID should be legalized in a particular community depends on the actual contingent situation in that particular community, viz., whether the legalization of MAID would actually lead to an unacceptable level of abuse.[67] For this reason, I do not claim—though Fan mistakenly thought that I did—that MAID should be legalized in China. However, I claim that Fan's arguments—in particular his arguments grounded in filial piety, the role of the physician, and the consequences of legalizing MAID—should not persuade us into thinking that MAID should be criminalized.[68]

Second, I am not against Confucian ethics *per se*. (There are different versions. It depends on what it says because some version is premised upon a kind of universalism.) I am only against Fan's use of relativism to support his Confucian conclusion, and also his arguments against MAID that claim to rely upon Confucian premises.

---

to show that most students that I came across said they were in favor of the legalization or decriminalization of MAID and, more important, that they were subject to the sway of their instructors.

[67] If a society is not governed by the rule of law, and if there is rampant corruption, then one is justified to worry that abuse would occur. If, on the other hand, a society is well-ordered and governed by the rule of law, then one need not worry as much about abuse.

[68] I have discussed the argument from Hippocratic Oath in Chap. 4, and the likely consequences of legalizing MAID in Chap. 1.

(D) Further Reflections

Confucian Ethics vs. Confucian Public Reason

17. When reading Fan's accusation that I have misinterpreted him,[69] I was reminded of a true legal tale. A defendant was acquitted of a serious criminal offense by the court. But the judge told the defendant: "You are acquitted of the criminal offense. But you have only yourself to blame for casting the shadow of suspicion onto yourself."

If indeed I have misinterpreted Fan's view, this is only because he has not written clearly. For instance, he never mentioned the difference between Confucian ethics and Confucian public reason clearly or at all until Chap. 9. (In fact, the term "Confucian ethics" does not even appear in his Chaps. 3 and 6!) According to Fan in Chap. 9, Confucian ethics is universal and everyone in the world should be a Confucian. Yet, Confucian public reason is a form of "soft relativism" because "a suitable conception of public reason for a society must seek some social goals for that society, such as its stability, cohesion and peace, in addition to seeking the cognitive goal of moral truths." However, this restatement raises more questions than provides answers. First, in Chap. 6, Fan seems to be attacking moral universalism. If so, why should he now hold that twenty-first-century liberals, for instance, should subscribe to Confucian ethics (instead of liberalism)? If Confucian ethics is universal, why shouldn't Confucian public reason also be universal? Why would realizing certain social goals turn Confucian public reason into a form of "soft relativism"? These questions, and in particular the relation between Confucian ethics and Confucian public reason, are left unanswered.

Are Universal Basic Rights Western?—Fan's Puzzlement

18. According to Fan, my view that he seems to reject liberal or Western elements from his theory of public reason is "puzzling" because he has

[69] See Chap. 9 of this volume.

"explicitly integrated a Rawlsian list of human rights into" his theory. But Fan endorses Rawls' understanding that this list of human rights are neither liberal, nor Western:

> A list of basic liberties and rights to be covered by my Confucian conception of public reason includes two parts: one is universal human rights that should be granted by all societies, and another is special Confucian rights that should be enjoyed in East Asian societies. Universal human rights are a decent minimal list of basic liberties and moral rights that has been proposed by Rawls in his later work: it is "a special class of urgent rights" such as "liberty (but not equal liberty)", which does not represent a full-blown liberal democratic conception of rights (Rawls, *The Law of Peoples*, henceforth "*LP*," 79). Instead, it is "a proper subset of the rights" similarly possessed by citizens in a liberal constitutional democratic regime and the members of a decent hierarchical society (*LP* 81). This list is as follows:
>
>> Among the human rights are the right to life (to the means of subsistence and security); to liberty (to freedom from slavery, serfdom, and forced occupation, and to a sufficient measure of liberty of conscience to ensure freedom of religion and thought); and to formal equality as expressed by the rules of natural justice (that is, that similar cases be treated similarly). (*LP* 65)
>
> *These rights, as Rawls stresses and I agree, "cannot be rejected as peculiarly liberal or special to the Western tradition" (The Law of Peoples, 65). [footnote omitted]. They do not "belong to certain kinds of political institutions," liberal democratic or other (LP 79). Instead, they set a limit to the pluralism among peoples, and constitute a necessary condition of the decency of any political institution* (*LP* 80). Although the Confucian tradition has never proposed this concept of universal human rights, I think such rights can be worked out from within the Confucian tradition because they are implicit, if not explicit, in the fundamental implications of the basic Confucian virtues. This is to say, they are not only needed to protect legitimate individual interests in Confucian-influenced societies, but can also be consistently developed out of Confucian virtue-based tradition. In fact, in my previous work I have shown that such rights can be derived from the moral requirements of the basic Confucian virtues, such as *ren* (humanity) and *yi* (appropriateness), regarding how individuals

> should treat one another as well as how they should be treated by their government. Accordingly, they should be accepted and upheld by contemporary Confucian-influenced societies for pursuing comprehensively good life, such as *zhishan* I mentioned before (Ruiping Fan, "A Confucian Conception of Public Reason and Bioethics," Chap. 3 of the present volume; emphasis added).

This passage shows that, according to Rawls, this list of basic human rights are "universal rights" (*LP* 65) and *hence neither liberal nor Western*. Second, it shows that *Fan agrees with Rawls*. Note that Rawls uses the terms "liberal tradition" and "Western tradition" as meaning the same thing.[70] Third, Fan himself said that these basic human rights "can be worked out from within the Confucian tradition because they are implicit, if not explicit, in the fundamental implications of the basic Confucian virtues." I believe this is another reason for which Fan said that these rights are not "peculiarly liberal or special to the Western tradition."

However, Fan is now saying that although he rejects liberalism, he does not reject "Western" views because he subscribes to Rawls' list of basic human rights.[71] Apart from the point that Fan agrees with Rawls that these basic human rights are neither liberal nor Western, Fan even argues that "such rights can be derived from the moral requirements of the basic Confucian virtues."[72]

Yet, to argue that my reasoning is puzzling, Fan is now arguing that the basic human rights on Rawls' list are Western. How so? First, Fan would be inconsistent because he says (in Chap. 3) that the list of basic human liberties and rights are neither liberal nor Western. Second, I can think of no other reason why, on Fan's view, this list of basic human rights is Western, except perhaps that Rawls is a Westerner. (Would this list of rights become Western because Rawls is a Westerner?) But if this reconstructed reasoning was indeed Fan's reasoning, let us remind Fan that Rawls is also a liberal. (Does this fact make the list of rights liberal?) The answers to these

[70] Also note that "liberal democracy" is also known as "Western democracy." See https://en.wikipedia.org/wiki/Liberal_democracy#Liberal_democracies_around_the_world.

[71] See Chap. 9 of this volume.

[72] See Fan, Chap. 3 of this volume. Also note that "liberal democracy" is also known as "Western democracy."

rhetorical questions in parentheses are no. Surely, not everything said by Rawls is Western. Otherwise, if Rawls says "2+2=4" to a child, say, this would become a piece of Western theory or view? That would be absurd.

Can Confucianism Cohere with Liberal Ideas or Basic Human Rights?

19. One reason why I believed that Fan does not want his theory to be tainted by anything liberal or Western—where both terms have the same referents—is that in his Chap. 3 he wrote:

> [N]o fundamental reason or value contained in the Confucian conception of public reason should be directly inserted into it from another tradition (such as the liberal social-democratic tradition).

Fan does not explain why.

Fan criticizes Joseph Chan's and Sungmoon Kim's versions of Confucianism on the grounds that these versions are mixed with liberal (or Western?) elements. Fan criticizes Joseph Chan's version of Confucianism on the grounds that if Chan adopts the Rawlsian idea of civility and appeals to Confucian virtues that are self-standing, viz., without the comprehensive doctrines, then the resulting Confucianism is no longer Confucianism. Chan could reply that the self-standing Confucian virtues (such as *ren*, *yi*, *li*, *xiao*, *he*) are still Confucian virtues. They do not become something else (e.g., Aristotelian virtues) by becoming self-standing.[73] So I do not see the force of Fan's objection.

More important, the objection raised by Fan—as to why Chan's Confucianism is Confucianism—is a nominal one. The correct response is: why care about the label?

Fan raises two objections against Kim's version of Confucianism. According to Fan's interpretation, Kim argues that Confucianism should

[73] The four cardinal virtues for Aristotle are temperance, bravery, justice, and *phronesis*. Some translate *phronesis* as "practical wisdom," whereas others (e.g., Terence Irwin) translate it as "prudence."

be espoused with democracy and the idea that all citizens are equal. Fan rejects Kim's view on the grounds that it clashes with the Confucian view that "the Confucian virtues are unequally possessed by individual citizens," and hence "Confucian public reason may not accept that everyone has a moral right to vote." Kim can respond by saying that no one who argues in favor of democracy would hold that every individual citizen possesses the same degree of virtue, and hence that the idea of democracy is compatible with the idea that citizens do not possess virtues to the same extent. Fan supports meritocracy. If by meritocracy is meant the view that those who have higher or greater merits should have more responsibility or obligation, then meritocracy is compatible with democracy. In fact, democracy (meaning one-person-one-vote), as the best way to express the General Will, may also be the best way to select a president while minimizing risks pertaining to totalitarianism.

Fan's assumption that democracy and meritocracy are mutually exclusive is unwarranted. He argues:

> Confucianism can take universal suffrage as a useful instrumental/empirical mechanism: if this mechanism turns out to be helpful to the people's happiness or flourishing in a particular society, it should be adopted; to the contrary, if it is not useful or is even detrimental to people's good lives, it should not be adopted. Accordingly, one person one vote may or may not become a political or legal right in a society, but it is not a fundamental human or moral right in the Confucian view.

Can someone espouse Confucianism while also holding the view that universal suffrage is intrinsically important? Fan thinks no, but his argument is not persuasvie. Fan's argument that because universal suffrage can be justified only contingently (because it might be "a useful instrumental/empirical mechanism"), it cannot be "a fundamental human or moral right" on any reasonable Confucian view is problematic.[74] This is because his conclusion is a *non sequitur*. People of various political stances, in particular the rule-utilitarians, can certainly argue that because (1) universal suffrage is good instrumentally (to a contemporary society beyond a certain level of

[74] Because it is obvious that pre-twentieth-century Confucianism does not allow universal suffrage, Fan must mean any reasonable version of Confucianism in the twenty-first-century.

wealth and whose citizens are literate), therefore (2) its citizens should have a fundamental or moral right to universal suffrage.

If the way in which Fan argues against Chan and Kim were acceptable, we could also argue that his version of Confucianism is not Confucianism. Pre-twentieth-century Confucianism does not recognize rights of any sort, with the possible exception of the emperor's right to govern. The list of universal human rights (viz., rights to life, to liberty, and to formal equality) which Fan adopts as part of his version of Confucianism was unknown to, and not recognized in, pre-twentieth-century Confucianism. Fan's revised version therefore clashes with pre-twentieth-century Confucianism. On the other hand, if Fan wishes to claim his own version as Confucianism, even though it clashes with pre-twentieth-century Confucianism, then to be consistent he must also allow Chan and Kim to claim their versions as Confucianism too.

Does Confucianism Stabilize Absolute Monarchy?

20. In Chap. 4, I wrote that Confucianism supported or stabilized Absolute Monarchy or feudalism in China. (I understand "Absolute Monarchy" and "feudalism" as having the same referent.) Fan says that it was Legalism, not Confucianism, that defended Absolute Monarchy. In the history of Chinese philosophy, it was the Legalists who attempted to give strong support to the idea of absolute monarchy by proposing *shih* (power or authority), *fa* (law or authority), and *shu* ("statecraft"). But did the Legalists succeed? The answer is no, because they ignored the rites (*li*). Moreover, does it follow, even if Legalism purported to defend feudalism, that Confucianism did not support or stabilize Absolute Monarchy? The answer is that it does not follow. Let me explain why.

The proposition that Confucianism or Legalism stabilized feudalism (or Absolute Monarchy) can be understood in two different ways. The first way is to understand "X stabilizes Y" as saying that X contributes to the stability of Y (but it is an open question whether or not this contribution is sufficient to insure that Y remains intact); here "stabilize" is used as an "attempt-verb." Another way to understand "X stabilizes Y" is to understand X as a sufficient condition for the continued existence of Y; here, "stabilize" is understood as a "success-verb" (see Raymond Geuss, *The Idea of A Critical Theory*, 15-16).

When I said that Confucianism stabilized feudalism or Absolute Monarchy in China, I had in mind that Confucianism was *successful* in stabilizing feudalism (or Absolute Monarchy). In fact, I subscribe to the following functional explanation, namely, that feudalism selected Confucianism as the ruling ideology because Confucianism could (successfully) stabilize feudalism. Second, more important, such a functional explanation is meant to explain something that actually happened in history. If we understand "stabilize" as a "success-verb," then Legalism failed spectacularly as a means to stabilize feudalism. With the exception of the Qin Dynasty (221 to 206 BC), which lasted for merely 15 years, Legalism was never accepted as a ruling ideology in China. On the contrary, Emperor Wu of the Han Dynasty (202 BC-220 AD) "dismissed a hundred schools of thought and respected Confucianism only." The statesman who did most to entrench Confucianism as the ruling ideology was Dong Zhongshu (or Tung Chung-shu) (董仲舒), a high-ranking minister and influential adviser of the Emperor Wu of Han (ruling from 141 to 87 BC). Dong's most important work was *Luxuriant Dew of the Spring and Autumn Annals* (《春秋繁露》), in which he frequently emphasized the power of the emperor.[75]

Since the Han Dynasty until early Twentieth Century, only Confucianism was enshrined as the ruling ideology by the government. Scholars who aspired to become officials had to sit public examinations, where they must display their knowledge and understanding of Confuciani classics. These texts advocate ideas governing the "five cardinal relationships," namely that between monarch and subject, father and son, husband and wife, elder and younger brothers, and between friends as well as the "three *kang*" (see Chap. 4, "Liberalism or Western Theory," of this volume). In this way, Confucianism imposed hierarchical submission and promoted political conservatism.[76]

Confucianism advocated emperors to "rule over [the people] with dignity and they will be reverent; treat them with kindness and they will do their utmost; raise the good and instruct those who are backward and they will be filled with enthusiasm" (Confucius, *The Analect*, tr. D. C. Lau,

---

[75] I thank Kwong-loi Shun for helpful discussion on this point.

[76] See Fung Yu-lan, *A Short History of Chinese Philosophy*, Chap. 17; Fung Yu-lan, *History of Chinese Philosophy*, Vol. II, Chap. 2; Shihlin Ema Fu, "The Economic Morality of Leadership: The Confucian Ethics that Affected Emperor Kangxi of the Qing Dynasty," MA thesis, p. 15; https://citeseerx.ist.psu.edu/viewdoc/download?doi=10.1.1.940.509&rep=rep1&type=pdf.

Book 2, Chap. 20). Confucianism also emphasized the importance of rites (*li*) and moral propriety.

Confucianism has a quasi-religious character. It maintained that the emperor's rule was legitimated by the Mandate of Heaven, and that natural disasters were signs that Heaven was angry at the emperor's poor governance. To say that "Confucianism *stabilized* feudalism (or absolute monarchy)" is to say that "Confucianism was selected as the ruling ideology because it could (successfully) stabilize the economic structure of feudalism." This is to propose that such a functional explanation is true or justified (just as it is a functional explanation that "birds have hollow bones because hollow bones facilitate flight").[77] To actually make such a functional explanation would not be a simple task, but one which many people (I myself included) believe can plausibly be made. I shall not attempt such a task here, except to say that such an explanation would involve social stability made possible by political conservatism, hierarchical submission, rites and moral propriety, all of which Confucianism promoted through the mechanism of the public examination, as well as the quasi-religious character of Confucianism. It would also involve how such stability enabled feudalism to develop the prevailing level of technology in an agrarian mode of production.

The Point of Critical Morality

21. Fan complains that it is misguided to think that one side has only positive morality whereas another side has positive morality and critical morality (see Chap. 9). A quick reply: it has happened in history that some people did evil things because they did not care about critical morality. Cases abound: the Holocaust, Stalin's Great Purge, the Cultural Revolution, and the "dirty war" in Argentina are obvious examples. The point of alluding to the idea of critical morality was not to say that some group occupies the moral high ground and can therefore condescend to others. Rather, the point is that positive morality does not justify anything unless it is backed up by normative reasons. It urges us to continually look for justification and moral reasons when we (as a person, a

[77] For an excellent discussion of functional explanation and its application in social explanation, see G. A. Cohen, *Karl Marx's Theory of History: A Defence*, Chap. 10. I would like to thank Chung-yi Cheng for a helpful discussion on whether Confucianism stabilized feudalism.

group, or a community) make an important decision. This is also the whole point of public reason: someone who makes a proposal must believe in good faith that those at the receiving end should not only understand the terms of our proposal but could also reasonably accept them. If we just have positive morality, but not critical morality, the dimension of normativity and justifiability is lacking.

22. In Chap. 3, Fan says that the Confucian view that he articulates and defends is a communitarian one, though he does not explain exactly what he means by the term "communitarian." Nevertheless, he seems to think that communitarian values are ones that are simply shared in a community. Thus understood, Fan's version of communitarianism would lead him to relativism. But such understanding of communitarianism is not only impoverished but also fails to do justice to communitarian scholarship.

First, in *Spheres of Justice*,[78] Michael Walzer (who is usually considered to be a communitarian) argues that the way in which goods are distributed in society is determined by their social meanings that are shared among members of the community. Yet, Walzer subsequently explains that social meanings must be *correctly* interpreted.[79]

Second, Hegel has been a source of inspiration to Charles Taylor and Michael Sandel (both of whom are usually considered to be communitarians).[80] Yet, Hegel holds that our views must be scrutinized and subject to rational reflection. As Allen Wood writes:

> Hegel's use of the term *Sittlichkeit*, which might be translated from ordinary German as "customary morality," which has often been interpreted as an endorsement of moral traditionalism, of the view that to do what is morally right, all I need to do is act in conformity with the accepted standards of my people and culture. It is true that Hegel regards objective and determinate moral standards as founded on the organization of a concrete social order, and that he regards some enlightenment moral theories as shallow and overly individualistic. But Hegel's conception of the ethical life

[78] *Spheres of Justice* (Oxford: Blackwell, 1983).

[79] See Michael Walzer, *Interpretation and Social Criticism* (Cambridge, MA: Harvard University Press, 1987).

[80] See Alexander Kaufman, "Hegel and the Ontological Critique of Liberalism," *The American Political Science Review 91: 4*, 1997, 807–817.

> of modern society (as presented in the *Philosophy of Right*) is not conservative or traditionalist in its orientation. Hegel insists that the institutions of the modern state have a claim on us only because they are *rational* (PR s.258R), and Hegel takes it to be the function of rational reflection to confirm what we do by custom and habit through insight and scientific cognition (PR s.147R). The "ethical" standpoint in Hegel is better interpreted as a certain type of critical reflection on existing social institutions than as a rejection of such reflection.[81]

Thus, Hegel would demand "critical morality" (though such morality must be contextualized), and would reject moral relativism.

Finally, even Alasdair MacIntyre (another communitarian), who holds that we must carry out moral reasoning within a tradition, affirms that some tradition is objectively better than another one insofar as the former can resolve an epistemic crisis that plagues the latter. We can now see that communitarian values should not be understood as values that are *simply* shared among members of a community. Otherwise, the Nazi view that inferior peoples should be annihilated could be the basis of communitarian values in Nazi Germany. Instead, communitarian values should be values that can stand rational or critical reflection.

(E) Reflection on Rawls

Scanlon's Contractualism[82]

23. In *Political Liberalism* and "The Idea of Public Reason Revisited," Rawls says that we should order the relevant political values when faced with an issue concerning "constitutional essentials or matters of basic

[81] Allen Wood, "Hegel's Ethics," in *The Cambridge Companion to Hegel*, ed. Frederick Beiser, Cambridge: Cambridge University Press, 1993, 224–5.

[82] Fan claims in Chap. 6 that Scanlon's contractualism—"an act, policy or law is permissible if and only if it can be justified to everyone affected by it"—is too abstract. I referred to this abbreviated version toward the end of Chap. 1. When I first introduced Scanlon's contractualism in Chap. 1, I wrote that, according to Scanlon's contractualism, "an act is impermissible if and only if 'its performance *under the circumstances* would be disallowed by any set of principles for the general regulation of behavior that no one could reasonably reject as a basis of informed, unforced general agreement'" (emphasis added). Is this too abstract? I do not think so because whether an act or practice is permissible depends on the circumstances.

justice." What political values does he have in mind? In discussing abortion, Rawls mentions three political values: the due respect of human life, the ordered reproduction of political society over time, and the equality of women as equal citizens (*PL* 243 *nf*).[83]

Rawls believes that to approach the problem of abortion, say, we should first reasonably rank in order the relevant political values—viz., the due respect of human life, the ordered reproduction of political society over time, and the equality of women as equal citizens—and come to a conclusion as to what is the right thing to do in the circumstances by reasonably balancing these values. He holds that a pregnant woman has the right to abortion in the first trimester, and that this is the only reasonable conclusion. Yet, he does not tell us how his reasoning goes.

If we look at the literature on abortion, none of the well-known arguments proceed in the way that Rawls suggests. In other words, none of the famous papers by Judith Thomson, Mary Ann Warren, and Don Marquis,[84] for instance, identify the political values and then make an ordering of these values. Instead, they proceed with plausible premises and draw a conclusion as to whether abortion should be permissible under particular circumstances. Especially clear is Thomson's "A Defense of Abortion," where she uses self-standing *arguments* (i.e., without any doctrines). For this reason, I suggest in Chap. 1 that we should incorporate Scanlon's

[83] There are other political values. Other political values related to the family are the equality of children as future citizens, the value of the family in securing the orderly production, and reproduction of society and of its culture from one generation to the next (*PL* 163–4). Elsewhere, he mentions the values of equal political and civil liberty; fair equality of opportunity; the values of economic reciprocity; and the social bases of mutual respect between citizen (*PL* 139). He later mentions as examples of political values those in the preamble of the US Constitution, namely, a more perfect union, justice, domestic tranquility, a common defense, the general welfare, and the blessings of liberty for ourselves and our posterity (*LP* 144). Closely related are other political values, such as equal basic liberties, equality of opportunity, ideals concerning the distribution of income and taxation, as well as effectiveness and efficiency (*LP* 144). Political values related to animals and nature are the good of preserving the natural order; to foster species of animals and plants for the sake of biological and medical knowledge; to protect the beauties of nature for purposes of public recreation and the pleasures of a deeper understanding of the world (*PL* 245). Finally, liberty of conscience (*LP* 151), freedom of association (*LP* 158), and the freedom of religion (*LP* 163), including the freedom to affirm no religion (*LP* 145) are political values. The "transcendent values" of salvation and eternal life—the *Visio Dei*—is, however, not a political value. This list of political values is not a complete one.

[84] These are Judith Thomson's "A Defense of Abortion," *Philosophy & Public Affairs* 1, no. 1 (Autumn, 1971), 47–66; Mary Ann Warren's "On the Moral and Legal Status of Abortion," *The Monist* 57, no. 4 (January 1973) 43–61; Don Marquis' "Why Abortion is Immoral," *Journal of Philosophy* 86, no. 4 (April 1989), 183–202.

contractualism into political liberalism because the *principles* he employs in argumentation are self-standing and the way Scanlon carries out his *arguments* is reasonable. However, have I read too little into Rawls' idea of ranking the political values in order? For one thing, he has included "ideals and principles, standards and values" (*LP* 144) and arguments in public justification. For another, Rawls has included as values of public reason:

> great values … expressed in the guidelines for public inquiry and in the steps taken to secure that such inquiry is free and public, informed and reasonable. These values include not only the appropriate use of the fundamental concepts of judgment, inference, and evidence, but also the virtues of reasonableness and fair-mindedness as shown in adhering to the criteria and procedures of commonsense knowledge and the methods and conclusions of science when not controversial, and in respecting the precepts governing reasonable political discussion. (*Restatement* 190)

Furthermore, Rawls strongly recommends Judith Thomson's article, "Abortion" (different from "A Defense of Abortion"), in which he thinks Thomson has used various political values of public reason.[85] The three values Thomson uses in "Abortion" have a certain degree of resemblance to Rawls' three political values on abortion mentioned above.[86] More important, Thomson also uses moral *arguments*, for example, in refuting various positions. But Rawls believes that the right way to arrive at a conclusion is by way of a reasonable balance of political values.

I myself would reject Rawls' approach. For me, the advantage of Scanlon's contractualism is that the way in which arguments proceed is not necessarily a matter of ordering political values relevant to a situation, and then reasonably balancing them. Rather it is to make arguments involving principles which everyone can reasonably be expected to accept.[87]

---

[85] See Judith Thomson, "Abortion," *Boston Review*, 20: 3 (Summer 1995).

[86] Thomson writes: "In sum, my case here against regulation of abortion rests on three ideas. First, restrictive regulation constrains women's liberty. Second, severe constraints on liberty may not be imposed in the name of considerations that the constrained are not unreasonable in rejecting. And third, the many women who reject the claim that the fetus has a right to life from the moment of conception are not unreasonable in doing so. All three ideas seem to me plausible."

[87] Nevertheless, one may ask whether Scanlon's contractualism amounts to the same approach as Rawls' suggested ordering and balancing relevant political values. I believe that they are different, and that Scanlon's approach is the right approach.

Constitutional Essentials

24. Rawls has clarified that if some issues "border" on constitutional essentials and are "politically divisive," then these issues ought to be discoursed in terms of public reasons in the public forum (*Restatement* 117). But why shouldn't the use of public reason be extended to important social issues that do not border on constitutional essentials? For instance, shouldn't questions about animal experimentation and killing nonhuman mammals for food, as well as questions about helping nations in dire need, be discussed in terms of public reasons, assuming that they do not border on any constitutional essential? This is a complex issue and deserves a much fuller discussion than I can give it here.[88]

Should All Metaphysics Be Excluded?

25. According to Rawls' theory of political liberalism, public justification should involve public reasons but not comprehensive doctrines. That means, comprehensive doctrines—even if they are very reasonable ones—should not figure in any discourse on constitutional essentials and matters of basic justice in the public forum. A doctrine, to be reasonable, has to be epistemically coherent and consistent.[89] It also has to accept "some form of the political argument for toleration" (*LP* 176).

It is not clear whether all religious, moral, metaphysical, and philosophical doctrines should be bracketed in such a public discourse, or whether only controversial ones are to be bracketed. Perhaps claims about the afterlife in any religion (e.g., Catholicism, Islam, Buddhism, Taoism, atheism) are by their very nature speculative or controversial. But this may not be true in moral or metaphysical views. In her article, "Abortion," Thomson not only alludes to (what Rawls calls) public values but also uses moral arguments, for example, against Ronald Dworkin's view on abortion. These moral arguments (which are what you might expect to

[88] For a view that public reasons should be used more broadly than Rawls approves, see Jonathan Quong, "The Scope of Public Reason," *Political Studies*, June 1, 2004; https://doi.org/10.1111/j.1467-9248.2004.00477.x

[89] See discussion earlier in this chapter.

find in an applied ethics article)—insofar as they are persuasive—can be seen as self-standing because they do not depend on any controversial religious, moral, or metaphysical doctrines. Yet, are there self-standing metaphysical views?

Consider the metaphysical view that a baby (or a fetus) is the result of the merging of a particular sperm and a particular ovum, and that if both sperm and ovum were different, then a different baby would be born.[90] Consequently, a woman who wants to have an abortion now (perhaps because she is a scholar seeking tenure and is simply too busy to have a baby), and plans to have a baby three years later, would be having a different baby. Whether this metaphysical view might figure in any moral or political debate, my claim is that this view seems to be self-standing because it is self-evident.[91] If I am correct, even though all religious doctrines should be bracketed (unless they can be stated in terms of public reason to fulfill the Proviso), we should perhaps bracket only those moral, metaphysical, or philosophical views that are controversial. To the extent that I am tentatively inclined to this position, I have come a little bit closer to the Natural Law Theory defended by Farrell and Tham (who hold that nothing should be bracketed), though we still remain miles apart.

Is Political Liberalism Simply the Product of Western Hegemony?

26. Finally, I come to the last issue for reflection: isn't political liberalism simply the ideology of the hegemony of advanced welfare capitalism in the West? And if this is imposed on China and other countries with a long history and rich cultural heritage, do they not have strong reasons to object?

The first thing to observe is that no one would say that scientific theories and technological invention from the West are ideologies of Western

---

[90] Saul Kripke defends this view in *Naming and Necessity* (Cambridge, MA: Harvard University Press, 1980), 111.

[91] Such a metaphysical claim is relevant to the Non-identity problem. See Derek Parfit, *Reasons and Persons* (Oxford: Oxford University Press, 1984), Part IV. This idea also figures prominently in Michael Lockwood, "When Does A Life Begins?" *Moral Dilemmas in Modern Medicine* (Oxford: Oxford University Press, 1985), 9–31.

hegemony. In order to make the objection, one must claim that normative theories (e.g., moral and political theories) are different from scientific theories in that the former are subjective, whereas the latter are objective. To support this claim, one might argue that there is a physical world to which scientific theories might correspond, and hence we can make sense of the statement that the general theory of relativity is true. However, one might continue, moral and political theories do not correspond to anything in reality. Consequently, either an error theory[92] or moral noncognitivism[93] is adopted.[94]

I myself would defend moral realism, the theory that there are moral truths. On this view, whether a moral judgment is true or not depends on the best reasons for or against the judgment. And reasons, at least those that are persuasive, are real or objective. I do not have the space to elaborate on moral realism. Suffice it to say that I adopt the theory of moral realism articulated by T. M. Scanlon and Thomas Nagel.[95]

Because I embrace moral realism and the "universality of reason" (see Sect. 12 of this chapter), I hold that moral and political reasons that apply in one case (or one community) must also apply in another relevantly similar case (or in another community where the circumstances are relevantly similar). Thus, if bribery is wrong in one community, it must also be wrong in another community, if the situations are relevantly similar.

---

[92] According to this view, first-order moral judgments—for example, abortion is (im)permissible—are meaningful, but they are all false. See J. L. Mackie, *Ethics: Inventing Right and Wrong*, chap. 1, esp. section 9.

[93] This is the theory that moral propositions are like expression of emotion, and hence are meaningless and not capable of being true or false. The classic statements were made by A. J. Ayer, *Language, Truth, and Logic*, 1936, and also Charles Stevenson, *Facts and Values*, 1963.

[94] It is sometimes attributed to Marx the view that the ruling ideology arising from an economic base is false in the pejorative sense. Marx's attitude to philosophy is clear in this passage: "One has to 'leave philosophy aside'..., one has to leap out of it and devote oneself like an ordinary man to the investigation of actuality, for which there exists also an enormous amount of literary material, unknown, of course, to the philosophers.... Philosophy stands to the investigation of reality as masturbation to sexual love" (Marx and Engels, *The German Ideology*, ed. C. J. Arthur, 103). See also Richard W. Miller, *Analyzing Marx*, chaps. 1 and 2.

[95] See T. M. Scanlon, *Being Realistic about Reasons*, esp. chap. 2; Thomas Nagel, *The Last Word*, chap. 6.

I do not believe that the reason for which Rawls wrote *Political Liberalism* was to apply his political liberalism to a different kind of society (one that is not a constitutional liberal democracy), such as China. As stated at the beginning of this chapter, Rawls is concerned with stability of a liberal society for the right reason, and also the legitimacy of laws. He tries to solve two internal problems. His theory is a normative rather than a descriptive one. When he proceeds to deal with the two problems in constitutional liberal democracy, where reasonable and rational citizens are supposed to have the two moral powers, he is not saying that this ideal corresponds to the reality of the United States or of any other country. Rather, he is making some normative assumptions in order to launch his normative inquiry.

Although Rawls is not interested in applying political liberalism to a society that does not practice constitutional liberal democracy, such as China, that should not stop *us* from reflecting on this issue. On one view, the political system (as part of the basic structure) in a society is merely an *instrument* for developing the productive forces in that society. Given a certain level of technological development, one country can choose constitutional liberal democracy, whereas another country may opt for a more authoritarian system in which democracy and freedom are more restricted. One can argue that Singapore, for instance, would not have been so successful had its government practiced constitutional liberal democracy.[96] Although Rawls would consider Singapore as falling short of being a just society,[97] he would regard its government as "non-liberal but decent" (*LP* 3) since, among other things, Singapore is non-aggressive.[98]

[96] In "A Brief History of Elections in Singapore," Thum Ping Tjin says that "Singapore has never had an election to parliament (and its predecessors) which has been free and fair and with a universal franchise" (https://newnaratif.com/research/a-brief-history-of-elections-in-singapore/; March 19, 2020).

[97] I am assuming that Singapore fails to satisfy the first principle of justice by failing to safeguard citizens' right to a genuinely fair voting system.

[98] There might be some truth to the "instrumental" theory that there is more than one kind of employment practice. For instance, in Japan, an employee in a big corporation is basically guaranteed a job until retirement. But in the West, promotion, demotion, and dismissal of an employee depends on her performance as well as the finances of the employer. It might be possible to argue that both practices are equally functional for capitalism, despite how different they are.

Another view, on the other hand, says that a political system is not merely an instrument for developing the economy. It should safeguard citizens' basic liberties and rights, and fair equality of opportunity, and should also provide an adequate all-purpose means to enable effective use of their freedoms, whether by installing a roughly egalitarian economy (e.g., as prescribed by the Difference Principle), or by the provision of a basic income, or by pursuing a middle course. Rawls proposes that there are universal basic rights: to life, to liberty, and to formal equality—which apply to all societies (*LP* 65). Imagine two similar societies, one being Singapore and another a society like Singapore except that it is a liberal democracy. If they have comparable economic achievements, then the liberal democracy is more just and equal and is a well-ordered society.[99] It is therefore also a better society. What this means is that a political system can be assessed with respect to how it safeguards basic liberties, rights, and opportunity, as well as how it honors distributive justice. Because the three basic rights are universal (as held by Rawls and also Fan), there can be cross-national comparisons.

Fan may want to claim that there can be no cross-national comparison of the political system between, say, the United States and China, because these countries have very different cultures. It is plain that "culture" is a vague term because a culture comprises different components. Thus, "culture" could refer to (1) customs, (2) people's moral (and political) beliefs and practices, or (3) the political system in place.

In the first sense, culture refers to customs: Americans celebrate Thanksgiving and give family members presents at Christmas, whereas the Chinese commemorate the Dragonboat Festival and adults give red pockets (containing money) to children at Chinese New Year. It would not make any sense to ask the Chinese to celebrate the Thanksgiving any more than urging the Americans to celebrate the Dragonboat Festival.

Second, "culture" could mean moral (and political) beliefs and practices. According to moral relativism, whether the act or practice of a group (e.g., a nation) is right or not is relative to the moral (and political)

---

[99] A society is well-ordered if and only if (1) everyone accepts, and knows everyone accepts, the same principles of justice; (2) its basic structure is publicly known (or with good reasons believed) to satisfy these principles; (3) its citizens have a sense of justice and comply with the society's basic institutions, which they regard as just. See *PL* 35.

beliefs shared by members of this group. I have argued in Chap. 4 that merely shared moral beliefs ("positive morality") by themselves do not justify anything. Whether a practice is justified depends on the reasons for or against it. Different practices in different communities can be justified by reasoning arising from their different circumstances; they are not justified *merely* by their respective beliefs that their own practices are justified. Thus, as I said in Chap. 4, it makes perfect sense that adult children should support their parents financially in China because its social security system is "inadequate," "incomplete," and "limited,"[100] whereas in the United States retirees can claim social security benefit and hence do not need their children's support.

Third, the political system in place might also fall under the term "culture." We might say that Communist countries have a different political culture than liberal democracies. Does it mean that the different political systems in these countries are justified in virtue of the fact that their peoples hold different beliefs? Again, moral contextualism helps to answer this question. Whether political practices (including the political system) are justified depends on reasons arising from the circumstances. Although liberal democracy works well for many Western countries today, it may not work for very poor or "burdened societies" (*LP* 90, 106). Thus, when China began to move away from its Maoist centralized economy in 1978, but was still quite poor and backward, it was understandable and probably acceptable for its government to impose restrictions on liberty of conscience and freedom of religion and thought.[101] But in 2020 China's economy is the second largest in the world, and is the largest if based on purchasing power parity. Under today's very different circumstances, it is difficult to justify glossing over people's freedom of religion[102] and thought, and generally the liberty of conscience.[103] Yet, it is rather widely

[100] See Li Jiange, et al., "Social Security Reform in China: Issues and Options," January 27, 2005. (https://economics.mit.edu/files/691).

[101] In *Theory,* Rawls argues that basic liberties and rights should not be sacrificed for economic progress. I take a slightly different view here and allow some tradeoff of basic liberties for economic reward only if the country is in a very poor, unfavorable, or "burdened" condition.

[102] See Lee Edwards, "Is China Totalitarian?" *The Heritage Foundation*, February 26, 2020.

[103] The liberty of conscience should include the right to criticize the government, if the criticism is fair and conducted in good faith.

believed (outside of China) that China compares unfavorably with at least constitutional liberal democracies in the areas of (1) democracy, (2) liberty of conscience including freedom of religion and thought, and (3) toleration of fair comments. In addition, it is increasingly difficult to justify existing income disparities among different groups of people, and different regions, in the country.[104]

But why should anyone insist that China ought to practice liberal democracy? As I argued earlier, Singapore sacrifices people's right to *genuinely* fair election. In the case of China, political power is highly concentrated in the Communist Party, in general, and the political politburo, in particular.

An authoritarian political system stands in stark contrast to the ideal of liberal democracy. A check-and-balance system can reduce the risk of political domination. This is shown by the fact that had there been no checks and balances in the American political system, President Donald Trump would probably still stay in power, despite having lost the 2020 US Presidential election.[105]

My point is a structural one. It has to do with the structural relation between power, corruption, and the lack of checks and balances.[106] (It is not a statement about individuals.) An authoritarian political system tends to restrict liberty of conscience, and freedom of religion and thought. In order to claim that liberal democracy might be good for the United States, but it is unsuitable for China, as Fan claims in Chap. 3, there has to be a persuasive argument for this claim. A relativistic claim on an abstract level would not suffice.

---

[104] According to Deng Xiaoping in 1986, "we allow some people and some regions get rich first, the general principle of common prosperity" (http://www.fengqi.sh.cn/en/news/knowledge9.html). Presumably this was a somewhat temporary measure for a transitional phase that should not last for 40 years. Premier Li Keqiang of the People's Republic of China recently said that 600 million people in China still receive one thousand Yuan or less per month.

[105] Another possibility is that there would be no election if political power is concentrated only in the hands of the President.

[106] See Stephen K. Ma, "Reform Corruption: A Discussion on China's Current Development," *Pacific Affairs*, vol. 62, no. 1 (Spring, 1989), 40–52 (https://www.jstor.org/stable/2760263?seq=1#metadata_info_tab_contents); Yukon Huang, "The Truth about Chinese Corruption," *Diplomat*, Carnegie Endowment for International Peace, May 29, 2015 (https://carnegieendowment.org/2015/05/29/truth-about-chinese-corruption-pub-60265); Bao Tong, "How Deng Xiaoping Helped Create a Corrupt China," *New York Times*, June 3, 2015 (https://www.nytimes.com/2015/06/04/opinion/bao-tong-how-deng-xiaoping-helped-createa-corrupt-china.html).

8

Further Reflections

Dominic Farrell LC and Joseph Tham LC

In our earlier replies, we outlined both where we agreed and disagreed with Hon-Lam Li's Rawlsian conception of public reason and its bearing on medically assisted suicide and Ruiping Fan's Confucian approach to these issues. Similarly, both Li and Fan have presented similar assessments of our initial paper and each other's. Here we take stock of their replies and reflect on the value and results of this exchange.

1 Reflections on Li's Replies

Li's replies to Chap. 2 are divided into six sections and, by our count, contain nine main objections. We shall give a brief assessment of these objections, mainly with the aim of clearing up possible misunderstandings. We do so, not to have the last word, but to lay the groundwork for further constructive discussion on these matters. Two general

D. Farrell LC (✉) • J. Tham LC
Pontifical Athenaeum Regina Apostolorum, Rome, Italy
e-mail: Dfarrell@upra.org; jtham@legionaries.org

H.-L. Li, M. Campbell (eds.), *Public Reason and Bioethics*,
https://doi.org/10.1007/978-3-030-61170-5_8

observations are in order, though. First, Li's objections tend to focus on peripheral aspects, both real and imagined, of our original contribution rather than the main argument. Many of his objections are concerned with confuting natural law ethics, our interpretation of it, or some of its normative claims. This is a legitimate line of argument since we offered an overview of the natural law tradition (NLT). Nevertheless, Chap. 2 did not set out to give a full-scale defense of the NLT as a theory of normative ethics and political philosophy. The main aim was to explain why NLT conceives public reason as truth-directed inquiry of a political variety, how public reasoning of this kind works, and why it is more consistent than political liberalism when it comes to legislating on bioethical issues. It is more consistent, we argued, because it requires that one address rather than bracket and arbitrarily second-guess the second-order questions necessarily involved in such legislation. Regrettably, Li does not consider the central point of our chapter with the detail that we would have expected and desired. Instead he deems this approach unsuitable and argues that it involves the disqualification and exclusion of people who hold views at odds with Catholic moral teaching. This leads us to a second general observation. At several points, Li misunderstands and so misrepresents various of our own arguments or those of natural law ethics in general.

First, Li raises in quick succession a series of objections against natural law ethics or our account of it (*Human* Telos *and Common Good*). The general tenor of these objections is that natural law ethics is based on the mistaken view that there is morally normative teleology inherent in human nature and actions. It is not possible to deal adequately with each of his specific objections here. Several are addressed in detail in the literature on natural law cited in the notes of Chap. 2. It is worth noting, though, that, Li generally misunderstands the natural law theories which he cites or critiques.

For example, his Kantian objection to ethical naturalism involves a fallacy of equivocation. It uses "human nature" in a different sense than ethical naturalists usually do. According to this Kantian objection, human nature is not a reliable standard for moral judgments because people are not naturally inclined to the same things, let alone moral virtue. Here human nature is equated with an individual's propensities *qua* individual.

When ethical naturalism speaks of "nature," on the other hand, it is usually referring to those ends to which an individual is directed *qua* member of a natural kind. It does not refer to an individual's peculiar psychology or set of motivations, but to an Aristotelian categorial, real essentialism, and natural teleology.

Similarly, Li does not provide an accurate description of certain aspects of the "new natural law theory" (a theory with which we do not fully agree). He notes that one of its main representatives, John Finnis, rejects the view that "natural functions or tendencies are moral standards and ought to guide deliberation and choice." Li concludes that Finnis does not believe that humans have natural ends. He goes on to note how the new natural law theory rejects the view that humans have a single natural end. Curiously, in making this last point, Li cites an article whose title signals that the authors intend to show, in typical natural law fashion, that humans have ultimate ends. In that article, the authors defend the view that integral human fulfillment and basic human goods are ultimate ends of persons, albeit in different senses. Rather, they argue that we have a single ultimate end, but it does not constitute a single ultimate reason for acting or principle of practical reason. Instead, there are various incommensurable basic human goods (life, friendship, marriage, play, knowledge of the truth, religion), each of which is an irreducible reason for acting or principle of practical reason.[1] The new natural law theory, therefore, does not reject natural teleology and the rootedness of human good in human nature. Rather, as we indicated Chap. 2 (note 13), it holds that we come to know the principles of practical reason, not by deriving them inferentially from an observation of the teleology present in the world and human nature, but by understanding in a non-inferential manner that we are directed toward various basic goods.[2] This is what Finnis is getting at when he rejects the proposition, "natural functions or tendencies are moral standards and ought to guide deliberation and choice." Li, however, mistakes this for proof that even some natural law theorists disavow natural teleology.

[1] Germain Grisez, Joseph Boyle, and John Finnis, "Practical Principles, Moral Truth, and Ultimate Ends," *American Journal of Jurisprudence* 32, no. 1 (1987): 99-151.

[2] John Finnis, *Natural Law and Natural Rights*, 2nd ed. (Oxford: Oxford University Press, 1980, 2011), 33-34.

Li also misconstrues aspects of Finnis' account of marriage and sexual morality. He attributes to Finnis the view that it is wrong for a married couple to engage in sexual intercourse for pleasure. Finnis holds that it is not wrong for them to do so. Rather, he argues that it is wrong to have sexual intercourse when pleasure alone is sought and at the expense of the good of marriage.[3]

Li also attributes to Finnis the view that "an act is wrong unless it actualizes a common good" and notes that there are acts, such as one-player games, that are good without actualizing a common good. Indeed, Li is right that there are acts which do not actualize a "common good." However, there is also a sense in which it would be correct to say that an act is wrong unless it actualizes a common good. Writings in the natural law tradition, including our earlier contribution, do not always adequately specify what they mean by "common good." In fact, a good can be common by way of predication ("X is F," "Y is F," "Z is F"), finality ("F is the common end of the association of X, Y, and Z"), or distribution ("F is divided up between X, Y, and Z").[4] The basic human goods are common by way of predication: there is a sense in which they are a good for each and every human.[5] Some goods, however, such as that of marriage and political society, are common by way of finality. They are the specific end of an association and can only be achieved through joint cooperation. Furthermore, they are not common by way of distribution, but belong to all members of the association and cannot be divided up. Finnis' point is that sexuality, unlike a game and many other acts, is directed to a particular kind of interpersonal relation and so to a good that is common by way of finality.

These are several regards in which Li's objections to natural law ethics in general are based on an inaccurate or inexact account of the contested position. As a result, they do not land where they need to.

---

[3] John Finnis, "Marriage: A Basic and Exigent Good," *The Monist* 91, no. 3/4 (2008): 391-92.

[4] Gregory Froelich, "The Equivocal Status of *bonum commune*," *The New Scholasticism* 63, no. 1 (1989). John Finnis, *Natural Law and Natural Rights*, 168.

[5] Finnis, "Marriage: A Basic and Exigent Good," 389. Here, Finnis also suggests that they may even be common by way of finality, to some extent, insofar as they generally can only be attained with the help of others.

Second, Li objects that, contrary to what we propose, moral disagreements are intractable and cannot be resolved through a dialectical engagement between members of alternative traditions who are committed to the requirements of rational inquiry (*Alasdair MacIntyre*). However, his assessment is unsatisfactory. On the one hand, he believes that, contrary to what we outlined, MacIntyre takes agreements of this kind, such as that between utilitarianism and Thomist, to be intractable. MacIntyre's position is more nuanced. MacIntyre believes that it is possible to show through argumentation whether a tradition is right about some matter, but not by appeal to shared, neutral standards. Instead, the two traditions need to engage one another to determine which one has the resources to not only resolve its own problems, but also to identify and explain the failures that the other has in resolving its own problems, given its own principles.[6] MacIntyre does not believe, though, that one tradition possesses arguments that will be compelling to each member of a rival tradition. Instead, the dialectical engagement should focus on principles rather than the positions it draws from them.[7] Even so, Li deems such a project infeasible because, unlike science, it cannot test rival theories through common and well-defined method of experimentation. This, we suggest, is not a compelling objection. Whereas the sciences are primarily theoretical in nature and regard the deterministic causal processes in nature, public reason is an exercise of practical reason and regards action. It has far more in common with the reasoning and methods of the social rather than the natural sciences. Indeed, the sort of reasoning that we find in policy making and law not only bears on many of the same subjects as moral and political philosophy, such as human dignity, rights, justice, the nature, and various kind of actions, but employs very similar methods of analysis and argumentation.

Third, Li claims that the moral absolutism that we defend is incompatible with the contextualism inherent in any concrete action and in Catholic moral teaching (*Alasdair MacIntyre*). By moral contextualism, we mean, and Li appears to have in mind, the view that φ-ing, a specific

[6] Alasdair Chalmers MacIntyre, "Intractable Moral Disagreements," in *Intractable Disputes About the Natural Law: Alasdair MacIntyre and Critics*, ed. Lawrence Cunningham, (South Bend: University of Notre Dame Press, 2009), 32-36.

[7] MacIntyre, "Intractable Moral Disagreements," 51-52.

kind of intentional action, is neither morally good nor bad in and of itself but only good or bad under a specific set of conditions or circumstances.

On the one hand, Li points to how Aristotelians and Thomists believe that humans perform particular kinds of intentional action, many of which are either good or bad in and of themselves, and that we can indicate their moral quality in exceptionless, universally binding rules. He then proposes contextualism as a plausible way of accounting for how universal rules or principles apply to the particularity of any concrete action. Nevertheless, merely asserting the plausibility of contextualism does not disprove the anti-contextualist view. Indeed, Li does not examine the details of the natural law tradition's characteristic action theory, its explanation of how the various kinds of intentional action are determined by their inherent end-directedness rather than their circumstantial features, nor show this action theory to be internally inconsistent.[8]

This failure to address the natural law tradition's anti-contexualist and non-consequentialist action theory, which shares much in common with Kant's, may also explain why Li believes that our absolutist account of natural law is out of kilter with the Catholic Church's official moral teaching. In his view, the latter is committed to a contextualist rather than an absolutist approach to moral norms because it allows for certain exceptions to its prohibition to abortion.[9] Even if this were the case, it would not be relevant to a critique of the argument of Chap. 2. It would be relevant if this were a volume on theology, concerned with assessing the soundness of our scholarship on Catholic doctrine. However, we took care to distinguish the natural law tradition from Catholic belief and present it in its own terms.

At any rate, Li misrepresents Catholic moral teaching. First, he asserts that "the Roman Catholic Church approves of abortion if and when the fetus is implanted in the mother's fallopian tube, or if the mother suffers

[8] Support for a contextualist take on natural law can be found in Jonathan Crowe, *Natural Law and the Nature of Law* (New York: Cambridge University Press, 2019).

[9] Catholic theologians who defend proportionalism (or revisionism) do support a contextualist conception of norms. See, Christopher Robert Kaczor, *Proportionalism: For and Against*, (Milwaukee: Marquette University Press, 2000); Kaczor, *Proportionalism and the Natural Law Tradition* (Washington, D.C.: Catholic University of America Press, 2002).

from cervical cancer." In support of this statement, he does not cite any official documents of papal magisterium, but two webpages. The second of these webpages, put together by unnamed compilators, inaccurately characterizes the cases it mentions as "exceptions" to the Catholic Church's prohibition of abortion. The first webpage, though, consists of an article written by a scholar of Catholic bioethics, Tadeusz Pacholczyk. However, contrary to what Li states, at no part of the article does Pacholczyk claim that the Church approves of abortion in the case of an ectopic pregnancy. Instead, he considers three medical treatments that are used for a life-endangering ectopic pregnancy: the use of methotrexate, a salpingostomy, and a salpingectomy. Moreover, he deems that only the last one is morally acceptable *because* it alone does not constitute an act of direct abortion and so is consistent with the Church's teaching on abortion. He presents his own considered opinion because there is no specific official Catholic teaching on which treatments for life-threatening ectopic pregnancies are licit. Rather, the general teaching, which he clearly has in mind, is "that direct abortion, that is, abortion willed as an end or as a means, always constitutes a grave moral disorder, since it is the deliberate killing of an innocent human being... No circumstance, no purpose, no law whatsoever can ever make licit an act which is intrinsically illicit" (John Paul II, *Evangelium vitae* 62).

This means that, whenever the Church approves of the removal of the cancerous womb of a pregnant woman or a salpingectomy performed to remedy a life-threatening ectopic pregnancy, it does not take itself to be authorizing acts of abortion or exceptions to a moral rule. Rather, it takes each treatment to constitute a different and morally good kind of intentional action, albeit one that will have the foreseen but unintended effect of causing the death of the unborn child. Further on, Li correctly notes that the Catholic Church allows certain medical interventions that will result in death of the fetus on the grounds of such double-effect reasoning. Nevertheless, he wrongly concludes that it thereby allows for certain exceptions to its general prohibition of abortion.

Due to this misreading, Li believes that the Catholic Church's official moral teaching is committed to contextualism. This is hard to square with the preceding quote from *Evangelium vitae*. After declaring abortion to be a grave moral disorder, John Paul II states a general principle of

Catholic moral teaching: that no circumstance can make licit an act which is intrinsically illicit. He explains this in greater detail in another major encyclical (*Veritatis splendor* 78-80), whereas the *Catechism of the Catholic Church* (n. 1754), drawing no doubt on Aquinas (*Summa theologiae* I-II, q. 18, a. 10), summarizes the matter as follows.

> The *circumstances*, including the consequences, are secondary elements of a moral act. They contribute to increasing or diminishing the moral goodness or evil of human acts (for example, the amount of a theft). They can also diminish or increase the agent's responsibility (such as acting out of a fear of death). Circumstances of themselves cannot change the moral quality of acts themselves; they can make neither good nor right an action that is in itself evil.

None of this proves that we are right. It only means that Li has misunderstood and misrepresented the Catholic Church's official moral teaching and cannot adduce its purported anti-contextualism as proof of the unsoundness of the natural law tradition's action theory.

Fourth, Li claims that we do not allow for reasonable pluralism (*Objections and Replies*). He notes that whereas, foundational disagreements that are both reasonable and irresolvable should not be possible for our account of natural law ethics, Rawls insists that, given the burdens of judgment, such disagreements do exist. Our position is more nuanced. We largely agree with Rawls' notion and classification of the burdens of judgment, but not with the conclusion that he draws. We believe that moral disagreements over foundational issues are resolvable in principle, but often not in practice. Moreover, we acknowledge that such disagreement exists and that this constitutes a problem for the natural law tradition. Drawing on Alasdair MacIntyre, we then considered how NLT can resolve this issue.

Fifth, Li claims that, contrary to our reading, "Rawls says that being reasonable is not an epistemological idea (though it has epistemological elements). It is rather part of the idea of democratic citizenship" (*Objections and Replies*). However, in Chap. 2 we reproduce the very passage in which Rawls describes reasonableness as a political rather than an epistemological ideal (*Political Liberalism*, henceforth PL, 62). Of course, our interpretation of Rawls' political liberalism and assessment of its coherence can be called into question. Still, we have not overlooked that he treats reasonableness as a political rather than an epistemological ideal.

Nor have we claimed that Rawls is committed to epistemic abstinence. Rather, we objected that his conception of reasonableness is circular. He identifies reasonableness with a commitment to political liberalism but rules out the possibility of assessing this commitment on external, self-standing grounds. We then argued that the circularity cannot be resolved by appealing to epistemic abstinence and that Rawls, despite his efforts, may end up proposing a comprehensive liberalism, albeit one of a more pragmatist bent.

In this regard, Li proposes the interesting case of the Good Samaritan. He notes that everyone deems the Samaritan's action praiseworthy, regardless of which ground or comprehensive doctrine motivates his decision to succor the traveler who has been left for dead by his assailants. What matters then is that people are committed to fairness, equality, and freedom, not the conflicting grounds underlying this common commitment. What matters is the overlapping consensus, whether that which exists between rival traditions of normative ethics or the various groups within a liberal democracy.

Significantly, this sort of overlapping consensus may only extend to the most general forms of social cooperation and support. It is likely to break down when it comes to more specific moral and legal norms. Whereas everyone is likely to agree that murder is wrong, people disagree over the morality and politics of abortion and medically assisted suicide. This is probably why Rawls wisely restricts the scope of public reason to the fundamental constitutional arrangements of a liberal democracy. It is questionable, therefore, whether the case of the Good Samaritan provides good support for applying the political liberal's conception of public reason when it comes to legislating on bioethical issues. More importantly, it does not account for how laws are framed and applied. Although the consequentialist, contractualist, Kantian, virtue-ethicist, or proponent of natural law all approve of the Samaritan's compassionate altruism, each approves of it on different grounds, as Rawls recognizes. Indeed, each approves of it on grounds that may be very different from those of the Samaritan and which he may even find objectionable and offensive. Now this can explain why each will have a reason to commit to laws that fit in with this overlapping consensus. Nevertheless, the Samaritan is a private person, not a political authority. Laws, however, are framed with very specific values, reasons, and comprehensive doctrines in mind. This is evident, for example, from the majority and dissenting opinions of the judges of the US Supreme Court. The Good Samaritan scenario explains why adherents of different comprehensive doctrines have reason to commit to the laws of a decent society, at least its fundamental ones, but not how laws are made. True, Rawls argues that the laws of liberal democracy are legitimate if and only if they are acceptable to all reasonable persons. However, as we tried to show in the final section of Chap. 2, actual laws are not rooted in an overlapping consensus of comprehensive doctrines but in substantive positions concerning the second-order questions involved in that law.[10]

Sixth, Li claims that, because we accept classical natural law ethics but not political liberalism, by construing public reason as truth-directed inquiry, we not only believe that a society's law should conform with the classical natural law ethics, but also, since this is simply Catholic moral teaching, that all members of society should agree with the underlying comprehensive doctrine, Catholicism *in toto* (*A General Objection to NLT*). Once again, Li misrepresents our position. He conflates public reason as truth-directed inquiry with classical natural law ethics, and the latter with the Catholic faith. In Chap. 2, we distinguished the three. On the one hand, we explained why the natural law tradition is not to be conflated with either Catholic moral teaching or, more broadly, the Catholic faith. The natural law tradition does not appeal to the Gospel but to ethical naturalism, and so the same sort of considerations we find in, say, Aristotle. A Catholic *qua* Catholic will believe abortion or medically assisted suicide to be wrong out of faith and for reasons specific to the Christian faith. However, the natural law tradition believes that abortion or medically assisted suicides are wrong on different grounds. On the one hand, it believes that a human can only become a good human by choosing those goods to which we are directed by our nature. On the

[10] Of course, the law can be applied impartially to people holding different comprehensive doctrines so long as it bears on one's actions rather than on one's beliefs and motivations.

other hand, it believes that abortion and medically assisted suicide constitute intentional actions that are contrary to one of these goods: caring for human persons. Furthermore, the natural law tradition is a variety of moral and political philosophy rather than Christian theology. It is concerned with the question of the good life and which kind of society is best suited to it, not with explaining and justifying the Christian faith. Consequently, public reason in the natural law tradition is restricted to determining what is necessary for the common good of political society and relies on strictly philosophical analysis and argument.

So, while we do believe that a society's law should conform with natural law, we do not believe that these are just if and only if they are predicated upon the acceptance of the Catholic faith. Our general point is that a society's laws should be morally good. They should respect human dignity, natural rights, and basic human goods. In this regard, we are no different from a consequentialist, such as Peter Singer. Whereas a consequentialist will argue that her version of consequentialism provides the suitable framework for designing and revising the law, we believe that a more compelling case can be made for the natural law tradition. Suppose the consequentialist believes and argues that a liberal democracy is not only the best kind of constitution, from a utilitarian standpoint, but also that it should follow consequentialism in designing and revising its laws. In making this argument, she is advocating, not political liberalism, but one comprehensive doctrine as the proper framework for determining the proper constitution, its institutions, and its eventual laws on bioethical issues. Should she make this argument in the public square and political institutions, she is engaging in public reasoning, construed broadly. Furthermore, if she makes this argument in an intellectually honest way, aiming to convince by rational argument, engage rival traditions, and ready to follow the better case, then she is not only committed to public reason as a truth-directed inquiry but is also implicitly proposing such a conception of public reason. At the same time, she believes that his proposed comprehensive doctrine is the right one and that anyone who gives due consideration to his arguments has reason to agree, even if in practice many will either misunderstand them or decline to give them due consideration. The only difference between the consequentialist and us, in this case, is that we bring the natural law tradition rather than classical utilitarianism to the table. If a consequentialist can share the same conception of public reason as we do, then our proposed account of public reason cannot and should not be conflated with natural law ethics.

Seventh, Li objects that we do not provide a complete and fully satisfactory defense of moral absolutes in general or of the thesis that the direct killing of an innocent person is always wrong (*Missing Links*). That is correct. We have not given a complete defense of either because that was not our aim in Chap. 2. Nevertheless, our account may not be as deficient as he makes out. We did illustrate, albeit obliquely, that there are grounds to believe that an intentional action is of a certain kind on account of its inherent teleology or directedness rather than its purely circumstantial features or the agent's ulterior motives. As we noted in dealing with the third objection, Li never really comes to grips with this kind of an anti-contextualist action theory. He also objects to our statement of the view that medically assisted suicide (MAS) is wrong because it consists of the direct killing of an innocent person. First, he claims, without any further explanation, that "whereas euthanasia is killing, medically assisted suicide is assisted suicide." Drawing this contrast makes sense if "killing" denotes a wrongful action. If so, Li is right to distinguish euthanasia from MAS. The former is broader than the latter because it can be applied without the patient's consent. However, we used "killing" to designate a certain kind of intentional action which has both licit and illicit forms. Our point was that, whether right or wrong, MAS consists in providing medical expertise to help a certain class of patient kill themselves (*sui-cidium*). We do not see anything objectionable in calling it an act of killing in this morally neutral sense or as a statement of fact. Second, Li objects that killing an innocent person is not wrong in and of itself but only when there is no reason for killing the innocent person or it is not in his or her interest to do so. To show that it is not wrong in and of itself, he cites possible exceptions to the rule. These are interesting and merit further discussion. Nevertheless, the traditional view on the immorality of directly killing an innocent person can accommodate these cases. Killing in legitimate self-defense a morally innocent but life-threatening aggressor is not murder. On the one hand, it is not a direct killing. The aggressor's death is preterintentional. On the other hand, notwithstanding our potentially misleading talk of an "unjust aggressor," an aggressor is not an innocent person. The aggressor may be morally innocent. Perhaps the aggressor is not compos mentis or has been drugged against her will. Nevertheless, the aggressor is not entirely innocent. The aggressor is committing a life-threatening aggression. Morally responsible or not, an aggressor is not non-nocive (*in-nocens*) but doing harm (*nocens*). Here the operative word is "doing." Some degree of sentient action is involved, even when the aggressor is drunk, drugged, or simply not compos mentis. However, someone who has been blown off the top of a skyscraper and will squash a few oblivious pedestrians upon impact is not doing anything but is subject to external forces. For this reason, like the fat potholer or the fetus in a life-threatening ectopic pregnancy, the person falling from the top of a skyscraper is not an aggressor or *nocens*. There is a difference between doing harm and causing it without doing anything. For these reasons, the traditional view does have an account of why killing a morally innocent but actual aggressor is a case of killing in legitimate self-defense; aborting a human fetus, even in a life-threatening ectopic pregnancy, is not. The former is an aggressor; the latter is not.

Eighth, in making the previous objection, Li claims that, contrary to what we maintain, no reasonable person believes that health care agents should be forced to provide, against their conscience and moral convictions, a treatment that is contrary to one's conscience (*Missing Links*). To be exact, we noted that a certain class of conflict could be avoided if public authorities, notwithstanding the depenalization or legalization of a controversial medical procedure, such as abortion, refrained from making its provision mandatory whenever an eligible applicant requests it. Li opposes—commendably, in our view—policies that would oblige health care agents to act against their conscience and provide certain treatments. However, policies and judicial rulings of this kind do exist. To cite just one example, the Ontario Court of Appeal has upheld regulations, issued by the College of Physicians and Surgeons of Ontario, that would require physicians who refuse on grounds of conscience or religion to carry out medically assisted suicide ("medical assistance in dying") to provide the patient with an effective referral for another doctor or health care provider.[11] This makes the conscientious objectors complicit in an act that they deem wrong. Furthermore, some bioethicists are opposed to conscientious objection for health care agents and approve of rulings and policies of this kind.[12]

Ninth, the fifth section of Li's replies (*Hippocratic Oath*) is based on a misreading. According to Li, we argued "that a physician has sworn the Hippocratic Oath that she would not kill any patient." He goes on to claim that we are thereby committed to the belief that one should follow rules that one has promised to follow, regardless of their validity.

Actually, we argued that MAS does not have the character of a medical act because, regardless of the laudable motives with which it is performed, it is, *qua* intentional action, directed toward the killing of a patient rather than the restoration of health. We then noted that this idea has a venerable pedigree. It is operative in the *original* version of the Hippocratic Oath. We singled out the original version to indicate that we were not referring to the subsequent versions of the oath nor one that doctors might currently make. Moreover, we did not claim to have made a knock-down argument against alternative conceptions of the medical act. Indeed, we alluded to some possible counterarguments and left the matter there. Rather, our remarks on the Hippocratic Oath were part of an argument on how, to be morally justified, the legislation of MAS needs to engage in a deeper inquiry into the essential characteristics of medicine. At no point did we mention the promises that doctors currently make or the rules that they promise to follow. Rather, in interpreting the sense of the original version of the Hippocratic Oath, we pointed to the logical sequence between some of its parts (pledges).

Nor did we claim that we should follow rules because we have promised to do so. Rather, we support the alternative position that Li outlines

---

[11] Christian Medical and Dental Society of Canada v. College of Physicians and Surgeons of Ontario, 2019 ONCA 393.

[12] Julian Savulescu and Udo Schuklenk, "Doctors have no right to refuse medical assistance in dying, abortion or contraception," *Bioethics* 31, no. 3 (2017): 162-170; Udo Schuklenk and Ricardo Smalling, "Why medical professionals have no moral claim to conscientious objection accommodation in liberal democracies," *Journal of Medical Ethics* 43, no. 4 (2017): 234-240; Alberto Giubilini, "Objection to conscience: an argument against conscience exemptions in healthcare," *Bioethics* 31, no. 5 (2017): 400-408.

These articles cite various rulings against conscience exemptions for health care agents.

and endorses: we should follow rules because they are right. After all, the fundamental premise of NLT is that one should do that which is inherently good, rather than follow mere convention.

Hopefully, this assessment of Li's objections and clarification of the misunderstood points can allow for a more careful consideration of our mutual positions and for further fruitful engagement.

2 Reflections on Fan's Reply

Fan agrees with the central claim of our chapter and believes that it is consistent with his reconstructionist Confucian account of public reason. He argues though that public reason should not be reduced to political deliberation informed by truth-directed inquiry. Rather, public reason needs to be construed more broadly as involving a commitment to "non-cognitive values" such as social cohesion, stability, and peace. Moreover, he argues that the cognitive element should enjoy greater weight in the discussion of policy; the non-cognitive ones should be prevalent in the justification of policy. Inquiry into second-order questions has its place in the discussion of which legislation should be adopted. However, in legislating, government should prioritize the non-cognitive elements. As a result, it should not base its legislation on controversial positions regarding such second-order questions but on the society's dominant beliefs on rights and virtue.

In Chap. 5, we have already outlined why we believe that truth-directed inquiry nonetheless should have a certain priority over the tradition that informs a country's public reasoning. Here, we shall consider Fan's point about the legislative body's need or duty to prioritize what he calls non-cognitive values.

Fan, if we have understood him correctly, is making two distinct points.

First, in referring to non-cognitive values, he is stressing that social values and bonds cannot be reduced to thought. We agree. Nevertheless, perhaps it is inaccurate to characterize these values as non-cognitive. Generally, the distinction between the cognitive and non-cognitive, as it is understood in moral and political philosophy, mainly in metaethics, is used to demarcate moral judgments that are truth-apt from those that are not. Classifying social cohesion, peace, and stability as non-cognitive values suggests that members of society value them for utterly subjective rather than objective reasons, and that the political discourse by which they commend the values and propose suitable policies is projectivist and manipulative. This is not what Fan means since he is committed to moral realism. However, if our judgments about the value and priority of social cohesion, peace, and stability are truth-apt, then they are also potential objects of truth-directed inquiry, and have been such at some stage or other of the history of political thought.

We propose that, the relevant distinction is not between cognitive and non-cognitive values, but one's cognitive and affective engagement with the goods, laws, and institutions of society. We take affectivity to encompass not only emotional but also volitional responses, such as choices and action. Moreover, just as any animal's concrete emotional or behavioral response is occasioned by its perception of the situation, so too is our emotional attachment to social goods and action as citizens distinct from our cognitive processes but always informed by them. It is inseparable from our underlying practical and public reasoning about them. Social goods are cognitive, therefore, because it is only through reason that we perceive, value, reassess, arrange, and pursue them. However, they are not merely objects of thought and theoretical inquiry but goods. As goods, they are objects of emotional attachment and political action. This is what Fan is getting at, if we have understood him correctly, when he proposes the distinction between cognitive and non-cognitive values.

In proposing the prioritization of non-cognitive over cognitive values, on the other hand, he is arguing that the legislature should base laws on society's dominant morality for the sake of social cohesion, peace, and stability. Some social goods—cohesion, peace, and stability— not only have priority over the others but are only secured if they are grounded in reasons acceptable to the main constituency of society. As a result, they need to be grounded in the dominant morality, on the condition that, as Fan specified in Chap. 3, they are consistent with the requirements of a decent society.[13]

[13] We made a similar point when we noted that some fundamental social goods, such as peace, enjoy lexical priority over others in such a way that one has reason to commit to a society's institutions even when they are not perfect or even unjust in certain regards.

No doubt, legislative bodies often follow such a logic or simply the lead of the dominant morality. However, Fan is not merely describing what goes on but proposing a normative principle. It is not clear whether he is proposing the prioritization of these values and the dominant morality as a sensible prudential norm or a categorical one. Either way, he does not seem to account for how public opinion changes and, with it, the dominant morality. This raises a problem that does not arise for our proposed conception, where public reason is an ongoing inquiry in which all have a right to participate. To address the problem, Fan needs to explain what he proposes for a traditionally Confucian East Asian country which reaches a point where Confucianism has ceased to be the dominant moral tradition. In this case, should Confucianism also cease to inform its laws since it no longer acts as the source of the values which underpin social cohesion, stability, and peace?

In this scenario, it is likely that committed Confucians will wish to convince others of the truth of their values. To do so, they will need to engage in truth-directed inquiry with members of rival traditions. They will need to have writers and artists who translate Confucianism into compelling works. They will also need to gradually influence the culture through communities and institutions which successfully embody and transmit their values. In this case, however, they will be engaging in the sort of public reasoning that we outlined, or rather outlined in part.

We can only claim to have outlined it in part because, as this last example illustrates, much meaningful public reasoning takes place in the broader culture, outside legislative bodies, courts, and lobbies. Indeed, just as the visible part of an iceberg is supported by the hidden ninety percent that lies underwater, the reasoning of politicians, judges, and lobbyists is often a reflection of the much broader discussion that is going on in universities and the media, communities, and social circles.

3 Closing Reflections

Some progress has been made. There is, we believe, extensive agreement between our own analysis of medically assisted suicide and that of Fan. There is even a significant agreement over the need for truth-directed inquiry in public reason and the place of Confucianism in East Asian policymaking, although Fan gives priority to what he takes to be non-cognitive sources of social agreement. On the other hand, while our exchange with Hon-Lam Li has brought to light some issues that each side must address, we continue to hold opposing positions on the nature of public reason and on medically assisted suicide.

It is only to be expected that none of the three parties has come to full agreement with any of the others. A volume of this kind can only get the debate underway, lay out the various positions, clear up some potential misunderstandings, and identify a few of the more promising lines of inquiry. More importantly, it should thereby stimulate both writers and readers to a greater understanding of alternative views and willingness to engage them with intellectual honesty.

Significantly, virtually all the contributions have broached John Rawls' account of public reason. This is a testimony to how much he has shaped contemporary reflection on the fundamental principles of political morality proper to a pluralistic society. This is mainly due to the strength of his work. It may also be due in part to some of the philosophy profession's current characteristics. Most faculty teaching philosophy in North America and Europe espouse, maybe not Rawls' exact positions, but the same general left-of-center liberal principles and assumptions that underlie his theories. For this reason, conferences and monographs which, like this volume, bring alternative traditions of ethics and political philosophy into the conversation perform an increasingly important function. By facilitating a greater understanding of various comprehensive doctrines, both those held by a significant part of the populace and those prevalent in the academy, they broaden our intellectual horizons. Without them, we are at a greater risk of becoming inadvertent captives of intellectual complacency or bias and, in the process, less inclined to give those with whom we disagree a fair hearing, more likely to be unduly dismissive toward them. Healthy academic conversations of this kind, therefore, can play a small but important role in preventing or remedying polarization within society.

In his classic defense of liberalism, John Stewart Mill insists on the need of precisely this kind of exercise. He notes that one "who knows only his own side of the case, knows little of that" and so does not have adequate grounds for holding what may be otherwise sound commitments. Hence, Mill argues, it is imperative to engage with people who are committed to opposing moral views and present them in the most persuasive way. Moreover, "if opponents of all important truths do not exist, it is indispensable to imagine them, and supply them with the strongest arguments which the most skillful devil's advocate can conjure up."[14]

On the other hand, conversations of the kind found in this volume constitute but one expression of a broader and more fundamental intellectual pursuit: a liberal education. Indeed, traditionally it has been believed that a liberal education is critical to the success of public reasoning because it aims at forming judicious citizens.

Of course, not everyone who boasts of a liberal education exemplifies its fruits, and many people who cannot boast of one are wise and thoughtful. By and large, though, it can improve the quality of public discourse.

The aim of a liberal education, as Leo Strauss notes, is to form oneself as a well-rounded person. To achieve this end, it adopts the strategy of learning from the great minds. In practice, this consists of the careful, lifelong study of the books that they have left behind. The books studied will differ somewhat from one region to another—say in the West and East Asia—mainly because one's language limits the range of cultures one can study with profit. Nevertheless, the study of a certain canon leads to critical reflection rather than indoctrination. It involves engagement with the wide variety of contrasting ideas held and articulated by the great thinkers and writers. Hence, the careful reading of their works requires that we bring them into dialogue with one another. It also promotes a greater intellectual humility and circumspection. Besides making one less

[14] John Stuart Mill, *On Liberty*, in *Essays on Politics and Society*, ed. John M. Robson, Collected Works of John Stuart Mill, vol. 18 (Toronto: University of Toronto Press—Routledge & Kegan Paul, 1977), 245.

susceptible to demagoguery, these qualities render one a more judicious and thoughtful citizen.[15]

This volume has put these principles into practice. To examine how a pluralist or secular society should legistlate on bioethical matters, it has drawn on Confucius and Mencius, Aristotle and Aquinas, Kant and Mill. In so doing, it reminds us of the value and importance of a liberal education. So, while this book debates public reason and bioethics, on a deeper level it reminds us of the importance of pursuing and promoting liberal education.

[15] Leo Strauss, "What is Liberal Education?," in *An Introduction to Political Philosophy: Ten Essays*, ed. Hilail Gildin, (Detroit: Wayne State University Press, 1989).

9

Further Reflections

Ruiping Fan

Hon-Lam Li could be forgiven for misunderstanding my view. While I am constructing a Confucian conception of public reason rather than formulating Confucian ethics in my chapter, I could have explicitly pointed out their differences. Confucian ethics is by no means moral relativism. Whereas I do hold that "Chinese should continue to be Confucians," it is not "because China has been practicing Confucianism for centuries" as Li mistakes in his Reply to me, but because Confucian ethics teaches true virtue (*de*) that all human beings should learn and practice. That is, it discloses the moral truth for every human being. For the same reason, any people, no matter whether they have been practicing Confucianism or not, should ethically become Confucians. It is only remarkable that Confucianism never supports using force to convert people into Confucianism. Instead, genuine moral power, in Confucian belief, lies in peaceful, virtuous exemplars.

R. Fan (✉)
Department of Public Policy, City University of Hong Kong,
Kowloon, Hong Kong
e-mail: safan@cityu.edu.hk

H.-L. Li, M. Campbell (eds.), *Public Reason and Bioethics*,
https://doi.org/10.1007/978-3-030-61170-5_9

On the other hand, "public reason" is primarily a political concept. Although a conception of public reason should, in my view, contain a universal moral core and have significant moral implications, it is not only for intellectual inquiry to discover moral truths. Instead, a suitable conception of public reason for a society must seek some social goals for that society, such as its stability, cohesion, and peace, in addition to seeking the cognitive goal of moral truths. Accordingly, distinct ethical resources of that society, together with a universal moral core, must be integrated into the suitable public reason conception for that society to direct its public discussion and policy making so as to pursue both moral truths and social goals. In the case of my Confucian conception of public reason for East Asian societies, a minimal idea of human rights is upheld (as it should be accepted universally in any public reason conception), and certain Confucian ethical codes are maintained (as they may not be accepted into other public reason conceptions for non-Confucian societies). This is to say, my Confucian conception of public reason does imply a "soft" relativism: what is a normatively right conception of public reason for a society should be evaluated with respect to the actual moral standards adopted in that society. This relativism is soft because it insists the same moral core of human rights for all public reason conceptions. In my view, unless one renounces any social goal for a public reason conception, this minimal relativism is as inevitable as it is reasonable.

This version of relativism is like the so-called agent's-group relativism, not appraiser's-group relativism, in the distinction that David Lyons has famously made. For instance, under my Confucian conception of public reason, a decision that medically assisted suicide (MAS) should NOT be legalized in East Asian societies is right because it accords with the norms of East Asian people, whereas the decision that MAS should be legalized in contemporary Western societies (under the Rawlsian liberal conception of public reason) is right because it accords with the norms of contemporary Western people. This version of relativism does not validate conflicting moral judgments for a group of people, either East Asians or Westerns, so that it is not incoherent. In a nutshell, to face pluralism in East Asia, my Confucian conception of public reason is constructed by selecting only those Confucian virtues and their central requirements that are still generally practiced and accepted in the region. It is not to include the entire content of Confucianism that I hold to be morally true. This is similar to the strategy that Rawls takes to build his liberal conception of public reason for modern Western democratic societies by drawing on their overlapping consensus. However, while both the West and the East face pluralism, the substance of their respective pluralism differs, since there are distinct religious and moral traditions embedded in each of them. It is misguided to assume that one side has only "positive moralities" while the other side has both "positive" and "critical moralities." In fact, the contemporary world bears witness to different "critical moralities" in addition to incompatible "positive moralities" between the East and the West, just as inside the West people disagree not only on comprehensive accounts of the good life, but also on the theories of rightness or justice. Although all people can be encouraged to engage in critical reflection on their respective moralities, either Confucian, or liberal, or natural law morality, and to back up "the objectively right morality" that they hold with reasons, principles, and arguments, it is still unlikely that they will all arrive at the same "critical morality."

However, neither soft relativism nor value pluralism should block Eastern and Western people from learning from each other. In my case, I have accepted a Rawlsian list of human rights into my Confucian conception, although it is Western and liberal in origin. I do not think Joseph Chan's or Sungmoon Kim's Confucian account as a suitable Confucian conception of public reason for East Asian societies not "because they…are mixed with Western or liberal theory" as Li asserts, but because I offer the specific moral reasons in my chapter that have been ignored by Li. In Chan's case, he wants to promote both "the rich insights and ethical resources" of Confucianism and the requirement of liberal "civility" in a modern Confucian-influenced society, whereas the latter requires that Confucian values be promoted only in a piecemeal way so as not to damage "civility." One problem with this "civility" requirement is that it makes his entire Confucian account impossible as Kim argues: his piecemeal Confucian values to be promoted will hardly be Confucian if they are detached from the comprehensive doctrine of Confucianism. In addition, I added an argument that this "civility" requirement is unnecessary for East Asia: when a comprehensive Confucian moral doctrine (such as that of filial piety in the sense that adult children have a moral obligation to take care of their elderly parents) is generally practiced and accepted in East Asia, it is not necessary to set up the "civility" requirement and refrain from promoting or drawing on them in making public policy. Such "promoting" or "drawing on" will not harm (but will, instead, contribute to) the stability, solidarity, or peace of any Confucian-influenced East Asian society, notwithstanding that leading liberal scholars (such as Jane English and Norman Daniels) and Western societies hold that children do not have such a moral obligation. In short, my conception does not accept this "civility" requirement not because it is Western or liberal, but because it is unnecessary and unbeneficial for East Asia.

Kim is the first Confucian scholar to construct an insightful Confucian conception of public reason for East Asian societies. However, he finds it necessary to integrate the democratic value of public equality into his conception. In so doing he has, in my view, overlooked two nonegalitarian Confucian structural features that should continue to be cherished in contemporary East Asian societies. One is a notion of political inequality due to the reason that virtues are unequally possessed by individuals through their own efforts so that only virtuous or meritorious persons should be selected to undertake political responsibility and lead society (but what methods should be taken to select them is a different issue that I have no space to address in my chapter or here). Another is a notion of opportunity inequality arising from the practice of the Confucian principle of differentiated and graded love with the existence of the family: given that the family holds a fundamental place in Confucian society, opportunity equality should only be pursued as appropriately qualified (namely family-based) equality, and certain reasonable opportunity inequalities among individuals generated by the existence of the family must be allowed and even respected. I can add that these nonegalitarian moral features should be defended not only because they do not go against any fundamental sense of human moral equality, such as the equality of human nobility (*gui*) or dignity, but because they are instrumental to the virtue-based familist idea of the good life that is still largely embraced by East Asian people. In short, in building my Confucian conception of public reason, I have learned a great deal from my Confucian friends Kim and Chan. Whatever disagreement remains between them and me, it must be due to different substantive moral concerns rather than any narrow-minded identification of value "pedigree" as Li suspects.

In reflection, it remains puzzling to me why Li asserts that "Fan holds that a Confucian theory should not include anything liberal or Western" when I have explicitly integrated a Rawlsian list of human rights into my conception. It is similarly puzzling why he wants to suppose that I think "Western theory" is bad. It is even more puzzling why he comes to "suppose that by 'Western theory,' which Fan thinks is bad, is liberalism and not Marxism." For the last supposition, Li picks out a few Confucian citations to exhibit that "Confucianism stresses the supremacy of men, not women" and contends that the relative equality of the genders has been held by Chinese Marxism but not Confucianism. I agree it is important to explore the Confucian view of the genders, but that is not the task of this book. Actually immense well-focused, critical literatures on the Confucian view of the genders have been generated by contemporary Confucian and non-Confucian scholars. Whatever the view is, it must be ill-served by offering a few highly selective, out-of-context citations to outline the view as Li does in his Reply. A twenty-first-century liberal scholar should do better than that. Isn't it illogical to go such a sophisticated way to "suppose" that a Confucian scholar thinks "liberalism" is the "bad" "Western theory" when he has clearly integrated a Rawlsian idea of human rights into his conception?

In his Reply to me Li contends that a Confucian should support medically assisted suicide (MAS). He states that "[a]s far as I know, [Fan's] view on MAS would clash with most Confucians' view." I have searched recent literatures and found no reliable survey about Confucian or Confucian-influenced attitudes toward MAS that can be used to support Li's statement. In a footnote he mentions that "I have personally talked to various Confucian scholars in Hong Kong and abroad." I am aware that some Confucian scholars support MAS, others oppose it, and I have no evidence about which side is a majority opinion. Accordingly, his claim on "most Confucians' view" based on his personal contacts is not indicative. More important, what is the proper Confucian view of MAS should ultimately be determined by fundamental Confucian values and commitments. Nevertheless, in making his personal investigation and statement, Li has cared about Confucians' view in general and the

Confucian virtue of filial piety in particular regarding the issue of MAS. I welcome this move he has made.

Indeed, from Li's vantage point, it is the "most Confucians' view" that "a filial son should respect his parent's earnest decision to choose MAS if she is terminally ill, going through unbearable suffering, and desperately wants to die." However, although nobody has reliable evidence to show a majority Confucian opinion on this issue, his claim is fatally flawed in terms of the Confucian virtue of filial piety. First, contemporary palliative care and pain management have earned great accomplishments. If done appropriately they do not leave a patient to "unbearable suffering," even if they cannot get rid of every bit of suffering or pain for the patient. Leaving his ill mother to "going through unbearable suffering" without looking for a skillful palliative care doctor and effective pain-relievers that are already available in contemporary society to make her suffering lessened and bearable is hard to constitute as a "filial son" in the Confucian moral sense. Moreover, in Confucian tradition, if his mother "desperately wants to die" in a medical situation, a filial son must examine himself to see if he has done anything inappropriate or insufficient so as to lead to his mother's "desperate" want for death. He can always improve himself in caring for her, materially, psychologically, or spiritually, to accompany her to live out the last stage of life rather than support her to commit suicide. Finally, even when his mother is terminally ill (which usually means she will survive no more than six months), a filial son's respect for her preference for MAS should still be balanced against his respect for the value of her life. Her life matters as a noble (*gui*) or dignified human life, which is especially important to him, since he, as her adult child, undertakes a special moral obligation to take care of her life according to the Confucian teaching of filial piety. Her preference for committing suicide for terminating her suffering in this case should not be taken as a righteous decision because, for one thing, the suffering can be medically reduced. It is the Confucian teaching (e.g., in the *Classic of Filial Piety*) that a filial son must remonstrate with the parent rather than comply with her unrighteous decision so that she would not sink into an unrighteous deed. In short, contrary to Li's understanding, a filial son should not support his parent's decision to choose MAS.

When I read Li's comments such as "Marx's idea of the dictatorship of the proletariat certainly coheres with Confucianism, which defended absolute monarchy for thousands of years," I was dismayed. For it is evident and widely held that it was Legalism, not Confucianism, which "defended absolute monarchy for thousands of years" in the history of China.[1]

Dominic Farrell LC and Joseph Tham LC in their "Replies to Li and Fan" rightly remind everyone that the "fact of reasonable pluralism," as Rawls describes it, does not exist. Pluralism goes deep. It is not the case that people disagree only on matters of religion and morality, but not, on due reflection, over the principles of justice. The fact of the matter is, as they show, there are conflicting theories of the right as there are different religious and moral doctrines. Accordingly, as they argue, there is no reason to rule out considerations concerning the good when it comes to public reasoning. Instead, public reason, for them, is only consistent if it does not bracket comprehensive doctrines but engages in truth-directed enquiry into second-order questions about human nature or metaphysical convictions.

I think they are right to emphasize the including of a truth-directed, epistemological ideal into the scope of proper public reasoning. They may also be right that from their natural law tradition perspective, there is a universally valid form of public reason to be approximated by every region of the world. However, there should be a difference between a normative ethics and a conception of public reason, as I draw in the beginning. While the former holds a purely epistemological goal to pursue the moral truth, the latter contains only part of all normative reasons for a society (facing pluralism) to pursue some social goals (such as society's peace, solidarity, and cohesion) as well as for the moral truth. In this case, I don't object to their way of marking my approach as "historicist" (which I think roughly amounts to the "soft relativism" in my terminology) as long as they do not keep a determinist sense of history in the term.

[1] It is worth noting that for another example of a debate between a liberal and a Confucian scholar, readers can see the Appendix to my book *Reconstructionist Confucianism*, which is a record of a dialogue between Andrew Brennan and myself. Although Brennan does not affirm the Confucian values I uphold, his approach takes the history of Chinese schools of thought sufficiently seriously.

When the central moral values and commitments of a moral tradition (either Confucianism, the Natural Law Tradition, or Liberalism) are embodied in the way of life of the people in a society and are largely taken to be true by the people, the people have a right to develop their particular conception of public reason based on that tradition to deal with the pluralism they face in their society, although that tradition is taken to be morally not true or less true than another tradition by the people of another tradition living in another society. Thus, every conception of public reason, whether it is Confucian, liberal, or the natural-law-based, is a loaded rather than neutral conception. And every people are inevitably culturally-and-historically conditioned in their way of pursuing moral truths or "tradition-transcendent standards."

At this point I should accentuate my above-mentioned belief on the importance of cross-tradition dialogue and mutual learning (and criticizing) for truth-directed inquiry. People from any tradition, including those from the liberal tradition, should be epistemologically modest in listening to the voices of other traditions. I will continue my reflection on the challenge raised by Farrell and Tham regarding the necessity to explain "where the robust moral cognitivist standpoint ends and the weak moral cognitivist one of the historicist approach begins" for my Confucian conception of public reason (as well as for any other conception at that). I think my suggestion made in my Replies to them is useful: we should distinguish the task of policy discussion from the task of policy justification under a conception of public reason. For policy discussion, the cognitive goal should dominate. Every participant in public discussion and deliberation should have a right and be encouraged to pursue moral truths based on their comprehensive religious and moral convictions, along with their engaging in critical moral reflections and conducting cross-tradition conversations. However, after a policy is made, the government should avoid appealing to controversial doctrines to justify the policy. This latter strategy is recommended primarily for meeting the social goals of peace, cohesion, and solidarity. In this way both moral truths and social goals can hopefully be sought in a society through its established conception of public reason.

Part II

Conceptual and Historical Background

10

Kant's Conception of Public Reason

Terence Hua Tai

1 Introduction: Kantian Autonomy and Public Reason

The idea of personal autonomy has played a fundamental role in contemporary moral and political philosophy. While 'autonomy' was a term initially applied in ancient Greek to a city-state that *rules itself*, Kant used it to characterize the will of a person who leads her life by following principles she imposes upon herself through reason. Under Kant's influence, many contemporary liberals appeal to personal autonomy as one of the key concepts that structure their theories, though they in fact understand it differently not only than Kant did, but also among themselves, so much so that it may be more accurate to view them as advocating different *conceptions* that purport to explicate (hopefully) the same *concept* of

T. H. Tai (✉)
Department of Philosophy, National Chung Cheng University,
Chiayi County, Taiwan

H.-L. Li, M. Campbell (eds.), *Public Reason and Bioethics*,
https://doi.org/10.1007/978-3-030-61170-5_10

(personal) autonomy.[1] John Rawls, who rests his political liberalism on a "political conception" of autonomy as a feature to be accorded to "free and equal citizens" in a constitutional democracy, distinguishes his conception from Kant's, holding that the latter belongs to "a comprehensive moral view in which the ideal of autonomy has a regulative role for all of life," and that "[t]his makes it incompatible with the political liberalism of justice as fairness."[2] According to Rawls, a political conception of autonomy must be able to serve as "a public basis of justification,"[3] which he thinks Kant's conception cannot. But this does not mean that Rawls would reject the latter conception. Rather, he leaves it open that "[a] comprehensive liberalism based on [Kant's] ideal of autonomy may … belong to a reasonable overlapping consensus that endorses a political conception, justice as fairness among them."[4] What we find in Rawls is a cautious treatment of Kantian autonomy that attempts to shun it for the sake of justifying a political conception of justice from the distinctive and, for some scholars, questionable moral theory and metaphysics in which Kant has embedded it.

The same trend of seeking theoretical resources in Kantian autonomy, though in a revisionary way, has also taken place in bioethics. Tom Beauchamp and James Childress, the arch-principlists in bioethics, advocate the principle of "respect for autonomy" as one of their famous four principles of biomedical ethics (the other three being those of beneficence, non-maleficence, and justice). They are keen to make clear what the principle is aimed at protecting. According to them, autonomy has been held by some theorists to pertain to "the abilities, skills, or traits of the autonomous persons, which include capacities of self-governance such as understanding, reasoning, deliberating, managing, and

[1] See Gerald Dworkin, *The Theory and Practice of Autonomy* (Cambridge University Press, 1988). The distinction between a concept and different conceptions falling under it was first made by H. L. A. Hart, *The Concept of Law* (Clarendon Press, 1961). John Rawls later adopted it in holding his theory of justice to offer a conception of justice as fairness; hence the title of his classic opus *A Theory of Justice* (Harvard University Press, 1971), henceforth "*Theory*."

[2] John Rawls, *Political Liberalism* (Columbia University Press, 2005, expanded edition), (henceforth "*PL*"), 98.

[3] Rawls, PL 99.

[4] Rawls, PL 98–9.

independent choosing."[5] Beauchamp and Childress avoid conceiving it this way, which has led to postulation of a capacity for so-called *second-order* reflection whereby an autonomous person is supposed to determine whether ordinary, *first-order* desires and preferences in her are genuinely her own or belong to her "true self."[6] As they see it, linking autonomy to such a capacity is problematic in various ways, one of which has to do with their need as bioethicists to find "a way for ordinary persons to qualify as deserving respect for their autonomy even when they have not reflected on their preferences at a higher level."[7] To make the principle practicable for bioethicists, they propose a "three-condition theory" that focuses on autonomous *choices*: at the minimum, we make such choices just in case we "act (1) intentionally, (2) with understanding, and (3) without controlling influences that determine [our] action."[8] Moreover, while some principlists in bioethics tend to assign the *highest* priority to the principle of respect for autonomy over other ethical principles, Beauchamp and Childress avoid doing so, thereby distancing themselves from what may seem to be a controversial Kantian moral outlook with which to begin bioethical theorizing.

Contemporary conceptions of autonomy have been criticized for their individualistic or egoistic implications by communitarian and feminist philosophers. Onora O'Neill, a renowned Kant scholar and bioethicist, adds additional fuel to this line of criticism when she asks,

> [C]ould the very conceptions of autonomy and of respecting autonomy, that have been at the heart of so many policies for regulating medicine, science and biotechnology, threaten the maintenance and creation of trust?

---

[5] See Tom Beauchamp and James Childress, *Principles of Biomedical Ethics* (Oxford University Press, 7th ed., 2013), (henceforth "*Principles*"), 102.

[6] See Dworkin, *The Theory and Practice of Autonomy*, 20, where he suggests, more fully than I have depicted here, that autonomy may be "conceived of as a second-order capacity of persons to reflect critically upon their first-order preferences, desires, wishes, and so forth and the capacity to accept or attempt to change these in light of higher-order preferences and values." See also Beauchamp and Childress, *Principles*, 102–3, where they refer to theories that postulate the second-order capacity in question as "split-level" ones. The idea of such a capacity is owed to Harry Frankfurt's classic essay, "Freedom of the Will and the Concept of a Person," collected in Frankfurt, *The Importance of What We Care About* (Cambridge University Press, 1988).

[7] Beauchamp and Childress, *Principles*, 103.

[8] Beauchamp and Childress, *Principles*, 104.

> Is loss of trust perhaps the price of increasing autonomy? Must we choose between respect for autonomy and relations of trust?[9]

O'Neill goes to great length to press these questions and eventually concludes,

> By themselves … conceptions of individual autonomy cannot provide a sufficient and convincing starting point for bioethics, or even for medical ethics. They may encourage ethically questionable forms of individualism and self-expression and may heighten rather than reduce public mistrust in medicine, science and biotechnology.[10]

She goes on to argue that contemporary conceptions of autonomy, including the one advocated by Beauchamp and Childress, are *un-*Kantian, and that not only is a Kantian conception of autonomy *not* individualistic, but it would also, if adhered to by bioethicists, enable them to avoid having to "choose between respect for autonomy and relations of trust."

More specifically, according to O'Neill,

> [Contemporary conceptions of autonomy] range from existentialist or quasi-existentialist conceptions … that identify it with mere, sheer independence in choosing and acting, to a wide range of positions that see autonomy as independence in choosing and acting that meets the demands of one or another conception of rationality.[11]

She is here dividing those conceptions into two groups: one group takes autonomy to be "mere, sheer independence," as exemplified by one who chooses and acts "lawlessly," that is, without relying on any principle at all; whereas the other group takes it to be "rational autonomy," as exemplified by one who chooses and acts according to principles that, though seemingly "lawlike," are not necessarily *universal*, that is, valid for

---

[9] Onora O'Neill, *Autonomy and Trust in Bioethics* (Cambridge University Press, 2002), (henceforth "*Autonomy*"), 16.

[10] O'Neill, *Autonomy*, 73.

[11] Onora O'Neill, *Constructing Authorities: Reason, Politics and Interpretation in Kant's Philosophy* (Cambridge University Press, 2015), (henceforth "*Authorities*"), 137.

everyone.[12] In spite of this difference, both groups are taken by O'Neill to "view choosing, hence autonomous choosing, in markedly individualistic terms."[13] Hence, her conclusion cited above that all these conceptions "may encourage ethically questionable forms of individualism and self-expression and may heighten rather than reduce public mistrust in medicine, science and biotechnology." On the other hand, referring to Kantian autonomy as "principled autonomy," O'Neill holds that it is inseparable from "public reason" (or, in Kant's own term, the "public use of reason"). As she puts it, "the conception of principled autonomy that [Kant] proposes and defends requires adoption of principles that combine *lawlike form* and *universal scope*, so are fit to be offered as [public] reasons to all others."[14] So interpreted, Kantian autonomy is held by O'Neill to enhance public trust to the extent that it yields lawlike and universal principles or norms that can be offered as *reasons acceptable to all* for public policies.

Although O'Neill's claim that Kantian autonomy is not individualistic and must be understood in terms of public reason seems to me defensible on the whole and illuminating, I think there is still something lacking in it. I will argue that a full account of Kantian autonomy must invoke not only Kant's ideas of "lawlike form," "universal scope," and public reason, as O'Neill does, but also, as she does *not*, two other and no less essential ideas in Kant's moral and political philosophy, namely: "humanity" (or our "rational nature"), and the domain of "right" (*Recht*) as opposed to that of morality. In Sect. 2, I show that humanity plays an indispensable role in the autonomous use of reason to choose principles that we can regard as morally binding on everyone, and that O'Neill has in fact tacitly invoked Kant's idea of humanity without singling it out in her account of Kantian autonomy. In Sect. 3, I argue that, when explaining how Kantian autonomy is conducive to trust in public policies that require coercive measures for their implementation, O'Neill fails to consider the autonomous use of reason *in the domain of right*, where it is also called by Kant the public use of reason. While O'Neill interprets Kantian autonomy as,

[12] O'Neill, *Authorities*, 125–8.

[13] O'Neill, *Authorities*, 127.

[14] O'Neill, *Authorities*, 138.

in her words, "the idea of acting on principles that we can will as universal laws,"[15] I suggest that the "principles" she refers to in this idea had better be taken, for the purpose of connecting Kantian autonomy with public trust, as including "positive laws" legislated by "external" authorities in the domain of right.

The overall plan of the following Sects. 4–7 is to explore how the public use one can make of one's reason according to Kant can be interpreted as justifying one in treating positive laws, which are by themselves laws of *heteronomy*, as if they were laws of *autonomy*. Section 4 explains the public use of reason as it figures in Kant's proposal made in "What Is Enlightenment?" to rulers in the world for "public enlightenment." Section 5 further explains the public use of reason in terms of what Kant speaks of as three "enlightened" "maxims of reason," which appear mainly in *Anthropology from a Pragmatic Point of View* and *Critique of the Power of Judgment*. Section 5 also unpacks the first of these maxims, that is, to "think for oneself," against the background of Kant's Doctrine of Right and, in particular, the so-called universal principle of right.

Section 6 introduces Kant's conception of members of a "state of right" (or civil society) as *equal* "partakers in happiness," according to which they are legally required to pursue happiness within the bounds delimited by positive laws. These laws must, according to Kant, themselves conform to the universal principle of right, which must in turn be grounded in the freedom human beings are supposed to have an "innate" right to by virtue of their humanity to do things in ways compatible with the like freedom of every other to do the same for the sake of pursuing happiness as each conceives it. This final appeal to humanity as what underlies freedom for all human beings is something O'Neill's account of Kantian autonomy passes over. Section 6 also shows that while all members of a state of right are equal "partakers in happiness," not everyone is qualified to enter into public reasoning about the moral legitimacy of positive laws unless one has "enlightened" oneself by way of the aforementioned "maxims of reason."

Section 7 argues that public reasoning plays a *double* function for Kant: It is "justificatory" in submitting positive laws to reason's "free and

[15] O'Neill, *Autonomy*, 89.

public examination,"[16] so that their legislation by an external authority can win respect and, therewith, public trust if they pass the test of such examination. It is also "exemplary" in serving to encourage what Kant calls the "unthinking masses" to enlighten themselves (which he takes them to have the potential to do), thereby making themselves eligible for participation in public reasoning. Section 8 concludes by recapping my criticism of O'Neill regarding the connection between Kantian autonomy and public reason.

A final word by way of introduction: I will try to show in this chapter that Kant can simply argue that positive laws could be so justified through the public use of reason that they could be thought of *as if* they were laws of autonomy. This is different from saying that positive laws can be made into *genuine* laws of autonomy. Kant would deny that they can be, because for him only moral laws can (and must) be laws of autonomy. Whether or not we can *really* legislate such laws to ourselves is a metaphysical question Kant must grapple with, and answer in the positive, regarding the possibility of morality, but he does not have to touch it in his Doctrine of Right. If so, Kant's conception of the public use of reason in the domain of right does not presuppose the kind of controversial and, for many, unintelligible metaphysical position that has been, rightly or wrongly, attributed to him.

2 Humanity as Ground for Laws of Autonomy

Let us consider autonomy as Kant expounds it in the *Groundwork for the Metaphysics of Morals*. He uses the Formula of Autonomy (FA, for short) together with the Formula of Universal Law (FUL) and the Formula of Humanity as End in Itself (FH) to express what he has identified in Section II of the *Groundwork* as the "supreme principle of morality" (and

[16] See *Critique of Pure Reason* (first edition 1781, second edition 1787), (henceforth "*CPR*"), trans. Paul Guyer and Allen W. Wood (The Cambridge Edition of the Works of Immanuel Kant, Cambridge University Press, 1997), A xi, footnote. Citations of CPR will follow the standard method of referring to the pagination of its first ("A") and second ("B") editions.

will justify it in Section III). Autonomy is at first said to be simply "the idea *of the will of every rational being as a will giving universal law.*"[17] Kant later states it in the form of a (so-called categorical) imperative as follows:

> *Act in accordance with that maxim which can at the same time make itself into a universal law.*[18]

Still later in Section II, we find a subsection headed "**Autonomy of the will** as the supreme principle of morality" (which suggests that the FA is *the* principle Kant looks for in the *Groundwork*), where the "principle of autonomy" is said to go as follows:

> Not to choose otherwise than so that the maxims of one's choice are at the same time comprehended with it in the same volition as universal law.[19]

Both imperatives cited above can be regarded as expressing the FA, the *third* formula. But it seems difficult to distinguish it from the *first* formula, the FUL, as follows:

> *Act only in accordance with that maxim through which you at the same time can will that it become a universal law.*[20]

Both the FUL and the FA are phrased in terms of "maxim" and "universal law." "Maxims" are for Kant *subjective* principles that guide the person who adopts them in making choices and acting accordingly. How, then, do the two formulas, the FUL and the FA, differ in regard to the relation of maxims (or subjective principles) to universal laws (which Kant also calls "objective principles")?

[17] *Groundwork for the Metaphysics of Morals* (1785), (henceforth "G"), trans. Allen W. Wood (Yale University Press, 2002): 4:431. Citations from Kant's works other than the *Critique of Pure Reason* will be by volume: page number of Kant's *gesammelte Schriften*, ed. Königlich-Preuβische Akademie der Wissenschaften, now Berlin-Brandenburgische Akademie der Wissenschaften (Berlin: Walter de Gruyter, 1900–).

[18] G 4:436.

[19] G 4:440.

[20] G 4:421.

To this question, Allen Wood has provided a helpful answer,[21] which may be put this way: Both the FUL and the FA seem to ask you to reflect on the same question as to whether a maxim of yours *can be willed to become a universal law*. But the self-reflective tasks you are to undertake in the two cases are in fact different. The FUL requires you to see whether your maxim can be, in Wood's words, "both thought and willed [without contradiction in conception and in willing, respectively] as a universal law"[22]; if it cannot be, it is impermissible for you to act on it; otherwise, it is permissible for you *either to act or not to act on it*. At least when your maxim proves *permissible*, you don't get from it, as yet, a universal law, which is supposed to be *binding* on everyone. On the other hand, the FA requires you to consider whether your maxim can "*make itself into a universal law*" or includes in itself the same volition with which you act on the maxim that it be a universal law. If the answer is yes, then this means that your maxim has acquired, through that "same volition," a property of making itself into a law, a property that Kant later in the *Critique of Practical Reason* calls "legislative form."[23] Thus, whereas the FUL specifies, by way of "thinking and willing without contradiction" as a test for the permissibility of maxims, a necessary (but *not* sufficient) condition under which a maxim can become a universal law, the FA shows that, in addition to meeting this condition, your maxim must *also* take on a "legislative form" so that it can *actually* become a universal law by virtue of the same volition with which you act on it.

What about the FH, the *second* formula that Kant holds to give the "ground," "matter," or "end" of the FA,[24] which, as mentioned earlier, is the "supreme principle of morality" Kant eventually settles on through the three formulas in question? The FH is stated as follows:

[21] See Allen W. Wood, *Formulas of the Moral Law* (Cambridge University Press, 2017), 68–9.

[22] Wood, *Formulas*, 68.

[23] *Critique of Practical Reason* (1788), (henceforth "CprR"), trans. Mary Gregor, in *Practical Philosophy* (The Cambridge Edition of the Works of Immanuel Kant, Cambridge University Press, 1996), 5:27.

[24] G 4:429 and 436.

> *Act so that you use humanity, as much in your own person as in the person of every other, always at the same time as end and never merely as means.*[25]

As used by Kant in the FH, 'humanity' (*Menschheit*) is a technical term for what all (normal) human beings have as their "rational nature" (*vernünftige Natur*), which involves whatever capacities we must avail ourselves of in *setting ends to ourselves through reason and then taking efficient means for their attainment*. That humanity ought to be treated as end in itself means that it has an "objective value" which gives everyone reason or "ground" for positively adopting the maxims-cum-laws to which the FA applies: the FH gives everyone reason to adopt such maxims independently *both* of (external) authorities other than one's own rational will *and* of her natural desires and preferences, however worth satisfying they are for the sake of her happiness; and to the extent that we can so adopt those maxims, our will is "autonomous" rather than "heteronomous." It is because humanity serves to ground the FA that Kant derives immediately from the FA (rather than from the FUL) a further formula, the Formula of the "Realm of Ends" (*Reich der Zwecke*):

> Act in accordance with maxims of a universally legislative member for a merely possible realm of ends.[26]

This formula directs our moral aspiration toward a "merely possible realm" that includes *both* its members, who should treat one another as ends in themselves, *and* the contingent ends which those members set to themselves and which, insofar as they are interpersonally in harmony, should be respected and, when necessary, assisted in their attainment by others.[27]

Now, in clarifying Kantian autonomy, O'Neill says,

> As Kant sees it, fundamental ethical principles should presuppose *merely* what it takes to be a principle at all. He therefore bases ethical reasoning on

[25] G 4:429.
[26] G 4:439.
[27] G 4:433.

the ideal of living by [subjective] principles that at least *could* be [objective] principles or laws for all, that have the *form of law*.[28]

As she goes on to argue,

The demand that we act only on [subjective] principles that we can view as [objective] principles for all is the basis for a rich and powerful account of ethical requirements.[29]

This is essentially the account O'Neill has to offer of Kantian autonomy. But, even though what she says in the first passage suggests that her account only needs to invoke the FUL, it turns out to require *more* when she later explains why the *subjective* "principle of coercion" cannot be made into a law (or objective principle) *for all*, as follows:

[The] capacities [of some persons] for action would be destroyed or undermined or bypassed by others' coercive action. The reasonably foreseeable result of anything approaching universal commitment to coercion would ensure that there could not be universally available effective means to coerce: universal coercion is therefore an incoherent project.[30]

It is supposed to be *only* the FUL test for the (moral) permissibility of maxims or subjective principles that is employed by O'Neill here. But, although the reason against the "principle of coercion" is taken by O'Neill to consist in its failure to pass the FUL test, her account of why coercion is to be rejected shows, on closer inspection, that the FH plays a part in the reason we have against the principle. For the "capacities for action" she speaks of belong obviously to our humanity. The same applies to O'Neill's account of why, in general, "[k]illing and coercing, injury and violence, manipulation and deception, torture and intimidation, enslaving and forced labour are all principles that *cannot be willed as universal laws*," namely that "any principle of action whose universal adoption

[28] O'Neill, *Autonomy*, 85, first emphasis added.

[29] O'Neill, *Autonomy*, 86.

[30] O'Neill, *Autonomy*, 87.

would destroy, damage or undermine *capacities for action* for some or for many cannot be willed as a universal law."[31]

For O'Neill, the FUL *itself*, which yields a purely "formal" test for the permissibility of maxims, suffices to give us "reason to reject a principle of coercion." Accordingly, her aim is to rely on the FUL alone for an account of Kantian autonomy as "the idea of acting on principles that we can will as universal laws." But, as pointed out earlier in this section along the line of Allen Wood, this FUL test only explains why it is impermissible to act on maxims that fail it, not why maxims can *positively* become universal laws that are binding on everyone. The fact that O'Neill's account of autonomy must nevertheless invoke capacities that belong to our humanity serves to show that this essential feature of ours as rational beings must be in play when it comes to deriving, as Kant later does in the *Metaphysics of Morals*, ethical duties that we are supposed to impose autonomously upon ourselves.

Wood reminds us that in the Doctrine of Right, Part I of the *Metaphysics of Morals*, it is from the FH that Kant derives our "innate" right to freedom by virtue of our humanity.[32] As Wood also indicates, in Part II, the Doctrine of Virtue, where Kant lays bare a system of ethical duties, the FH is "by a wide margin the formula of choice in justifying the system."[33] In agreement with Wood, I think the FH, or humanity as end in itself, does provide grounds for our adopting maxims that can make themselves into universal laws according to the FA. On the other hand, while Wood has noted the connection of the FH with the "innate" right to freedom, a right that underlies "the universal principle of right," I will go on to argue that humanity also plays a fundamental role in the moral justification of given positive laws through the public use of reason in the domain of right.

[31] O'Neill, *Autonomy*, 87–8, emphases added.

[32] See *Metaphysics of Morals* (1797), (henceforth "MM"), trans. Mary Gregor, in *Practical Philosophy* (The Cambridge Edition of the Works of Immanuel Kant), 6:237.

[33] Allen W. Wood, *Kant's Ethical Thought* (Cambridge University Press, 1999), 139.

3 Heteronomy and the Domain of Right

After explaining why "universal commitment to coercion" should be rejected, O'Neill immediately qualifies it by adding,

> It does not follow that all coercion is unjustified, for it may be that the best we can do if we are serious about rejecting coercion in so far as possible will still require some institutions that use certain, limited, regulated forms of coercion—for example, a police force, a taxation system. But these specific, limited, regulated uses of coercion would be justified only to the extent that they were indispensable elements of an underlying project of rejecting coercion and respecting other equally fundamental obligations.[34]

The "specific, limited, regulated uses of coercion" O'Neill speaks of involve punishment, and the institutions ("police force," "a taxation system," etc.) backed up by punishment as justifiable coercion must be legally established. Punishment of this kind is held by Kant in the Doctrine of Right to be "*[p]unishment by a court*—which is distinct from *natural punishment*, in which vice punishes itself and which the legislator does not take into account."[35] "Natural punishment" pertains to transgression of *moral* laws, even though in the domain of morality "vice" must be left for its own punishment because, in Kant's view, no one has a moral right to punish others (and, I take it, because what a wrongdoer violates are moral laws that he is supposed to legislate to *himself* in the first place). By contrast, "punishment by a court," or legal punishment, pertains to transgression of positive laws legislated in a state, where the court is at least legally justified in administering punishment. Since "the legislator does not take [natural punishment] into account," Kant's contrast between the two kinds of punishment suggests that the "regulated forms [or uses] of coercion" O'Neill holds to be justifiable in a Kantian fashion involve *legal* punishment. But when she says of these "regulated forms of coercion" that they are "indispensable elements of an underlying project of rejecting coercion and respecting other equally fundamental

[34] O'Neill, *Autonomy*, 87.
[35] MM 6:331.

obligations," it is unclear what she means by these "obligations." For one thing, are they *juridical* or *ethical* obligations (or duties)?

For Kant, ethical duties arise from ethical or *internal* lawgiving, and juridical ones from juridical or *external* lawgiving. As he puts it,

> That lawgiving which makes an action a duty and also makes this duty the incentive is *ethical*. But that lawgiving which does not include the incentive of duty in the law and so admits an incentive other than the idea of duty itself is *juridical*.[36]

The "underlying project" O'Neill alludes to may be thought of as a scheme of social cooperation that requires legislation of positive laws as means to ensuring a just social order. Positive laws fall under what Kant calls "external laws in general."[37] Some external laws "can be recognized as obligatory *a priori* by reason even without external lawgiving"; these are "external but *natural* laws."[38] But there are also external laws that "do not bind without actual external lawgiving (and so without it would not be laws)"[39]; these are positive laws. This "actual external lawgiving" is the juridical one that Kant explains alongside the ethical or internal one in the above passage, where he also distinguishes between ethical and juridical duties in a way that can be rephrased as follows: If an action is made an ethical duty through *internal* lawgiving, one must do it "from duty" (or do it because one recognizes that it is a duty); but if an action is made a juridical duty through *external* lawgiving, it does not matter whether one complies with the duty "from duty" or, say, from fear of legal punishment that would ensue if one did not comply.

O'Neill is eager to argue that public trust in "policies for regulating medicine, science and biotechnology" would be ensured if Kantian autonomy could play a part in the framing of those policies. But a gap remains open between the external lawgiving that generates public policies on the one hand and the internal lawgiving that yields lawlike and universal principles fit to be appealed to as reasons acceptable to all for

[36] MM 6:219.
[37] MM 6:224.
[38] MM 6:224.
[39] MM 6:224.

those policies on the other. As far as I can see, a crucial missing step in O'Neill's account of Kantian autonomy is to fill in the gap with what Kant has to say about the "public use of one's reason," to which I will now turn.

4 The Public Use of Reason

In his popular essay "An Answer to the Question: What Is Enlightenment?" Kant distinguishes the public use of reason from its "private" use when arguing that enlightenment, as a process (or progress) he takes himself to witness in the historical period of human development in which he was situated, requires nothing more than "freedom to make *public use* of one's reason in all matters."[40] He begins the essay by declaring,

> *Enlightenment is the human being's emergence from his self-incurred minority. Minority* is inability to make use of one's own understanding without direction from another. This minority is self-incurred when its cause lies not in lack of understanding but in lack of resolution and courage to use it without direction from another. *Sapere aude!* Have courage to make use of your own understanding! is thus the motto of enlightenment.[41]

As Kant uses it here, the word 'minority' (*Unmündigkeit*) means the quality of being a "minor" who leads a life under the direction or tutelage of others. This quality may be contrasted with that of "*being one's own master*," the latter being used by Kant in another essay, "On the Common Saying: That May be Correct in Theory but It is of No Use in Practice," to characterize a "citizen" as one who has "the right to vote in this legislation [i.e., the *original contract*, or the basic law]" and "*serves* no one other than the commonwealth."[42]

---

[40] "An Answer to the Question: What is Enlightenment?" (1784), (henceforth "WE"), in *Practical Philosophy*, trans. Mary Gregor, (The Cambridge Edition of the Works of Immanuel Kant), 8:36.

[41] WE 8:35.

[42] "On the Common Saying: That May Be Correct in Theory, but It is of No Use in Practice" (1793), (henceforth "Common Saying"), in *Practical Philosophy*, trans. Mary Gregor (The Cambridge edition of the Works of Immanuel Kant), 8:295.

It is worth noting that Kant is not concerned with *sporadic* enlightenment of some individuals or other, but with enlightenment of a "public." The latter is said to be not only possible but also "inevitable, if only [the public] is left its freedom," for

> there will always be a few independent thinkers, even among the established guardians of the great masses, who, after having themselves cast off the yoke of minority, will disseminate the spirit of a rational valuing of one's own worth and of the calling of each individual to think for himself.[43]

While this "spirit"[44] of "independent thinkers" will spread among people, "[the] public can achieve enlightenment only slowly." Further, Kant firmly believes that public enlightenment, though itself a kind of revolution or reform,[45] ought not to be brought about by any violent revolution against a ruler. Nor does Kant think it *could* be brought about in this destructive way:

> A revolution may well bring about a falling off of personal despotism and of avaricious or tyrannical oppression, but never *a true reform in one's way of thinking*; instead new prejudices will serve just as well old ones to harness the great unthinking masses.[46]

By "a true reform in one's way of thinking," Kant means what he elsewhere refers to as a kind of thinking that has three components, the first and foremost of which is "thinking for oneself" as an "**unprejudiced** way

[43] WE 8:36.

[44] By "spirit" Kant means "animating principle in the human being"; see his *Anthropology from a Pragmatic Point of View* (1798), (henceforth "*Anthropology*"), trans. Robert Louden, in *Anthropology, History, and Education* (The Cambridge Edition of the Works of Immanuel Kant, Cambridge University Press, 2007), 7:225.

[45] As Kant says in the *Anthropology*, "The most important revolution from within the human being is 'his exit from his self-incurred minority,'" 7:228, where he is referring to enlightenment as "a true reform in one's way of thinking" and takes it to have already occurred in mathematics and physics; see the passage cited immediately below, and Preface to the second edition of the CPR, B x-xi. So here we find *two* "revolutions" in the way of thinking, one being what Kant intends to promote in proposing a project of public enlightenment, the other being what he has discerned in the historical development of the sciences. We will see how Kant connects the two as we proceed.

[46] WE 8:36.

of thinking."[47] With this component (to be explained in the next section), the "true reform in one's way of thinking" will not, as Kant thinks a subversive revolution will, result in merely replacing old "prejudices" with new ones and having the "great unthinking masses" remain in their "self-incurred minority."

Kant is then ready to propose what he thinks is a feasible as well as morally permissible way to get such a "reform" started. As he writes,

> But I hear from all sides the cry: *Do not argue!* The officer says: Do not argue but drill! The tax official: Do not argue but pay! The clergyman: Do not argue but believe! (Only one ruler in the world says: *Argue* as much as you will and about whatever you will, *but obey!*)[48]

The parenthetical remark is made apparently as a tribute to the then ruler Frederick the Great, with a view to underscoring the feasibility of the proposal Kant is about to make. Kant's examples of the officer, the tax official, and the clergyman are used to make an important distinction in which his proposal is couched between the "private" use of reason and its "public" use. The first use is called "private" in a special sense:

> What I call the private use of reason is that which one may make of it in a certain *civil* post or office with which he is entrusted.[49]

To contrast this with the second use, Kant says,

> The *public use* of one's reason must always be free…; the *private use* of one's reason may, however, often be very narrowly restricted without this particularly hindering the progress of enlightenment. But by the public use of one's own reason I understand that use which someone makes of it *as a scholar* before the entire public of *the world of readers*.[50]

[47] *Critique of the Power of Judgment* (1790), (henceforth "CPJ"), trans. Paul Guyer (The Cambridge Edition of the Works of Immanuel Kant, Cambridge University Press, 2000), 5:294.

[48] WE 8:36–7.

[49] WE 8:37.

[50] WE 8:37.

A public reasoner is supposed to be someone who, as a "scholar," expresses his or her thoughts through writing "before the entire public of *the world of readers*." But this talk about "scholar/writer" is in fact no more than a metaphorical way of speaking that Kant models on the independent thinkers he has found to be already working in the sciences. Leaving this aside for the moment, the officer, the tax official, and the clergyman may be said to be the protagonists in Kant's illustration of how a project for public enlightenment can be promoted.

Recall that Kant has just held public enlightenment to be something both possible and inevitable because "there will always be a few independent thinkers, even among the established guardians of the great masses." In addition to these "guardians" as possible independent thinkers, the protagonists, who are "in a certain *civil* post or office," seem to be intended by Kant to be other possible examples of such thinkers. What Kant goes on to say may accordingly be read as suggesting the following likelihood to the "guardians": While those protagonists must play their role "merely *passively*, so as to be directed by the government … to the public ends," each can, "in his capacity of a scholar who by his writings addresses a public in the proper sense of the word, … certainly argue without thereby harming the affairs assigned to him in part as a *passive* member."[51] The protagonists can play both a passive, restricted role in the "private" use of reason (a use tied to their "post or office"), and an active, free one in its "public" use, even though they ought not to play them *simultaneously*, for this would be detrimental to the "public ends" assigned to them.

Kant then elaborates on how the protagonists can serve the two roles properly if the "guardians" or rulers are willing to promote public enlightenment:

> [I]t would be ruinous if an officer, receiving an order from his superiors, wanted while on duty to engage openly in subtle reasoning about its appropriateness or utility; he must obey [and command his soldiers to obey]. But he cannot fairly be prevented, as a scholar, from making remarks about errors in military service and from putting these before his public for appraisal.

[51] WE 8:37, emphases added.

> [A] citizen does not act against the duty of a citizen when, as a scholar, he publicly expresses his thoughts about the inappropriateness or even injustice of such [tax] decrees.
>
> So too, a clergyman is bound to deliver his discourse to the pupils in his catechism class and to his congregation in accordance with the creed of the church he serves. … But as a scholar he has complete freedom and is even called up to communicate to the public all his carefully examined and well-intentioned thoughts about what is erroneous in that creed and his suggestions for a better arrangement of the religious and ecclesiastical body.[52]

Although, as it stands, the second passage does not involve the tax official himself, it applies to him if he is, when off duty, willing to reason publicly with the tax-payer about the alleged "inappropriateness or even injustice" in tax laws. All three passages may then be read as showing how public enlightenment can get off the ground with a ruler's entrusted subordinates. They are also intended to address a fundamental political concern that Kant expresses shortly as follows:

> The touchstone of whatever can be decided upon as law for a people lies in the question: whether a people *could impose such a law upon itself*.[53]

If what Kant speaks of as "law for a people" is meant to encompass positive laws, then we can infer something of great importance for his conception of public reason from how the three protagonists are supposed to play their "active" part in Kant's project for public enlightenment.

As indicated in Sect. 3, positive laws in the domain of right are by themselves laws of heteronomy insofar as they must be established and enforced by external authorities. But if "a people *could impose such [laws] upon itself*" through freely reached recognition of these laws as morally legitimate, then they could be thought of as if they were laws of autonomy according to the "touchstone" cited above. For Kant, the recognition of positive laws as morally legitimate must be reached freely by members of a state of right who have enlightened themselves through their own efforts to unfetter the "yoke of minority." Moreover, these

52 WE 8:37–8.

53 WE 8:39, emphasis added.

members cannot be *forced* into enlightenment, which requires *freedom* to think for oneself.

Elsewhere, in the essay "Toward Perpetual Peace," Kant speaks of "[a] good moral education of a people [that] is to be expected from a good state constitution."[54] Given the (albeit metaphorical) scholar/writer relation of an "independent thinker" to his or her addressees, it seems safe to say that Kant's project for public enlightenment is unavoidably part of moral education, since one of its aims is to see to it that, even though positive laws are in essence heteronomous, members of a state—under the "animating"[55] influence of independent thinkers critically assessing those laws *and* in keeping with the "touchstone" phrased in the subjunctive mood above—may be able to think of those laws as if they had imposed them upon themselves.

But Kant should be realistic about his project: not only will public enlightenment be "slow," as he already concedes, but generational alteration will also make it an *unending* task. So should he be realistic about the "touchstone" as a test for the moral legitimacy of positive laws. The test should not be taken to require that *all* members of a state be able to think of positive laws in their state as if they could impose them upon themselves. And Kant does understand the "touchstone" in a realistic way. As he stresses near the end of "What Is Enlightenment?":

> [E]ven with respect to his [a ruler's] *legislation* there is no danger in allowing his subjects to make *public* use of their own reason and to publish to the world their thoughts about a better way of formulating it, even with candid criticism of that already given; we have a shining example of this, in which no monarch has yet surpassed the one [i.e., Frederick the Great] whom we honor.[56]

Leaving aside Kant's repeated assurance to rulers that a project for public enlightenment does not pose any threat of defiance against their authority, his point in the essay is that, as long as *some* members of the state have

---

[54] "Toward Perpetual Peace" (1795), in *Practical Philosophy*, trans. Mary Gregor, (The Cambridge Edition of the Works of Immanuel Kant), 8:366.

[55] See footnote 44 above.

[56] WE 8:41.

enlightened themselves, the ruler of the state would do better, both morally and to the benefit of his ruling power, if these independent thinkers were allowed to make public use of their reason to enable him to better formulate positive laws or reformulate some existing ones, so that these laws can have a greater claim to being thought of as if they were laws of autonomy for all citizens, *whether enlightened or not.*

Now, to return to O'Neill's account of Kantian autonomy as "the idea of acting on [subjective] principles that we can will as universal laws": given her aim to show that Kantian autonomy is conducive to public trust in "policies for regulating medicine, science and biotechnology," and given that these policies must be enforced by "institutions that use certain, limited, regulated forms of coercion," she should, I think, place Kantian autonomy in the context of positive laws. In this case, the idea she speaks of above may be rendered in terms of Kant's "touchstone" as the idea of "acting on [positive laws] that we [could] will as universal laws." But this is not yet adequate for Kantian autonomy in regard to positive laws. If I am right, the "we" involved must, as Kant hopes members of a state could be through moral education, qualify as *public reasoners* who can "think for themselves" while at the same time rationally valuing one's (and every other's) worth as end in itself, according to what Kant refers to as the "spirit" of independent thinkers. I will now explain this qualification by considering Kant's so-called maxims of reason that are supposed to make a public reasoner out of any folk.

5 The "Enlightened" Way of Thinking

Although Kant advocates public enlightenment in reference to matters of *positive laws* and *religion*, he argues, as we have seen, that enlightenment requires nothing other than "freedom to make public use of one's reason *in all matters*."[57] Kant mentions some other matters when he reminds rulers in "What Is Enlightenment?" that they (at least the enlightened ones

[57] This is quite unlike what Rawls has to say about the scope of public reason as he conceives it. For him, the use of public reason is restricted to issues over "constitutional essentials and matters of basic justice"; see Rawls, PL, Lecture VI, Section 4: "The Content of Public Reason."

among them) "have no interest in playing guardian over their subjects with respect to *the arts and sciences*."[58] Kant has in mind independent thinkers in the community of *scholars* (taken literally) who have been permitted to make *theoretical* use of their reason *publicly*. For Kant has already held in the *Critique of Pure Reason* that "a well-grounded way of thinking" prevails in such sciences as mathematics and natural science.[59] He hopes the same can take place in metaphysics so that it can progress on the "secure path of a science."[60] Without explaining what he means by this way of thinking, Kant declares, "Our age is the genuine age of **criticism** to which everything must submit."[61] Then he laments,

> **Religion** through its **holiness** and **legislation** through its **majesty** commonly seek to exempt themselves from it [criticism]. But in this way they excite a just skepticism against themselves, and cannot lay claim to that unfeigned respect that reason grants only to that which has been able to withstand its free and public examination.

Apparently, in Kant's view, external authorities in religion and legislation have been wielded as what must never be challenged, but they should be submitted to "free and public examination" for determination of whether they are rationally so grounded that they can acquire "unfeigned" or genuine respect.

Although Kant thinks the authority of rulers can be regarded as morally justified through what he conceives of as an "original," "hypothetical" social contract,[62] the contract does not justify them in declaring that whatever particular laws they legislate can be exempted from "free and public examination." Near the end of the *Critique of Pure Reason* Kant even speaks of such examination as "[that] which has no dictatorial authority, but whose claim is never anything more than the agreement of

---

[58] WE 8:41, emphasis added.

[59] See CPR, A xi, footnote.

[60] CPR, B xv.

[61] CPR, A xi, footnote.

[62] See MM 6:315–6, and "Perpetual Peace": 8:289.

free citizens, each of whom must be able to express his reservations, indeed even his *veto*, without holding back."[63]

In "What Is Enlightenment?" as we have seen, the freedom of such examination comes to be regarded as the freedom to make public use of one's reason, the only thing that Kant holds enlightenment to demand. He argues in the long remark last cited from this essay in Sect. 4 above that legislation, even if based on morally legitimate authorities, may leave room for "thoughts about a better way of formulating it, even with candid criticism of that already given." But these thoughts may seem *parochial*, as they pertain to laws in a particular state located at a particular place and time. Why, then, must a public reasoner express them as "a member [not only] of a whole commonwealth, [but also] even of the society of *citizens of the world*"?[64] And how? A straightforward answer to it (though a rough-and-ready one at this junction, as we shall see in Sect. 7) lies in the "way of thinking" Kant's proposal for public enlightenment is meant to encourage.

In the *Anthropology from a Pragmatic Point of View*, Kant contrasts "egoism" with "pluralism" by describing the latter as "the way of thinking in which one is not [as in egoism] concerned with oneself as the whole world, but rather regards and conducts oneself *as a mere citizen of the world*."[65] Calling the latter "pluralism of reason," Allen Wood criticizes Jürgen Habermas and his followers for neglecting it and thereby misunderstanding Kant's philosophy when they hold it to be "monological" or "solipsistic."[66] Thus, when Kant says in "What Is Enlightenment?" that there are independent thinkers who have instilled in themselves "the spirit of a rational valuing of one's own worth and of the calling of each individual to think for oneself," he is in fact appealing to this "spirit" to advocate a "pluralistic" standpoint of reason, which is essential to what it is to be an enlightened thinker.

---

[63] CPR, A 738–9/B 766–7.

[64] WE 8:37, emphasis added.

[65] *Anthropology*, 7:130, emphasis added.

[66] See Wood, *Kant's Ethical Thought*: 302, and 404 note 18, where Wood mentions that this point has been "well argued" by Burkhard Tuschling, "*Rationis Societas*: Remarks on Kant and Hegel," in *Kant's Philosophy of Religion Reconsidered*, ed. P. Rossi and M. Wreens (Indiana University Press, 1991).

Further, later in the *Anthropology* Kant holds there to be three "maxims" that are "unalterable commands" for enlightened thinkers. "To think *for oneself*" comes first, and the next two are "To think oneself (in communication with human beings) into the place of every *other person*," and "Always to think *consistently* with *oneself*."[67] They are also listed in the *Critique of the Power of Judgment*, where the first is called "the maxim of the **unprejudiced** way of thinking," the second, that of the "**broad-minded** way," and the third, that of the "**consistent** way."[68] Elsewhere, they are termed "enlightened" "maxims of reason."[69] While the third maxim is in any case a requirement of reason in both its theoretical and practical uses, the first and second are especially key to understanding the enlightened "way of thinking" with which one must appraise positive laws for their moral legitimacy.

In the *Anthropology*, the first maxim is said to be "negative (*nullius addictus iurare in verba Magistri* [nobody is forced to follow the words of the master]), the principle of *freedom from constraint*"[70]; in the third *Critique*, it is said to be "the maxim of a reason that is never **passive**," where "the tendency toward [passivity], hence toward heteronomy of reason, is called **prejudice**."[71] For Kant, in following the first maxim one must consider oneself as *negatively* free from constraint; and by "constraint" he apparently means "prejudice" as a tendency toward passivity, which implies one's submission to the "heteronomy of reason." But what does Kant mean by the "heteronomy of reason"? Recall that violent revolutions are held by Kant to end up in new "prejudices" replacing old ones, without helping people to extricate themselves from the "yoke of minority." I suggest that, as Kant understands them in this connection, "prejudices" have to do with the *ethos* created by the workings of governments. If so, it seems plausible to interpret the "heteronomy of reason" as what results from a ruler's external legislation of positive laws, which are by themselves laws of heteronomy. It also seems plausible to interpret Kant

[67] *Anthropology*, 7:228.

[68] CPJ, 5:294.

[69] See Guyer's translation of the *Critique of the Power of Judgment*, 379, translator's note 13.

[70] *Anthropology*, 7:228.

[71] CPJ, 5:294.

as alluding to the "heteronomy of reason" when he holds that "if reason will not subject itself to the laws it gives itself, it has to bow under the yoke of *laws given by another*."[72] Thus, to the extent that we subject ourselves to the "heteronomy of reason" by unthinkingly obeying externally given laws in the domain of right, our reason is far from autonomous.

Even so, Kant would insist that it is still in some sense *possible* for our reason to become autonomous, as may be seen from his saying that

> reason can use the mechanism of nature, through self-seeking inclinations that naturally counteract one another externally…, as a means to make room for its own end, the rule of right.[73]

The heteronomy of reason originates in the ruler's use of the "mechanism of nature" to establish a realm of positive laws that enable people to constrain one another *externally* through their "self-seeking inclinations." Even though these laws count as laws of heteronomy, they do not exclude the possibility of our obeying them as if they were what we had (autonomously) legislated to ourselves. For one thing, these laws are supposed to constrain only the external behavior of people toward one another, not their choices and ways of thinking. This limitation on positive laws comes from Kant's so-called "end [of reason]" which these laws can serve as means—or, in other words, an end that the ruler's heteronomous use of reason on his subjects in legislation ought to aim at. Kant identifies this end in the above passage as "the rule of right," which comes to be "the universal principle of right" in the Doctrine of Right, as follows:

> An action is right if it can coexist with everyone's freedom according to a universal law, or if on its maxim the freedom of choice of each can coexist with everyone's freedom in accordance with a universal law.[74]

[72] "What Does It Mean to Orient Oneself in Thinking?", trans. Allen W. Wood, in *Religion and Rational Theology*, (The Cambridge Edition of the Works of Immanuel Kant, Cambridge University Press, 1996), 8:145, emphasis added.

[73] "Perpetual Peace": 8:366–7.

[74] MM 6:230.

Being "right" in this sense is used specifically to characterize an action if its maxim is compatible with "everyone's freedom," which Kant later explains this way:

> *Freedom* (independence from being constrained by another's choice), insofar as it can coexist with the freedom of every other in accordance with a universal law, is the only original right belonging to every man by virtue of his humanity.[75]

This freedom is supposed to be the only "innate," "original," or "natural" right that human beings possess by virtue of their humanity as end in itself,[76] and in the domain of right, it may be called "external (hence rightful) freedom"[77] to do things that one sees fit for the pursuit of ends that one sets to oneself through one's humanity. This freedom must be guaranteed to every member of the state insofar as its exercise remains within the bounds of positive laws grounded in the universal principle of right, even though these laws can only be promulgated and backed up with punitive force, in which consists the heteronomy of reason.

It is then understandable why Kant's *first* "maxim of reason" requires us to think *independently* of whatever "prejudices" we are wont to as a result of living under the heteronomy of reason. For only in the "**unprejudiced** way" can we think, according to the *second* maxim, from the standpoint of *every other person* (as a "citizen of the world," and not merely as a member of any given state), who is as irrevocably entitled as we are to the status of "innate personality" by which one "can never be treated merely as a means to the purposes of another or be put among the objects of rights to things."[78] The second maxim is therefore a requirement that one think in the "**broad-minded** way" or, as Kant also puts it, from the "**universal standpoint**,"[79] so that one must adopt a "pluralistic" standpoint from which, as we have seen Kant explain in the *Anthropology*, "[one] regards and conducts oneself *as a mere citizen of the world*." But there is

[75] MM 6:237.

[76] MM 6:237.

[77] "Perpetual Peace": 8:349–50, footnote.

[78] MM 6:331.

[79] CPJ, 5:295.

something deeper and more substantive about this second maxim than is imparted by these quoted phrases.

6 "Partakers in Happiness"

Kant once unpacked "freedom of every member of … [a civil] society as a human being" with a single "formula": "No one can coerce me to be happy in his way (as he thinks of the welfare of other human beings)."[80] While this freedom belongs to our only "innate" right mentioned above, the formula brings to the fore an important purpose served by this right: it permits people to seek after what they *themselves* conceive as a happy life, insofar as their pursuit of it does not infringe on the humanity-based freedom of others to do the same. Kant once spoke of members of a state as "partakers in happiness,"[81] where, needless to say, he must have meant happiness whose pursuit is "rightful" (in the sense understood according to the universal principle of right).

This notion of "rightful" (pursuit of) happiness can help us explain Kant's formulation of the second maxim in the *Anthropology* as "the positive principle of liberals who adapt the concepts [*Begriffen*] of others." It does not seem far-fetched to understand these "concepts" as particular conceptions that other "partakers" have of their own happiness. Moreover, by a "liberal" Kant seems to refer to one who follows the demand of the *first* maxim for the "**unprejudiced** way" of thinking that one be "one's own master," and who also follows the demand of the *second* maxim for thinking from the "**universal standpoint**" that every person be allowed to pursue happiness as he or she conceives it as long as it is rightful. Thus, given that Kant intends to advocate the standpoint of "pluralism" alongside his idea of enlightenment, the two maxims are two sides of the same coin.

Now, for Kant, as I have so far interpreted him, "free and public examination" of positive laws can be undertaken only by people who have enlightened themselves through commitment to the three maxims of

[80] "Common Saying": 8:290.

[81] CPrR, 5:37.

reason (Sect. 5): only *they* can be said to have internalized the "spirit" of independent thinkers (or enlightenment), namely, "a rational valuing of one's own worth and … the calling of each individual to think for himself" (Sect. 4). In my view, there is nothing repugnantly "elitist" here, given that all (normal) human beings should be regarded as endowed with the potential to become enlightened, and that all members of a society should treat one another as equal "partakers in [rightful] happiness."

7 The Double Function of Public Reasoners: Justificatory and Exemplary

For Kant, as mentioned in Sect. 4, "[a] good moral education of a people is to be expected from a good state constitution." This, as I suggested, is also what public enlightenment calls for, in which case the public use that independent thinkers can make of their reason may serve a double function for Kant, namely: first, rationally appraising positive laws in their particular state of right; second, making themselves *exemplars* in the eyes of those as yet in their "self-incurred minority" so that the latter can be encouraged and guided to free their thinking from the heteronomy of reason and to envisage themselves as having a will that is free and autonomous. The function served by public reasoners in the domain of right may then be said to be not only "justificatory" but also "exemplary."

The exemplary aspect of public reasoning may help us answer a question raised in Sect. 5, namely: Why must one make public use of reason as "a member [not only] of a whole commonwealth, [but also] even of the society of *citizens of the world*"? I have tried to answer it by introducing Kant's three maxims of reason, the second of which, in particular, requires that one "think oneself (in communication with human beings) into the place of *every other person*." But why must public reasoners in their examination of controversial positive laws consider themselves as "communicating" their thoughts about these laws to all "citizens of the world"? The public reasoners are, for one thing, supposed to be concerned primarily with particular positive laws in their own particular state, in which case it does not seem apt for them to do so with "citizens of the world" at large as their addressees.

In reply, one might point out that, insofar as they have internalized the "spirit" of independent thinkers, the public reasoners must think of themselves as if both they and their addressees were all members of a single "realm of ends" encompassing all human beings as persons, and therefore that the public reasoners must take themselves to be speaking to *all* human beings, at least for the sake of fulfilling their *exemplary* function. This may well be what Kant has to say, but I think he must still give priority to the justificatory function over the exemplary, for no public discourse on whether or not some controversial positive laws are justifiable can serve as a good example of such discourse if it is *not* effective. Only by insisting that the justificatory function should be effective can Kant defend himself against the pejorative comment by one of his contemporary critics on his project for public enlightenment that, as Allen Wood puts it, "the 'public use of reason' [is] merely a 'sumptuous dessert' to be enjoyed only after the private use of reason supplies one's 'daily bread.'"[82]

We have seen in Sect. 4 that Kant illustrates the justificatory function by indicating how the three protagonists in "What Is Enlightenment?" can argue as "scholars." As he portrays them, they must (when off duty) deal with different issues over particular legislation and religious practices by addressing concerns that range from "appropriateness or utility" to "errors" and "injustice" in positive laws or institutional arrangements of their particular state. Let us consider how public reasoners could deal with matters of *justice* in ways pertinent to their particular state of right while also preserving the exemplary function Kant intends them to serve for people qua "citizens of the world."

For Kant, members of a state of right ought to be regarded as "partakers in happiness" with equal right to their humanity-based freedom. The ruler ought, accordingly, to legislate and enforce positive laws designed not only to (1) protect this freedom against intrusion by others equally for all members of the state, but also to (2) allow equal room for their use of that freedom to achieve the best possible happiness that is rightful.

[82] Wood, *Kant's Ethical Thought*, 308, and 406, note 26.

Equality in (what we may call) the "negative" sense of (1) is involved when Kant says,

> [Members of a commonwealth] are … all equal to one another as subjects; for no one of them can coerce another except through public law (and its executor, the head of state), through which every other also resists him in like measure.[83]

On the other hand, Kant speaks of equality in (what we may call) the "positive" sense of (2) when he goes on to say (though in terms of "social ranks" rather than happiness, the former being, I take it, an essential means to the latter),

> From this idea of the equality of human beings as subjects … there also issues the following formula: Every member of a commonwealth must be allowed to attain any level of rank within it (that can belong to a subject) to which his talent, his industry and his good fortune [*Glück*] can take him….[84]

This formula would require a state of right to take active measures to correct social and economic conditions that stand in the way of equal opportunities among members of the state to strive for their rightful happiness.[85] Thus, (1) and (2) taken together would yield a full sense of the equality of members of a state as "partakers in happiness."

For Kant, as far as legislation of positive laws and public reasoning about their moral legitimacy are concerned, both legislators and public reasoners must adhere to what some contemporary political philosophers would uphold as a kind of liberal "neutrality," namely, neutrality that forbids writing into positive laws whatever particular conceptions of a

[83] "Common Saying": 8:292.

[84] "Common Saying": 8:292.

[85] It is important to make the distinction that I attribute to Kant between equality in the negative sense and equality in the positive sense. For if the former is emphasized to the exclusion of the latter, the result will be a reading of Kant as a laissez-faire libertarian, whereas he can be taken to allow a state of right to take redistributive measures if he does advocate equality in the positive sense in the way explained here. For these two readings of Kant's theory of a state of right, see Allen W. Wood, *Kantian Ethics* (Cambridge University Press, 2008), 194.

happy life are adopted by different members of a state.[86] But neutrality is not, and should not be, the only desideratum because it has to do only with equality in the *negative* sense. If people are to be deemed "partakers in happiness" and accorded equal freedom based on the only innate right they have by virtue of humanity, then in the moral justification or criticism of particular positive laws, complex factors other than neutrality must be taken into account together with contingent social circumstances and facts about human nature, so that one can estimate whether equality in its *full* sense is secured externally through those laws under examination. Such factors would have to include, for example, how various social groups or individuals disadvantaged for morally irrelevant reasons can manage to pursue happiness according to their particular conceptions of it, whether they have actually embraced conceptions of happiness that had been distorted or shaped by their hardship so that they could not but adjust the former to the latter, or whether the positive laws they are subjected to may have made it difficult for them to realize their more ambitious conceptions of happiness despite their efforts to do so in conformity to those laws.

Thus, it is reasonable to say that the public use of reason as expounded by Kant in "What Is Enlightenment?" must attend to, rather than abstract from, factors specific to a given state of right—factors, however, that Kant's protagonists should be good at enumerating as a result of their public service in civil post. Nevertheless, these factors must be spelt out by public reasoners in an "**unprejudiced** way" according to the first maxim of reason. This, plus the fact that public reasoners must rest their moral appraisal of particular positive laws on reasons stemming from the humanity-based freedom for all and from the universal principle of right that Kant takes to regulate all states of right, should be enough to make it an obligation for public reasoners with adequate background knowledge about the particular issues at hand to communicate their thoughts to "the society of *citizens of the world*."

[86] What is meant by "neutrality" admits of different versions; see Gerald Gaus, "The Moral Foundations of Liberal Neutrality," in *Contemporary Debates in Political Philosophy*, ed. Thomas Christiano and John Christman (Blackwell, 2009). I believe the kind of neutrality attributed to Kant here counts as one version of it.

8 Conclusion

O'Neill seems to have interpreted Kantian autonomy in reference to the public use of reason along a line that finds support in Kant's remark that "the power to judge autonomously—that is freely (according to principles of thought in general)—is called reason."[87] But Kant is actually speaking here of "principles of *thinking* [*Denkens*] in general," by which he may reasonably be interpreted as referring, perhaps not exclusively, to the three maxims of reason that we have discussed in Sects. 5 and 6. O'Neill did take these maxims into account,[88] but unfortunately did not put them in the context of the "heteronomy of reason" *both* as something inescapable in the establishment of a state of right running on coercive, positive laws, *and* as something that can be *overcome* by public reasoners in their *thinking* according to the "unalterable commands" of the three maxims.

Further, as I argued in Sect. 2, O'Neill seems to rely too much on Kant's requirement of objective principles that they be applicable in "law-like form" and "universal scope." This, to be sure, matches the "**universal standpoint**" signified by the *second* maxim, but it misses the demand of the *first* for the "**unprejudiced** way" of thinking that one be "one's own master" by virtue of one's humanity or rational nature, which O'Neill passes over, but which lies, as far as the domain of right is concerned, at the basis of our "innate" right to freedom and, therewith, of the universal principle of right. Humanity is an objective value to be attributed to *everyone* (regardless of how much or little it has developed in them, as long as they count as equal "partakers in happiness"); and it underpins the first maxim: To "think *for oneself*."

Humanity as an objective value brings us from the first maxim to the second: "To think oneself (in communication with human beings) into the place of every *other person*." It is by putting these two maxims together and joining them with Kant's notion of equal "partakers in [rightful] happiness" (Sect. 6), rather than simply by focusing on the "**universal**

[87] "The Conflict of the Faculties" (1798), trans. Mary Gregor and Robert Anchor, in *Religion and Rational Theology*, 7:27. The remark is cited by O'Neill, *Authorities*, 138.

[88] See Chapter 1 of O'Neill, *Authorities*, esp. 32.

standpoint" of the second maxim as O'Neill does, that we can discern how the pluralism that Kant brings out in the *Anthropology* can lead to a *complex* standpoint from which positive laws in a state of right must be submitted to "free and public examination" by reason, or to the public use that independent thinkers can make of their reason, as a test for their moral legitimacy (Sect. 7). Not only can legislation win "unfeigned respect" (and therewith, trust) from its subjects if controversial positive laws it promulgates can pass the test (Sect. 5), but these laws can in that case also pass the test of Kant's "touchstone" for their being reasonably thought of as if they, as positive laws that are by nature heteronomous, *could* be self-imposed by us as laws of autonomy.

11

Autonomy, Neutrality, and Perfectionism

Yingying Tang and Lei Zhong

1 Introduction

Many contemporary liberals maintain that the state should be *neutral* between particular, disputed conceptions of the good life. The doctrine of neutrality has attracted some prominent political philosophers in the 1970s–1980s[1] and gained a reviving interest in recent years.[2]

[1] See, e.g., John Rawls, *A Theory of Justice* (Cambridge: Harvard University Press, 1971) (henceforth "*Theory*"); Robert Nozick, *Anarchy, State and Utopia* (Oxford: Basil Blackwell, 1974); Charles Larmore, *Patterns of Moral Complexity* (Cambridge: Cambridge University Press, 1987); Thomas Nagel, "Moral Conflict and Political Legitimacy," *Philosophy and Public Affairs* 16 (1987): 215–40; Brue Ackerman, *Social Justice in the Liberal State* (New Haven: Yale University Press, 1988).

[2] See Gerald Gaus, "Liberal Neutrality: A Compelling and Radical Principle," in *Perfectionism and Neutrality*, ed. Steven Wall and George Klosko (Lanham: Rowman and Littlefield, 2003),

Y. Tang (✉)
Qingdao Institute of Humanities and Social Sciences, Shandong University, Qingdao, People's Republic of China
e-mail: tangyingying@sdu.edu.cn

L. Zhong
Department of Philosophy, The Chinese University of Hong Kong, Shatin, Hong Kong
e-mail: leizhong@cuhk.edu.hk

H.-L. Li, M. Campbell (eds.), *Public Reason and Bioethics*,
https://doi.org/10.1007/978-3-030-61170-5_11

There are different ways of defending the neutrality principle. For example, one can argue for state neutrality by adopting a skeptical stance on conceptions of the good life. According to this skeptical argument, the state ought to be neutral between different conceptions of the good life, because we have no genuine knowledge about what ways of life are good or bad.[3] Nevertheless, this skeptical argument is problematic, in the face of a troubling dilemma. On the one hand, if the skeptical argument presupposes a *universal* skepticism concerning the entirety of morality, then the skeptical approach to neutrality, which is supposed to establish a principle regarding political morality, would be self-contradictory. On the other hand, if the skeptical argument adopts a *local* skepticism, which only denies that we have moral knowledge about conceptions of the good life, then the epistemic distinction between the scope of the good life and the rest of the moral domain (e.g., the scope of justice) would be suspiciously arbitrary.[4]

Many neutralists rather defend the neutrality principle by appealing to the value of autonomy.[5] On this account, the state should not promote particular conceptions of the good life because such state actions will violate the citizens' autonomy.[6] An autonomy-based justification of neutrality seems to be very promising: it resorts to some basic ideas of a liberal society, and need not presuppose any form of moral skepticism.

---

pp. 137–65; Ronald Dworkin, *Justice for Hedgehogs* (Cambridge: Harvard University Press, 2011); Martha Nussbaum, "Perfectionist Liberalism and Political Liberalism," *Philosophy and Public Affairs* 39 (2011): 3–45; Jonathan Quong, *Liberalism Without Perfection* (Oxford: Oxford University Press, 2011); Alan Patten, *Equal Recognition: The Moral Foundations of Minority Rights* (Princeton: Princeton University Press, 2014).

[3] See Ackerman, *Social Justice in the Liberal State.*

[4] Rawls famously holds that while reasonable people can agree upon principles of justice, they may have irresolvable disagreements over conceptions of the good life. See John Rawls, *Political Liberalism* (New York: Columbia University Press, 1993) henceforth "PL." Regardless of whether this distinction that Rawls makes is tenable, Rawls is not a local skeptic about the good life—in other words, his view is not that we have no *knowledge* or *justified beliefs* about the good.

[5] See, e.g., Rawls, *Theory*; Dworkin, *Justice for Hedgehogs*; Patten, *Equal Recognition.*

[6] One would readily accept the principle of neutrality if one already endorses Rawlsian political liberalism. However, as many philosophers point out, political liberalism is deeply unsatisfactory. See, e.g., Fabian Wendt, "Rescuing Public Justification from Public Reason Liberalism," in *Oxford Studies in Political Philosophy*, vol. 5, ed. Steven Wall, Peter Vallentyne & David Sobel (Oxford: Oxford University Press, 2019), pp. 39–64. Thus, this chapter does not assume the doctrine of political liberalism. We'd rather like to consider a justification of neutrality that appeals to comprehensive values (such as autonomy).

However, in this chapter, we argue that the idea of autonomy cannot justify the neutrality principle, if that is taken to rule out *any* method of advancing disputed conceptions of the good life. Nevertheless, we argue, autonomy raises a serious challenge to *perfectionism*, according to which it is permissible for the state to use any means, even coercion, to promote disputed conceptions of the good life. The remainder of this chapter is structured as follows. In Sect. 2, we distinguish *justificatory* neutrality from *consequential* neutrality and restrict ourselves to a discussion of the former rather than the latter. In Sect. 3, we discuss four major ways that the state can adopt to advance conceptions of the good life: *coercion*, *reward*, *persuasion*, and *creation*. In Sect. 4, we analyze and combine two different but related conceptions of autonomy: autonomy as *coherence* and autonomy as *independence*. In Sect. 5, we put forward an autonomy-based argument, which aims to establish that a person's living a way of life is *manifestly* valuable only if she is pursuing it *autonomously*. In Sects. 6 and 7, we argue that while some methods of advancing particular conceptions of the good life may probably violate the citizens' autonomy, other methods are compatible with the value of autonomy.

2 Justificatory Neutrality vs. Consequential Neutrality

In the literature, there are various versions of the neutrality principle. Here let us offer a standard formulation of neutralism as follows:

> [Neutralism] The state should not aim to promote a disputed, substantive conception of the good life for its own sake by *any* political means.

Two points to note in the formulation. First, neutralism maintains that the state should never promote particular conceptions of the good life *because of* its presumed intrinsic values—in other words, neutralism is concerned with the neutrality of justification. Second, neutralism rejects not just certain political means (such as coercion) for promoting particular conceptions of the good life, but also any political methods of promoting the good. We will turn to the second point in the next section. Here let us discuss the first point.

Many who reject neutrality worry that no government can ever achieve neutrality, since most laws, policies, and political institutions would inevitably favor some conceptions of the good life while disfavoring many others. For example, liberal institutions inevitably have non-neutral consequences. Because freedom of speech and freedom of association allow different citizens to pursue their own ways of life, and not all ways of life are equally appealing, some will have more difficulty attracting adherents than others.[7] So, non-neutral consequences seem to be unavoidable.

If this pitfall of neutrality is so lethal and pervasive, isn't the principle of neutrality doomed from the start? But most neutralists do not think so, and their standard response is to distinguish between consequential neutrality and justificatory neutrality.[8] The first asserts that governments should not adopt laws and policies that *have the effect of* promoting any particular conceptions of the good life, whereas the second asserts only that governments ought not take actions *in order to* promote any such conceptions. Neutralists generally reject the neutrality of effect as impracticable, but endorse the neutrality of justification. For example, John Rawls writes:

> [Citizens should not] use the coercive apparatus of the state to win for themselves a greater liberty or larger distributive share on the grounds that their activities are of more intrinsic value.[9]

> Fairness to persons may be achieved by a well-ordered society even though all (admissible) conceptions of the good do not flourish equally and some hardly at all.[10]

---

[7] Will Kymlicka, "Liberal Individualism and Liberal Neutrality," *Ethics* 99 (1989): 883–905, at p. 884.

[8] Kymlicka, "Liberal Individualism and Liberal Neutrality."

[9] Rawls *Theory*, p. 329.

[10] John Rawls, "Fairness to Goodness," *Philosophical Review* 84 (1975): 536–54, at p. 544.

And Charles Larmore puts it this way:

> [N]eutrality is not meant to be one of outcome, but rather one of procedure. That is, political neutrality consists in a constraint on what factors can be invoked to justify a political decision. Such a decision can count as neutral only if it can be justified without appealing to the presumed intrinsic superiority of any particular conception of the good life.[11]

According to these philosophers, the principle of neutrality is a basis for evaluating the justification of laws and policies rather than a doctrine for evaluating laws/policies as such. Consider Jeremy Waldron's example. That a law against Sunday business would conform to the requirements of a Sabbatarian doctrine is not a good reason for having such a law; but that it is necessary to prevent employees from being overworked may be a legitimate reason. A legislation based on the second reason is neutral on the account of justificatory neutrality, even if the law has the consequence of benefiting Sabbatarianism over other religious groups.[12]

Some may raise this question: How could we know about the real intentions of lawmakers behind the policies? Governments and lawmakers can sometimes appeal to neutral reasons to defend the policies through which they actually intend to promote particular conceptions of the good life. So, the principle of neutrality seems to be insignificant because many political policies can be neutrally justified.

But this objection is off the point. The objection seems to be based on a confusion between private intentions and public reasons, i.e., reasons publically presented in the political decision-making. It is the latter rather than the former with which the neutrality principle is concerned. Moreover, it is not true that public policies can always be neutrally justified. If we refrain from citing perfectionist reasons to support certain policies, sometimes it is hard to appeal to other reasons to justify them.

---

[11] Larmore, *Patterns of Moral Complexity*, p. 44.

[12] Jeremy Waldron, "Legislation and Moral Neutrality," in his *Liberal Rights* (Cambridge: Cambridge University Press, 1993), pp. 143–67, at p. 150.

3 Four Methods of Promoting the Good

According to the doctrine of neutralism, the state should not adopt any means to advance particular conceptions of the good life. In this section, let us discuss the main methods that the state can use to promote the good. Generally, the government can induce its citizens to pursue favored conceptions of the good in four ways. First, the government can use coercion (including imprisonment, fines, and taxation) to promote or disfavor particular ways of life. Neutralists, and liberals in general, maintain that citizens should never be punished merely on the grounds that their actions are base or valueless. For example, Ronald Dworkin says:

> Government must be neutral in matters of personal morality, and must leave people free to live as they think best so long as they do not harm others.[13]

However, there are various noncoercive ways of promoting the good. The second important way is by providing rewards. As in the United States, a government can, for instance, offer tax deductions to encourage charitable and religious donations, and subsidize art, literature, and scholarship to make intellectual activities more attractive.[14]

The third way of inducing citizens to pursue some desired lifestyles is by persuasion. The state can use political speeches, official media, and other resources to recommend or disfavor particular conceptions of the good life. There are two such kinds of persuasion: a government can persuade its citizens to take special actions by providing authentic value-based reasons (call this *justification*); or a government can influence the

---

[13] Ronald Dworkin, "Neutrality, Equality and Liberalism," in *Liberalism Reconsidered*, ed. Douglas MacLean and Claudia Mills (Totowa: Rowman and Allanheld, 1983), pp. 1–11, at p. 1. Here Dworkin seems to be appealing to Mill's *harm principle*. As Mill puts it, "the only purpose for which power can be rightfully exercised over any member of a civilized community, against his will, is to prevent harm to others. His own good, either physical or moral, is not a sufficient warrant." See John Stuart Mill, *On Liberty* (Indianapolis: Bobbs-Merrill, 1956), p. 13. However, the harm principle, which only rules out coercion, fails to justify neutralism, as there are noncoercive ways of promoting the good.

[14] George Sher, *Beyond Neutrality: Perfectionism and Politics* (Cambridge: Cambridge University Press, 1997), p. 35.

citizens by systematic indoctrination of certain ideas and distortion of relevant information (call this *manipulation*).

Thus far, we have discussed only methods of influencing the citizens' choices among *existing* options, but a government can also provide its citizens with *new* options. Thus, a fourth way of promoting the good is to provide some sorts of cultural options by sustaining, for example, museums, theaters, and universities.[15]

Different political doctrines have different attitudes toward the state's means of promoting the good. As we mentioned earlier, neutralism rejects all the ways of promoting the good. On the other end of the spectrum is what we call *perfectionism*:

> [Perfectionism] The state's most important goal is to promote particular conceptions of the good life.[16]

Perfectionism has the implication that it is permissible or even obligatory for the state to adopt any of the aforementioned methods to promote particular conceptions of the good life. If promoting the good were the most important goal, this goal would always outweigh other values (e.g., autonomy). It thus follows that perfectionism allows any means of promoting the good, even coercion and manipulation, whenever the methods are effective.[17]

In this chapter, we aim to challenge both neutralism and perfectionism by a closer examination of autonomy. In later sections, we shall argue that which way of promoting the good is legitimate essentially depends on whether it violates autonomy. While some ways (e.g., coercion and manipulation) violate the citizens' autonomy, other ways (such as justification and creation) do not. Thus, an autonomy approach to political legitimacy would imply that neither neutralism nor perfectionism is acceptable.

---

[15] Sher, *Beyond Neutrality*, p. 36.

[16] The view that the state ought to promote some second-order, general values like autonomy is not a version of perfectionism in our sense. See, e.g., Joseph Raz, *The Morality of Freedom* (Oxford: Oxford University Press, 1986); Steven Wall, *Liberalism, Perfectionism and Restraint* (Cambridge: Cambridge University Press, 1998).

[17] According to this specific definition of 'perfectionism', a liberal version of perfectionism is conceptually incoherent. If 'perfectionism' is characterized as a weaker thesis that promoting particular conceptions of the good life is *one* goal that the state should pursue, perfectionism could be compatible with liberalism.

4 I-Autonomy and C-Autonomy

Although the notion of autonomy has received much attention in contemporary moral and political philosophy, there is little consensus about what it exactly means. We shall analyze and elucidate the concept of autonomy and then adopt a definition of autonomy that is, we hope, most consistent with the fundamental ideas of a liberal society.

We can reveal the central idea that underlies the concept of autonomy by appeal to the etymology of the term: *autos* (self) and *nomos* (rule). According to Gerald Dworkin,

> The term was first applied to the Greek city state. A city had *autonomia* when its citizens made their own laws, as opposed to being under the control of some external conquering power. There is then a natural extension to persons as being autonomous when their decisions and actions are their own; when they are self-determining.[18]

Intuitively, being self-determining or autonomous has two important aspects: (1) in the inner world, to be self-determining is to be governed by 'true self' rather than impulse, passion, or a 'false self'; (2) in the outer world, to be self-determining is, in some sense, to be independent of the wills of others (O'Neill 1992: 203). We can call the two aspects *autonomy as coherence* and *autonomy as independence* respectively (*c-autonomy* and *i-autonomy*, for short). Many philosophers only emphasize either one aspect or the other so as to generate different conceptions of autonomy. But for the purposes of this chapter, we believe that both of the two aspects are indispensable.

Consider i-autonomy first. Autonomy as independence does not mean that we are self-sufficient—inevitably, we are always influenced by, and dependent on, others in various ways. Here we can simply understand i-autonomy as independence from others' interference and domination. Respect for i-autonomy is extremely significant because it reflects the underlying Kantian principle of humanity that individuals are ends and

[18] Gerald Dworkin, *The Theory and Practice of Autonomy* (Cambridge: Cambridge University Press, 1988), pp. 12–13.

not merely means. If the government invades a citizen's i-autonomy, then her way of life is shaped heteronomously in accordance with the legal authority's conception of the good—she is subjected to the wills of others.

Now turn to c-autonomy. What does the expression 'to be governed by true self' mean? Some philosophers may hold a more substantive understanding,[19] but here we can just adopt a weaker conception in the context of social and political philosophy. As Gerald Dworkin puts it,

> [C-autonomy] is conceived of as a second-order capacity of persons to reflect critically upon their first-order preference, desires, wishes, and so forth and the capacity to accept or attempt to change these in light of higher-order preferences and values.[20]

It is helpful to analyze the concept of c-autonomy in terms of Harry Frankfurt's *first-order desires* and *second-order volitions*.[21] According to Frankfurt, first-order desires are simply desires to do or not to do something. Besides wanting to do this or that, people also want to have (or not to have) certain desires and preferences—that is, they also have *second-order desires*. Many animals appear to have first-order desires, but no animals other than humans have "the capacity for reflective self-evaluation that is manifested in the formation of second-order desires".[22] When someone has a second-order desire to want a certain first-order desire to be her *will*, namely, an effective desire that would move a person all the way to action, the second-order desire is considered as her second-order volition.[23] On Frankfurt's account, one enjoys freedom of the will when one is free to will what one wants to will—in other words, when, with respect to any of one's first-order desires, one can make that desire be one's will by wanting it to be so at the level of second-order volition. If a

[19] See, e.g., Immanuel Kant, *Foundations of Metaphysics of Morals*, trans. Lewis W. Beck (New York: Macmillan, 1959); Gary Watson, "Free Agency," *Journal of Philosophy* 72 (1975): 205–20.

[20] Dworkin, *The Theory and Practice of Autonomy*, p. 20.

[21] Harry Frankfurt, "Freedom of the Will and the Concept of a Person," *Journal of Philosophy* 68 (1971): 5–20.

[22] Frankfurt, "Freedom of the Will and the Concept of a Person," p. 7.

[23] Frankfurt, "Freedom of the Will and the Concept of a Person," p. 10.

person cannot make that desire her own will, she does not enjoy freedom of the will (as in the famous example of an *unwilling addict*).

When c-autonomy is applied in the context of choosing some particular conceptions of the good life, this capacity is similar to the *capacity for a conception of the good* in Rawls's terminology, that is, the ability to form, to revise, and rationally to pursue a conception of one's good life.[24] A person who enjoys c-autonomy does not always make true judgments or have reasonable preferences; c-autonomy only means the coherence of the deep self and the superficial self.[25]

At the end of this section, we wish to stress that knowledge of deep self is *not* always transparent to the agent herself, whereas Frankfurt seems to be silent on this issue. In our view, the agent does not always have knowledge about her second-order desires/volitions—that is, she sometimes does not know whether she wants to have a first-order desire or whether she wants a first-order desire to be her will. We can see this point more clearly in a case of conflicting desires. Consider, for example, a person who has the desire to smoke, but at the same time wants to quit for the sake of health. Suppose that her desire to smoke outweighs her desire to quit, and hence she keeps smoking. It may be the case that she wants the desire to smoke to be her will (and succeeds), or the case that she wants the desire to quit to be her will (but fails). Regardless of which case it actually is, the agent may have no knowledge or hold a mistaken belief about herself. It is groundless to say that people always know about their own second-order volitions. Moreover, as we will discuss later, the epistemic opacity of the deep self has significant implications in moral and political philosophy.

[24] Rawls, PL 19.

[25] C-autonomy may require that a person's second-order volitions causally influence her first-order desires in forming her wills, more than the mere coincidence of first-order desires and second-order volitions, as Frankfurt's example of a *willing addict* illustrates. In this chapter, we leave this issue open.

5 Value, Reason, and Autonomy

When a person lives an autonomous life, a life in which both i-autonomy and c-autonomy are sufficiently fulfilled, she is not only a rational agent "who can choose between options after evaluating relevant information," but also an agent "who can be a part creator of her own moral world" and whose "personal integrity and sense of dignity and self-respect are made concrete."[26] Robert Nozick holds a similar view:

> A person's shaping his life in accordance with some overall plan is his way of giving meaning to his life; only a being with the capacity to so shape his life can have or strive for a meaningful life.[27]

In this section, we shall argue that living an autonomous life is a significant condition under which the value of a way of life can be *manifested.* Our argument is presented as follows.

1. A person *S*'s living a way of life *L* (in accordance with *substantive* conceptions of the good) is *manifestly* valuable if and only if (i) *L* is *potentially* valuable; and (ii) *S* is pursuing *L for the reason that L is valuable.*[28]
2. *S* is pursuing *L* for this reason only if *S* is pursuing *L* autonomously.
3. Therefore, *S*'s living a way of life *L* is manifestly valuable only if *S* is pursuing *L* autonomously.

Consider Premise 1 first. Here we are making a distinction between *manifest values* and *potential values*—or between *instantiated values* and *values per se.* When we call a way of life potentially valuable, what we mean is that the way of life has some intrinsically valuable features so that the value would be instantiated if it is pursued for the right reasons, say, for the reason that *L* is valuable. Even if some way of life is (potentially) valuable, the value cannot be instantiated or manifested by a person's

[26] Raz, *The Morality of Freedom*, p. 154.

[27] Nozick, *Anarchy, State and Utopia*, p. 50.

[28] For our purposes, we confine ourselves to the discussion of cases in which a person is pursuing substantive conceptions of the good life—rather than cases in which one is promoting general, primary goods (such as health, life, and freedom).

activity if she pursues the way of life for illegitimate reasons. For example, if a person chooses to live a Confucian life not because she endorses Confucian values, but because she fears the state's punishment, the values of a Confucian life would not be instantiated on this occasion, even though a Confucian way of life itself is valuable. Xunzi, an ancient Chinese philosopher, makes a similar point when he talks about reasons for moral learning:

> Students in ancient times learned for their own sake, but the students of today learn for the sake of impressing others. Thus learning for self-improvement is the learning of a virtuous person, while learning for the approbation of others is the learning of a petty man.[29]

Moreover, that a way of life is potentially valuable is also a necessary condition on the manifestation of the value. If a way of life is not valuable per se—that is, it has no potentially valuable characteristics—then the value can never be manifested even though the way of life is pursued for the right reasons.

Our view is structurally analogous to Kant's famous thesis that an action has moral worth if and only if (1) the action is in fact morally right; and (2) the action is performed for the right reasons—for Kant, it is done from the sense of duty.[30] Even though some sorts of actions (e.g., keeping a premise) are morally right, they lack moral worth if performed for illegitimate reasons (e.g., for an egoistic reason). On the other hand, if an action is not morally right, the action would also lack moral worth even if it is performed from the sense of duty. Just as the moral worth of an action depends on whether the action is performed for the right reasons, so, we wish to emphasize, the value manifestation of a way of life also relies on whether it is pursued in the right sort of way.

We will discuss Premise 2 in more detail in the following section. Here let us touch on this point very briefly. There seems to be a close connection between living an autonomous life and pursuing a way of life for the

[29] Xunzi, "An Exhortation to Learning," in *Xunzi: The Complete Text*, trans. Eric L. Hutton (Princeton: Princeton University Press, 2014), pp. 1–8, at pp. 5–6. The translation is slightly modified here.

[30] Kant, *Foundations of Metaphysics of Morals*, pp. 397–401.

right reasons. Suppose that a person does not pursue some way of life autonomously—for example, she is forced to choose a way of life she dislikes. In this case, she is not living this way of life for value-based reasons; rather she chooses the way of life for "alien" reasons, say, the reason to avoid being punished.[31]

If the aforementioned argument is successful, it follows that autonomy is a necessary condition on pursuing a valuable life. If a person is not pursuing a way of life autonomously, *she* is not living a valuable life at all, even though the way of life *itself* is valuable. Thus, autonomy is a second-order good that places constraints on the realization of first-order values.

6 Two Autonomy Constraints

There are two kinds of cases in which a person's autonomy is violated: violations of i-autonomy and violations of c-autonomy. First, the government may use coercion and punishment to promote particular conceptions of the good life. In such circumstances, the citizens' i-autonomy is undermined; they are subject to the wills of others. A person whose i-autonomy is invaded has difficulty pursuing her own way of life and is often forced to choose another way of life that she does not reflectively endorse. The way of life that a person is coerced to choose is merely an instrumental good—a "trade-off" to avoid punishment—rather than an intrinsic value for her. So, she is not pursuing the way of life for the right reason, that is, the reason that the way of life is valuable. Consequently, the value of the way of life is not instantiated.

As Jeremy Waldron points out, coercion and punishment are simply useless for perfectionist purposes, "because a person's allegiance to a conception of the good life is a matter of her inner commitment, rather than of her external conduct."[32] Someone or the government can threaten a person to change her external conduct, but cannot force her to change her inner commitment. For example, listening to classical music may be a valuable activity, but we will not make a person's life better by coercing

[31] Sher, *Beyond Neutrality*.

[32] Jeremy Waldron, "Legislation and Moral Neutrality," p. 155.

her into going to classical music concerts if she does not do so autonomously. This point has also been noted by Kant:

> No one can coerce me to be happy in this way (as he thinks of the welfare of other human beings); instead each may seek his happiness in the way that seems good to him.[33]

Will Kymlicka writes in a similar vein:

> No life goes better by being led from the outside according to values the person doesn't endorse. My life only goes better if I'm leading it from the inside, according to my beliefs about value.[34]

Although coercion violates i-autonomy, it is unclear whether coercion also undermines c-autonomy. Coercion typically intervenes in the path from first-order desires to actions. Consider a person who originally wanted to do *x* but is coerced to do *y* instead. The original path from *wanting to do x* to *doing x* is now disconnected due to the presence of coercion. But coercion seems to leave intact the coherence of first-order desires and second-order volitions. When a person surrenders her money to avoid being shot by robbers, she does not suffer any discrepancy between first-order desires and second-order volitions (suppose that she wants to stay alive and also wants this first-order desire to constitute her will). Similarly, when a citizen chooses a way of life that she dislikes in order to avoid being punished by the government, the (first-order) desire to pursue that way of life still conforms to her second-order volitions. After all, she still wants to have the desire to lead that way of life, or even wants this desire to be her will.

Nevertheless, there are some methods of promoting the good that may violate c-autonomy. For example, a government can induce its citizens to pursue some ways of life by *manipulation*. The government can use systematic indoctrination or 'brainwashing' to impose on the citizens some

[33] Immanuel Kant, *Practical Philosophy*, trans. Mary Gregor (Cambridge: Cambridge University Press, 1996), p. 291.

[34] Will Kymlicka, *Liberalism, Community and Culture* (Oxford: Oxford University Press, 1989), p. 12.

particular conceptions of the good life. Or the government can distort and conceal relevant information to make the citizens accept the favored ways of life. Suppose that living the way of life that the government promotes does not accord with some citizens' deep selves. However, due to indoctrination and distortion, those citizens develop the (first-order) desire to live that way of life. In this case, their first-order desires come apart from their own second-order volitions; their c-autonomy is thus undermined. As we have discussed in Sect. 4, knowledge of second-order volition is not always transparent. Because people under manipulation do not know that their first-order desires are in fact inconsistent with their genuine second-order volitions, they have no feelings of frustration—this is part of what we mean by 'manipulation' and 'indoctrination'. When people's c-autonomy is violated, they do not pursue the ways of life for right reasons, that is, for the reason that the ways of life are valuable. As a result, the values of the ways of life are not realized.

It is less clear whether rewards—by which the government aims to promote some particular conceptions of the good life—would violate people's c-autonomy. We suspect that at least in some circumstances, rewards may distort the citizens' self-understandings. It is possible that although pursuing the way of life that the government recommends does not accord with a person's second-order volition, she is somehow attracted by the rewards and hence wants to pursue that way of life. In this case, the person's first-order desire does not reflect her deep self or second-order volitions. The government's rewards are functioning like bribes.[35] If a person pursues a way of life in order to get rewards, she does not respond to the potential values of this way of life.[36]

We admit that when natural conditions rather than artificial factors cause a person to pursue a way of life for illegitimate reasons, the values of the way of life are not instantiated either. Nevertheless, an obvious reason why we do not place emphasis on the obviation of natural obstacles is that unlike intentional interventions, natural restrictions are morally irrelevant—they do not "owe" anything to us. More specifically, for

[35] Jeremy Waldron, "Autonomy and Perfectionism in Raz's *Morality of Freedom*," *Southern California Law Review* 62 (1989): 1097–152.

[36] Unlike cases of coercion, neither manipulation nor reward seems to undermine i-autonomy, for the citizens' actions are still caused by their first-order desires.

the purposes of this chapter, we are only concerned with the government's violations of autonomy, whereas there are certainly other sorts of agents who induce people to develop some ways of life by invading autonomy.

7 Beyond Neutralism and Perfectionism

The government adopts some special methods in order to advance particular conceptions of the good life. But if these methods invade and undermine persons' autonomy, the values of these desired ways of life cannot be manifested. As we have pointed out, coercion, manipulation, and perhaps reward will undermine the citizens' autonomy in one way or other. Therefore, such state actions are simply irrational and self-defeating in that the means is ill-adapted to the end the government claims to be pursuing.[37] The autonomy constraint provides an explanation of the distinction "between what is needed to justify belief and what is needed to justify the employment of political power".[38] Even if some conceptions of the good life are justified and true, it does not follow that the government should promote them by any means—so understood, perfectionism is false.

But would every method of promoting the good undermine autonomy? We do not think so. A government can induce its citizens to pursue certain ways of life, for example, by advertising the genuine values of these ways of life and by creating institutions or social forms that make the favored ways of life feasible. In those cases, neither the citizens' i-autonomy nor their c-autonomy is undermined, and the values of the desired ways of life can still be manifested. The government does not encourage its citizens to respond to inappropriate reasons in the process of pursuing the ways of life. In the case of justification, the government only provides value-based reasons to which its citizens may respond; and in the case of creation, the government merely increases the options that

[37] Waldron, "Legislation and Moral Neutrality," p. 155.

[38] Nagel, "Moral Conflict and Political Legitimacy," p. 229.

the citizens may choose. The consideration of autonomy fails to rule out some methods of promoting the good, and neutralism is thus groundless.

It may be objected that justification and creation will also undermine some citizens' autonomy, though in an indirect way. Either in the case of justification or in the case of creation, the government has to use fiscal revenue to promote certain conceptions of the good life, but these funds are ultimately drawn from taxation, including the contributions of those who do not endorse the conceptions of the good that the government recommends.[39] According to some scholars, the government is basically seizing the dissenters' money for promoting the conceptions of the good life that they disfavor. In such cases, their autonomy is violated.

This objection seems to presuppose the principle that any public policy is legitimate only if it receives unanimous collective agreement. But this principle is too demanding. Let us temporarily switch to the domain of distributive justice. The government uses fiscal revenue (ultimately from taxation) to, for example, develop the economy, invest in public infrastructure, and provide social welfare. But if legitimacy requires unanimous agreement, most policies in distributive justice would be illegitimate, because they are not endorsed by everyone. And it is unhelpful to appeal to hypothetical consent. Just as fundamental principles of distributive justice can be endorsed in an ideal contractual situation, so the principles regarding the good life (such as the principle that the government should promote particular conceptions of the good in appropriate ways) could also be agreed upon in a suitable contractual situation.

Some philosophers may point out a disanalogy between a politics of the good and a politics of the just: while policies regarding distributive justice are concerned with the distribution of *primary* goods that are neutral to the citizens' particular life projects, but policies regarding the good life aim at promoting *substantive* conceptions of the good that not every one shares. Thus, as some philosophers argue, non-neutral state actions would be *disrespectful* for the citizens who do not endorse those conceptions.[40] Martha Nusbaum puts it this way:

---

[39] Gaus, "Liberal Neutrality: A Compelling and Radical Principle".

[40] See, e.g., Nussbaum, "Perfectionist Liberalism and Political Liberalism"; Quong, *Liberalism Without Perfection*.

> These doctrines are so deeply a part of people's search for the meaning of life that public governmental denigration of those doctrines puts those people at a disadvantage, suggesting that they are less worthy than other citizens, and, in effect, not treating them as fully equal ends in themselves.[41]

First of all, we would like to note that this reason for neutrality is no longer based on the value of autonomy, but instead on the idea of respect. This topic deserves a fuller discussion than we are able to provide it in this chapter. Here let us address this issue briefly. It is important to make a distinction between respect for persons and respect for their views, *regardless of how closely* the persons are attached to their views. Our capacity for a conception of the good, like our capacity for a sense of justice, implies that we sometimes make mistakes and we should revise our views in response to evidence and reasons.[42] Some state actions, such as coercion and manipulation, are *paternalistic* in the sense that the government assumes that people do not have adequate rational capacity to respond to relevant reasons. Hence those methods of promoting the good are disrespectful for the persons. But not every method of promoting the good conflicts with the idea of respect. As we mentioned earlier, a government can provide value-based reasons to which its citizens may respond, or increase the options that the citizens may choose. In the cases of justification and creation, the government shows respect for the persons' rational agency. Therefore, justifying neutrality on the ground of respect is unsuccessful.

8 Conclusion

By appeal to the notion of autonomy, we have argued that neither neutralism nor perfectionism is acceptable. Perfectionism is mistaken in that the state actions that violate people's autonomy (e.g., coercion and manipulation) would fail to realize the values of the desired ways of life. Neutralism is problematic in that some methods of promoting the good

[41] Nussbaum, "Perfectionist Liberalism and Political Liberalism," p. 22.

[42] Rawls, *Theory*; Raz, *The Morality of Freedom*.

(e.g., justification and creation) do not undermine autonomy and hence can be politically justified.

We want to end this chapter by pointing out a distinction between autonomy as a political *constraint* and autonomy as a political *goal.* Although the main purpose of this chapter is to argue that a government ought not to violate its citizens' autonomy in promoting disputed conceptions of the good life, we are sympathetic with the view that a government should *positively* induce the citizens to live an autonomous life.

It is a tricky issue, however, whether a government should promote its citizens' autonomy by *coercive* means—for example, whether the government should *force* some people to abstain from hard drugs (in such a case, the government invades those people's i-autonomy for the sake of promoting their c-autonomy). This could be an interesting topic for further investigation on another occasion. But this chapter, which is only concerned with the competition between autonomy and other values, leaves it open how to balance different aspects of autonomy.

12

What We Have Reason to Value: Human Capabilities and Public Reason

Nancy S. Jecker

1 Introduction

Living together in a community, people inevitably make all sorts of demands on one another. We ask not only that people abide by laws, but that they conduct themselves in ways that comply with social morality. We strive to reach agreement peacefully about how we ought to behave with one another despite the fact that members may hold different beliefs about personal morality and religion, the good life, and how society should be ordered. Notwithstanding these differences, most of the time, we are able to resolve differences by appealing to reason.

Public reason is the name sometimes given to the reasoning we invoke in public life to come to agreement about matters requiring social action, such as the design of public institutions, laws and policies potentially affecting all, and social values and priorities. Unlike *private reason*, which

N. S. Jecker (✉)
Department of Bioethics and Humanities, University of Washington School of Medicine, Seattle, WA, USA
e-mail: nsjecker@uw.edu

H.-L. Li, M. Campbell (eds.), *Public Reason and Bioethics*,
https://doi.org/10.1007/978-3-030-61170-5_12

is the individual reflection we engage in with people who share our values and way of life and identify with our moral or religious convictions, public reason extends beyond like-minded people; it rests on modest premises that can gain wide acceptance.

This chapter sets forth an interpretation of public reason that appeals to our central capabilities as human beings. I argue that appealing to central human capabilities and to the related idea of respect for threshold capabilities is the best way to understand public reason. My defense of this position advances stepwise. (1) First, I consider the central alternative to a capability account, which is to regard public reason as a matter of contracting based on voluntary choice and rational deliberation among equals. (2) Next, I describe central concerns with contract views and (3) show how a capability view can avoid them. (4) Lastly, I consider extending a capability account of public reason beyond national borders and address the objections that capability views of public reason apply only to Western constitutional democracies or that they embed exclusively Western values.

2 Public Reason as Contracting

Public reason is a way to ensure that the moral and political rules that govern common life are justifiable to all people to whom the rules apply. A well-known account, with roots in social contract theories of the seventeenth and eighteenth centuries, regards public reason as a process of contracting. According to this account, free and equal individuals pursuing their own interests agree to join together in a civil society and to be subject to its constraints. This kind of contractual account is found in the philosophies of Hobbes, Locke, and Rousseau, where it is used as a justification for the existence of a sovereign and the restrictions of the state. Kant endorses a version of social contract which is based on rational unanimity binding each legislator to "give his laws in such a way that they could have arisen from the united will of a whole people"[1] and stresses

[1] Immanuel Kant, "On the Common Saying: That Might be Correct in Theory, But it is of No Use in Practice" (1793), *The Cambridge Edition of the Works of Immanuel Kant: Practical Philosophy*, trans. Mary J. Gregor (Cambridge, UK: Cambridge University Press, 1999), pp. 296 and 297.

that "[a]ny rights and duties stemming from an original contract do so … because of the rightful relations embodied in the original contract,"[2] Taken together, the political philosophies of

> Hobbes, Locke, Rousseau and Kant brought to the fore a crucial problem for social life among autonomous moral agents under modern conditions: using their reason as well [as] could reasonably be expected, people will arrive at conflicting judgments about morality. The solution of the social contract theorists was public reason as expressed by the political umpire.[3]

This manner of reconciling individual differences requires each person to exercise reason and consent to a framework for social life. No individual's private morality dictates; instead, each retains autonomy and authorizes a moral framework for social life as normative for them.

Twentieth-century thinkers share and extend this conception, shifting from the justifying the state to justifying moral principles of justice. Gaus, for example, maintains that when we use our private reason, we disagree about what is right and good, but when we articulate public reason, we find areas of overlap between diverse moral viewpoints. The strategy gives expression to the idea, rooted in social contract theory, that a shared public morality must meet a condition of justifiable to each: "For each person, she only has one source of normative judgment—her overall set of evaluative standards"; therefore, "social morality must not make moral demands on a person that cannot be justified to her."[4] The requirement of justifiability to all legitimates social morality in the same way that it legitimates the state. The basis for this condition is the belief the authority of social morality must be morally based, and that morality first and foremost arises at the level of individuals. Gaus writes, "While no one's vision of the moral truth can dictate the social moral framework, the aim is for all to see the social moral framework as normative, given

[2] Frederick Rauscher, "Kant's Social and Political Philosophy," in *Stanford Encyclopedia of Philosophy*, ed. Edward N. Zalta, (2017). At: https://plato.stanford.edu/archives/spr2017/entries/kant-social-political/.

[3] Gerald Gaus, "On Being Inside Social Morality and Seeing it," *Criminal Law and Philosophy* 9 (2015): 141–153, at p. 142, henceforth "*BI*".

[4] *BI*, pp. 145, 146.

their individual understandings of moral truth."[5] This coordination of moral sensibility makes possible a social life and public morality.

Underlying these analyses is a twofold consensus. First, on fundamental matters of how we ought to live our lives, reasonable people will disagree, where "reasonable" means the process of thinking and conversing in good faith about something. Second, public life should not be based on disputed features of any single individual's private morality, but on a shared vision, which in a diverse society can only be a thin vision of the good, reflecting a narrow overlapping consensus. According to Larmore, social morality "must seek its principle in a minimal morality, which reasonable people can share despite their expectably divergent religious and ethical convictions."[6] The corresponding aim of public morality must be nothing more than "restraining the struggle for advantage and the violence of emotion…."[7]

A striking feature of this account of public morality is the commitment to respect persons by seeking to justify social morality to each individual. Rawls calls this requirement the liberal principle of legitimacy.[8] For Rawls, as for Kant, people possess an inviolability that makes it wrong to regard them as mere means to a public good. According to Kant, the requirement of justifiability to all reflects the worth and dignity of persons: "a human being regarded as a person, that is, as the subject of morally practical reason, is exalted above all price … as an end in himself he possesses a dignity by which he exacts respect for himself from all other beings in the world."[9] Broadly speaking, justifiability to all expresses the idea that just political principles are those that can be justified to all those who are subject to them. The intuitive idea is that it violates the dignity and worth of an individual to force them to act against their own reasons.

[5] *BI*, p. 143.

[6] Charles Larmore, "The Moral Basis of Political Liberalism," *Journal of Philosophy* 96, no. 12 (1999): 99–625, at p. 600, henceforth "*MB*."

[7] *MB*, p. 601.

[8] John Rawls, *Political Liberalism* (New York: Columbia University Press, 2005), henceforth "*PL*."

[9] Immanuel Kant, *The Metaphysics of Morals* (1797), in *Immanuel Kant Practical Philosophy*, trans and ed. Mary Gregor (Cambridge, UK: Cambridge University Press, 1996), at pp. 434–435.

3 Disability Critiques of Contracting

Yet critics of the social contract tradition note that the conception of dignity it offers is narrow in scope, applying only to people who possess rational agency. Someone incapable of having reasons for acting falls outside the scope of contracting. To further consider what has come to be known as "the disability objection," it is helpful to distinguish two strands within the social contract tradition: first a *contractarian* or Hobbesian strand and second, a *contractualist* or Kantian strand.[10] The contractarian stance leans toward narrow self-interest as a basis for bargaining and is associated with thinkers such as Hobbes. The contractualist stance leans opposite, away from narrow self-interest toward a broader requirement of justification before others as a requirement of deliberating and is associated with the tradition of Kant. It is relatively easier to make the case that people who lack the capacity for public reason, such as people with severe intellectual disabilities, fall outside the scope of hard bargaining and narrow self-interest associated with a contractarian interpretation, since there might be a perceived social burden associated with caring for people with lifelong dependencies. Yet even when self-interest is tempered by a requirement to justify oneself to others, as it is on the contractualist view, the social contract position remains vulnerable to the disability objection. For example, Rawls requires that contracting parties deliberate by drawing on "moral powers," such as the capacity to form and revise a conception of the good and to have a sense of justice that enables them to act on and apply fair terms of cooperation.[11] Critics, such as Young, argue that an implication of Rawls' view is that duties of justice cannot be owed directly to individuals with serious intellectual disabilities.[12] For Young, and others of her ilk,[13] members of society who do not meet the require-

[10] Elizabeth Ashford, Tim Mulgan, "Contractualism," in *Stanford Encyclopedia of Philosophy*, ed. Edward N. Zalta (2018), at: https://plato.stanford.edu/archives/sum2018/entries/contractualism/.

[11] *PL*, pp. 18–20.

[12] Iris Marion Young IM, "Rawls's Political Liberalism," *Journal of Political Philosophy* 3, no. 2 (1995): 181–190.

[13] Eva Feder Kittay, *Love's Labor*, 2nd Edition (New York: Routledge, 2019), henceforth "*LL*"; Martha C. Nussbaum, "Capabilities and Disabilities," *Philosophical Topics* 30, no. 2 (2002): 133–165, henceforth "*CD*."

ment of rational agency are nonetheless owed justice, even though the justifiability to all condition does not apply to them. Others note that this form of objection applies not just to people with lifelong disabilities, but also to all human beings during the usual course of the human life cycle, which brings with it periods of immaturity during infancy and early childhood and high rates of cognitive impairment and dependency during later life.[14] Other interpretations of the outlier problem hold that social contracting positions African Americans[15] and women[16] at distinct disadvantage, even if they are not excluded categorically.

Kittay elaborates the disability objection by arguing that a contractarian view would make an outlier of her daughter, Sesha, who is profoundly mentally and multiply disabled; Kittay claims Sesha has an inviolability founded on dignity that social contract traditions miss. Raising Sesha imparted a lesson in humility, Kittay notes, by leading her to realize that "what … I—thought was at the center of humanity, the capacity for thought, for reason, was not it, not it at all."[17] Kittay identifies several features, unrelated to the capacity for thought, as morally salient: the human way in which Sesha appreciates music; shows sensitivity to others; stands in social relationships; evinces a "strong clear sense of herself"; "is capable of great joy and great love"; and seems to remember and anticipate people, places, and music she has not heard for years.[18] Kittay reasons that Sesha's intellectual limitations might mean that she lacks the capacity to understand the distinct moments of her life as a meaningful whole, with a narrative-like unity, which would mean that Sesha's life might not have a certain richness that her own life has. Yet, she rejects the idea that this shows that Sesha or other individuals with intellectual impairment should be compared to non-human animals in worth or dignity. Although Sesha cannot express cognitive capacities, and shows no measurable IQ, "What Sesha can do she does as a *human* would do them,

---

[14] Nancy S. Jecker, *Ending Midlife Bias: New Values for Old Age* (New York: Oxford University Press, 2020), henceforth "*EMB*."

[15] Charles Mills C., *The Racial Contract* (Ithaca, NY: Cornell University Press, 1997).

[16] Carol Pateman, *The Sexual Contract* (Stanford, CA: Stanford University Press, 1989).

[17] *LL*, p. 160.

[18] Eva Feder Kittay, "At the Margins of Moral Personhood," *Ethics* 116, no. 1 (2005): 100–131, at pp. 127–129, henceforth "*MMP*."

though frequently imperfectly, but it is *humanly* imperfect."[19] Kittay's conclusion is that Sesha's life contains "an immeasurable amount of good."[20] Nussbaum says of cases like Sesha's that they show with "naked clarity the extent to which the very choice of a bargaining model biases the whole idea of the benefits of social cooperation."[21]

In reply, it might be claimed that some versions of social contract theory escape the disability objection. Scanlonian contractualism purports to do just that. According to this view, what makes an act wrong is that it can be reasonably rejected; specifically, "an act is wrong if its performance under the circumstances would be disallowed by any set of principles for the general regulation of behavior that no one could reasonably reject."[22] Scanlon goes on to say that human beings with significant disability deserve protection even when they themselves are incapable of having reasons to reject principles and unable to form even the more minimal "judgement sensitive attitudes," which are attitudes that may arise spontaneously, without reflection, self-consciousness, or any judgment concerning oneself or what is done to one.[23] The justification Scanlon offers us for duties to beings with significant intellectual limitations is that "the mere fact that a being is 'of human born' provides [us] a strong reason for according it the same status as other human beings."[24] He goes on to defend this claim against the charge of speciesism, arguing that "it is not prejudice to hold that our relation to these beings gives *us* reason to accept the requirement that our actions should be justifiable to them."[25] He adds, "the beings in question here are ones who are born to us or to others to whom we are bound by the requirements of justifiability. This tie of birth gives us good reason to want to treat them 'as human' despite their limited capacities."[26] The general idea seems to be that an individual like

[19] *MMP*, pp. 127–128, emphasis added.

[20] *MMP*, p. 120.

[21] *CD*, p. 152.

[22] T.M. Scanlon, *What We Owe to Each Other* (Cambridge, MA: Belknap Press of Harvard University Press, 2000), p. 153, henceforth "*WWO*."

[23] *WWO*, p. 23.

[24] *WWO*, p. 185.

[25] *WWO*, p. 185.

[26] *WWO*, p. 185.

Sesha can be wronged simply by virtue of how *others* are rationally disposed to relate to her. As Kumar notes, as a contractualist, Scanlon is committed to the "individual reasons restriction," which roughly holds that what is morally relevant to justifying principles that regulate individual relations are only those considerations that have a bearing upon the recognition of persons capable of rational self-government.[27] Therefore, those with intellectual impairments like Sesha's count morally only so far as they figure into the lives of rational self-governors.

Yet, a limitation of Scanlon's account in these passages is that he includes people with disability derivatively, namely, because they stand in a certain relation to beings who have a reason to object to their ill-treatment. For example, we should not inflict gratuitous harm on Sesha because others could reasonably object. If our overall theory is contractualist, it makes sense to give a contractualist analysis of why treating Sesha cruelly would be wrong. Contractualism, as Scanlon presents it, is an account of reasoning about moral principles which assumes a specific normative ideal of a person, namely, someone with the capacity for reason and rational self-government.[28] However, in contrast to Scanlon, we want to say that Sesha's value is not derivative. She herself deserves respect and she herself possesses an inherent worth and dignity. Her dignity holds irrespective of her relation to others. These reflections lead us to look elsewhere for an account of public reason.

4 Public Reason as Human Capabilities

One way of spelling out an alternative vision is to focus on common humanity, a focus which derives from a more general idea I call, "species integrity."[29] In contrast to the social contract tradition, which begins with

[27] Rahul Kumar, "Who Can Be Wronged?" *Philosophy and Public Affairs* 31, no. 2 (2003): 99–118, at p. 108.

[28] Rahul Kumar, "Reasonable Reasons in Contractualist Moral Argument," *Ethics* 114, no. 1 (2003): 6–37, at p. 9.

[29] *EMB*, pp. 29–51.

the choices of free and equal individuals, species integrity begins with the idea of a being who commands respect, not by virtue of our volition or choice, but by virtue of the kind of being they are. Darwall refers to this form of respect as "recognition," distinguishing it from "appraisal."[30] While appraisal respect is the product of an evaluation we make of another's merit, recognition respect indicates "deference, in the most basic sense of yielding: self-absorption and egocentric concerns give way to consideration of the object, one's motives or feelings submit to the object's reality, one is disposed to act in obedience to the object's demands."[31]

If we begin with recognition respect, rather than consent, the first step is to ask what kind of being is Sesha that she commands this kind of respect? What Kittay tells us is that Sesha is human, and her imperfections are "humanly imperfect, not canine perfect."[32] In other words, she is not to be compared with any other species because our response to her is partly owing to the fact that she is human, like us. Kittay's description makes evident that Sesha can do and be many, though not all, of the central things that human beings generally can do and be. A more complete list might be the following, adapted from Nussbaum[33] and defended at greater length elsewhere[34]:

Central Human Capabilities

1. Life: having an unfolding story or narrative of one's life;
2. Health: being able to have all or a cluster of the central capabilities at a threshold level;
3. Bodily Integrity: being able to use one's body to realize one's goals;
4. Senses, Imagination, and Thought: being able to imagine, think, and use the senses;
5. Emotions: being able to feel and express a range of human emotions;
6. Practical Reason: being able to reflect on and choose a plan of life;

30 Stephen Darwall, "Two Kinds of Respect," Ethics 88, no. 1 (1977): 36–49.

31 Robin S. Dillon, "Respect," in *Stanford Encyclopedia of Philosophy* ed. Edward N. Zalta (2018). At: https://plato.stanford.edu/archives/spr2018/entries/respect/.

32 *MMP*, p. 128.

33 Martha C. Nussbaum, *Creating Capabilities* (Cambridge, MA: Harvard University Press, 2011a).

34 *EMB*.

7. Affiliation: being able to live for and in relation to others;
8. Nature: being able to live in relation to nature and other species;
9. Play: being able to laugh, play, and recreate; and
10. Environment: being able to regulate the immediate physical environment.

Appealing to central human capabilities is an appeal to salient features of shared humanity. Such an appeal extends well beyond the purely procedural focus found in Hobbes, Locke, Rousseau, and Kant (and in our own day, Rawls). With the capability view, the opening question is, 'What kind of being is a human being?' Only after there is an answer to the first question does the view ask, 'What is minimally required to respect that being's central capabilities at a minimal threshold?' Finally, the capability view asks a third question, namely, 'What reasonable steps can society take to support a minimal threshold with respect to each capability?'

If we assume that the proposed list of central human capabilities is at least a plausible way to answer the first question, then we turn to the second question. Here, our answer is that a capability threshold for Sesha (or another person with similar disabilities) focuses on supporting her capacities to be physically, emotionally, and mentally healthy; have bodily integrity; exercise senses and imagination; feel and express human emotions; affiliate with others; go outdoors; laugh, play, and recreate; and exercise some measure of control over her physical environment. All of these capabilities are part of the central things that Sesha can do and be as a human being. Since Sesha lacks some central capabilities, such as the capability to have a narrative or reflect on a plan of life, respecting her dignity will not require things that would be required if she had these capabilities.

The last step in a capability analysis is pinpointing reasonable steps society must take to bring about a capability threshold. This is a matter of balancing the responsibilities to respect Sesha in the ways described among various groups, including Sesha's family; public/private institutions, such as schools, health, and social care systems; and various public services, such as public safety and security, education, and recreation. It also implies duties on the part of the long-term care facility where Sesha lives and government funding at various levels to reasonably accommodate people with disabilities like Sesha's.

The case of Sesha and the disability challenge more broadly indicates that although common humanity is a constant, the presence or absence of particular capabilities varies from one individual to the next. In addition, particular capabilities vary for the same individual over time, from birth to death. For example, human practical reason is something acquired gradually as the human brain develops. Nor are we born with the ability to move from place to place or be sexually intimate; if we acquire these abilities, they might be diminished later in life. These variations between and within people do not change the underlying fact that each of us is equally and fully human. When we speak of *respecting human dignity*, it is the underlying humanity we speak to. When we speak of *showing* respect for human dignity, there will be different things required for different people and for the same person over time. Since people display unequal abilities to convert resources into functioning and opportunities, some individuals will require more resources than others to attain the same capability threshold.

In response, someone might think it matters *why* a person requires more resources to reach the same capability level, claiming that if the need for more is due to choices the individual made, it is the individual, not society, who should be held accountable.[35] Others might question why inequalities should matter only up to the threshold, rather than beyond it.[36] In reply, following Casal, I interpret the capability view as comprised of two separate theses: a positive thesis, claiming "the importance of people living above a certain threshold" and a negative thesis, denying "the relevance of certain additional distributive requirements" after threshold sufficiency.[37] It is possible to embrace both theses or to endorse just one. For example, Shields adopts a hybrid position, appealing to what he calls, "the shift thesis," which holds that it is worse to be unequal below than it is to be unequal above the threshold level.[38] My

[35] Kasper Lippert-Rasmussen, *Luck Egalitarianism* (London: Bloomsbury Academic, 2015).

[36] Liam Shields, *Just Enough: Sufficiency as a Demand of Justice* (Edinburgh University Press, 2016).

[37] Paula Casal, "Why Sufficiency is Not Enough," *Ethics* 117, no. 2 (2007): 296–326, at p. 298.

[38] Liam Shields, *Just Enough: Sufficiency as a Demand of Justice* (Edinburgh: Edinburgh University Press, 2016), at p. 30.

emphasis throughout is the positive claim that we ought to reasonably support people at a capability threshold, leaving open questions about what additional principles, if any, justice entails.

If we accept a capability strategy for going about public reasoning, we are left with a more demanding account of what public reason requires. Public reason that appeals to central human capabilities requires supporting *each* central human capacity, not just the capacity for practical reason. Some charge that a capability view is overly demanding for this reason—it compels supporting people in a wide range of things they can do and be as human beings.[39] In reply, there are several features that temper the view's demandingness. First, as I interpret it, the capability view is life stage sensitive. For example, if a child is deprived of basic childhood opportunities, such as primary education, society should intercede, but it is not required to help a child reach their highest possible potential. Second, reasonable efforts do not demand exorbitant or useless efforts. For instance, Sesha will never be able to acquire certain kinds of functioning, such as literacy and numeracy, and it would be futile to attempt to provide these opportunities to her. Third, a pragmatic feature is built into the notion of what is "reasonable." For example, when the view is deployed in low resource settings, it might sanction moves in the right direction, while acknowledging that they fall short of the minimal threshold set by capability sufficiency.[40]

A final objection is that a capability account properly understood is not a way of engaging in public reason at all, but instead an outcome of public reason, that is, a preferred framework arrived out through a process of public reasoning. In response, like other accounts of public reason, capability views aim to tell us what is required in order to justify ourselves to others. Rather than using the social contract apparatus to meet this requirement, a capability view meets it by appealing to human dignity. While respecting human dignity demands respecting people's capacity for reason, it also requires respecting their other central capacities. Just as

[39] Laura Capitaine, Guido Pennings, Sigrid Sterckx, "Why Jecker's Capabilities Approach to Rationing Is Incapable of Containing Health Care Costs," *American Journal of Bioethics* 13, no. 8 (2013): 22–23, at p. 22.

[40] Nancy S. Jecker, "Age-Related Inequalities in Health and Healthcare," *Developing World Bioethics* 18 (2017): 144–155.

we cannot force people to do things that they reasonably oppose, we cannot ride roughshod over their other central capacities. The capability view agrees with social contract views that human dignity resides in our capacity for rational agency, but it adds to this the claim that human dignity also resides in the many other capacities that make us human. This widening of the scope of public reason does not rest on a comprehensive doctrine, but instead on a view of humanness that departs from the view consent-based theories employ.

5 Public Reasons Beyond Borders

So far I have defended a capability approach to public reason by arguing that it carries advantages over contracting by including respect for all the central human capabilities, not just the capability for practical reason. In this way, it affords a fuller picture of common humanity and offers a vision of human dignity that encompasses the range of differences that humans display. However, a further test is how these different renderings of public reason fare at a global level. How can public reason extend beyond national borders and guide deliberations between nations? Does contracting or capability serve us better at an international level?

Contracting

An initial way to extend a Rawlsian contractual account of the just state to a global level is by introducing of a two-stage contract.[41] At the first stage, an agreement is struck among individuals with free and equal power who agree to abide by certain principles of justice in the domestic sphere. At the second stage, an agreement is struck among independent states with equal power that come together on the world stage and agree to abide by an international Law of Nations, which includes conventions such as keeping treatise, honoring nations' self-determination, and waging wars justly.

[41] John Rawls, *Theory* (Cambridge, MA: Belknap Press of Harvard University Press, 1971).

Yet, this initial attempt faces obvious concerns related to the preconditions it sets. First, states are unequal in power, wealth, and development; second, their borders change over time; and third, they come to the bargaining table with preexisting dependencies. A subsequent attempt to adapt a Rawlsian view replaces the Law of Nations with a Law of Peoples, which applies exclusively to liberal democracies. Pursuing this approach, Rawls opines that

> Just as a citizen in a liberal society is to respect other persons' comprehensive religious, philosophical, and moral doctrines, provided they are pursued in accordance with a reasonable political conception of justice, so a liberal society is to respect other societies organized by comprehensive doctrines, provided their political and social institutions meet certain conditions that lead the society to adhere to a reasonable law of peoples.[42]

The Law of Peoples is more robust and demanding than the Law of Nations, in the sense that it accords all people human rights and assigns all nations duties to assist people living under unfavorable conditions, regardless of national origin. As Nussbaum rightly notes, this more expansive conception is not based on contracting, but goes beyond it, making a direct appeal to those who are part of the bargain to respect human rights.[43] To find a basis in public reason for the appeal to human rights made in the Law of Peoples requires looking beyond contracting to a philosophical conception of human dignity, a concept often regarded as the philosophical basis of human rights.[44]

Capabilities

A capability account is at home with the language of international human rights and can offer tools for making sense of such appeals. Simply put, to honor human rights at an international level is to make good on them

[42] John Rawls, "The Law of Peoples," *Critical Inquiry* 20 (1993): 36–68, p. 37.

[43] Martha C. Nussbaum, "Beyond the Social Contract," *Oxford Development Studies* 32, no. 1 (2004): 3–18.

[44] Jeremy J. Waldron, "Is Dignity the Foundation of Human Rights," in *Philosophical Foundations of Human Rights* ed. Rowan Cruft, Matthew S. Liao, Massimo Renzo (New York: Oxford University Press, 2015).

through reasonable measures to ensure an adequate threshold of the central human capabilities they refer to. As Nussbaum notes, capabilities can work *either* with a conception of human rights as their philosophical underpinning, or they can function directly, as a standalone framework.[45] Those who regard the language of rights as beset with controversy, or as not especially informative beyond appeals to dignity, are sometimes led down the latter path, shedding the language of rights and using capabilities as a free-standing ethical analysis. Since 1993, for example, the United Nations (U.N.) Development Programme has assessed the quality of life in nations of the world using the concept of people's capabilities or their abilities to do and to be certain things deemed valuable.[46] Appealing to threshold capabilities, the U.N. directs economic development in low- and middle-income nations toward creating conditions that enable and create choice with respect to a variety of functions, while at the same time leaving open the question of which functions people choose to realize:

> Someone who has access to adequate nutrition can always fast for religious reasons: but there is a great difference between fasting and starving. Someone who dislikes leisure and play and prefers a workaholic life can choose that life: but this is a huge difference between that chosen life and the 'double day' (working a full-day job and then doing all the child care and domestic labor) that stops millions of women the world over from choosing leisure activities that help to make their lives meaningful. One may also refuse to vote or participate in politics, as the Old Order Amish do—but it would be quite another thing, and an assault on their fundamental political equality, to deny them the *capability* of voting.[47]

In this way, a capability account articulates a set of enabling conditions which, together with the threshold level of adequacy, invites a conception of public reason and social morality that can be broadly embraced among nations.

[45] Martha C. Nussbaum, "Capabilities and Human Rights," *Fordham Law Review* 66, no. 2 (1997): 273–300.

[46] United Nations, Development Programme, *Human Development Report 1993* (Geneva: United Nations 1996).

[47] Martha C. Nussbaum, "Political Liberalism and Global Justice," *Journal of Global Ethics* 11, no. 1 (2015): 68–79, at p. 71.

Is a Capability View Individualistic or Western Biased?

However critics might worry that a capability approach is Western biased or presents a view of public reason that makes implicit assumptions that many societies would reject. One concern is that since a capability view rests on a particular list of central human capabilities, some societies might regard the chosen list as not reasonably reflective of their way of life. For example, a capability list might be perceived as Western biased if it were overly individualistic, focusing too much on the capabilities and functioning of individuals and not enough on the capabilities and functioning of groups. As a result, it may not sit well with collectivist leaning societies. Another worry is that a capability list might be biased in ways some societies would reject if it were perceived as overemphasizing rationality and underemphasizing affective qualities, such as caring and solidarity.

In reply, on any interpretation of a capability view, some leeway is given to the specification of a list. Rather than strive for a "culturally neutral list," the aim is instead to specify lists in accordance with the way of life and conditions of the society and to continuously adapt them as these conditions shift. On some accounts, such as Sen's, capability lists are left entirely open for each society to choose; thus, Sen opposes "a cemented list of capability, which is absolutely complete (nothing could be added to it) and totally fixed (it could not respond to public reasoning and the formation of social values)".[48] Yet, at the same time, Sen allows that some central human capabilities figure in every list.

On the specific account I defend, multiple lists are possible provided certain normative constraints are met.[49] First, a capability list must be *balanced* between individual and relational capabilities and between cognitive and noncognitive capabilities. For example, emotions, affiliation, and play are primarily relational and noncognitive, whereas practical reason and thought are primarily individualistic and cognitive. The balance requirement ensures an even-handed specification of dignity that

[48] Amartya Sen, "Capabilities, Lists, and Public Reason," *Feminist Economics* 10, no. 3 (2004): 77–80, at p. 78.

[49] *EMB.*

can gain traction among people with diverse moral orientations. Second, capability lists must be *provisional*. This allows not only for the possibility that new arguments or objections may come to light, but also that the conditions of human beings or the environments in which they live might change in ways that impact what human beings can do and be. For example, genetic modification or degradation of the environment, may transform some of the central things that people can do and be. Third, an adequate capability list must be *life stage sensitive,* since what individuals can do and be shifts over the life course. The ability to move from place to place or affiliate with others, for example, is different for an infant than it is for an adult; capability lists must reflect this, rather than attempting to be life stage neutral.

Critics might also raise the concern that a capability view is narrow in the same kind of way that contractarian and contractualist versions of social contract are, namely, relegating those who lack rational faculties to the margins of the theory. If capability lists rest, for example, on an overlapping consensus among people with diverse metaphysical and religious conceptions, as Nussbaum claims,[50] this presupposes that people have the requisite intellectual capability to have a metaphysical or religious worldview in the first place. How would Sesha or others with significant intellectual impairment partake in an overlapping consensus?

In reply, although justifiability to all is retained on some versions of the capability account, when it is retained it occurs at a later stage in the argument. The argument starts with human capabilities. It is the idea of what we can do and be as human beings and what it means to respect these capabilities at a threshold level that drives the argument and underlies the conception of human dignity. Elsewhere, I defend in greater detail the claim that every human being who possesses some (at least one) of the central capabilities has full and equal moral standing and is owed the respect that dignity demands.[51] If this is right, then over time, the fact that these ideas are stable (for the right reasons) among those who have rational capabilities adds force to their justification.

---

[50] Martha C. Nussbaum, *Frontiers of Justice* (Cambridge, UK: Cambridge University Press, 2006), at p. 70.

[51] *EMB.*

A different kind of worry is that the capability view applies only to nations with governments that are liberal constitutional democracies. For example, Sen has argued that philosophical reasoning about justice is always democratic.[52] It is partly for this reason that Sen leaves the creation of capability lists to a later stage, as an evaluative exercise to be carried out by democratic decision-bodies. Reasoning along similar lines, Crocker argues that the capability approach needs to be connected to the theory and practice of deliberative democracy.[53]

In reply, while some interpretations do link justice and democracy, not all do. Siding with Nussbaum, the approach defended in this chapter is not merely a stimulus for public debate but also a substantive account of obligations of states to respect human dignity and to widen the scope of obligations to all citizens, including people with disabilities, such as Sesha, who are permanently incapable of the rational agency required for democratic deliberation. This is not up for a vote; rather, it sets parameters on how the state might act. The view I defend might be dubbed a "blended view," since it is partly procedural and partly substantive. The procedural aspect refers to the 'open-ended' feature of capability lists, which can be specified in different ways in different contexts. The substantive aspect refers to the fixity of the requirement that threshold capabilities must be reasonably supported as a matter of justice.

It could be argued that to the extent that a capability view relies on procedural methods that are democratic, it reflects a Western democratic bias. However, this concern misses the mark. Although the institutional structure of the contemporary practice of democracy is largely a product of the West, participatory governance itself has "surfaced and resurfaced with some consistency in different parts of the world."[54] People's participation in public reason is not quintessentially Western; instead, as I have argued, it is among the central human capabilities that human beings everywhere share. As Sen notes, democracy "gives expression to a

[52] Amartya Sen, *The Idea of Justice* (Cambridge, MA: Belknap Press of Harvard University Press, 2011), henceforth "*IJ*."

[53] David A. Crocker, *Ethics of Global Development* (Cambridge: Cambridge University Press, 2008).

[54] *IJ*, p. 323.

tendency in social living that has a much longer and more widespread history."[55]

It could also be argued that public reason is objectionable in the opposite way, namely, it is antidemocratic. Harbermas, for example, raises the concern that public reason fixes the content of public reason prior to any actual democratic debate.[56] Quong puts the challenge this way: "how can citizens engage in public reasoning about what the principles of justice ought to be if justice is also meant to be the basis for the grounds for their adherence to the practice of public reason?"[57] In reply, we might say that at a deeper level, what public reason requires is a prior commitment to civility, that is, a *duty of civility* that requires people who engage in public reason to

> explain to one another how our important political positions are justifiable by reference to a reasonable political conception of justice, and to refrain from supporting positions when we believe they can only be justified by appeal to a religious doctrine, or some other comprehensive doctrine that we cannot reasonably expect everyone to endorse.[58]

While this duty is restrictive, it is not unduly restrictive, since the duty of civility applies only to individuals in their capacity as citizens, such as voting or expressing views publicly on essential matters of justice. In daily life within a family, friendship circle, or religious group, individuals are not generally constrained by it.

Another way of specifying the commitment assumed in public reason is suggested by Ebels-Duggan, who argues that when we invoke capabilities in public reason, we rely on a lenient interpretation of what it means to be reasonable, which holds that someone is *reasonable* provided they recognize the existence of reasonable disagreement and aim to

---

[55] *IJ*, p. 323.

[56] Jürgen Habermas, "Reconciliation Through the Public Use of Reason," *Journal of Philosophy* 92, no. 3 (1995): 109–131.

[57] Jonathan Quong, "On the Idea of Public Reason," In *A Companion to Rawls* ed. Jon Mandle, David A. Reidy (New York: John Wiley & Sons, 2014): 265–280, at p. 274, henceforth "*IPR*."

[58] *IPR*, p. 265.

cooperate.[59] They need not reach an overlapping consensus, yet they must share a spirit of cooperation and mutual respect. According to this version of a capability view, public reason extends to diverse societies around the world where we find a pledge to respect persons.[60] This approach diverges by a longshot from traditional social contracting, since it bears the mark of a broader exchange, one that engenders trust among committed participants.[61]

In defense of this line of reasoning, it could be argued that the social contract alternative requires similar kinds of assumptions. Hampton, for example, argues that a Hobbesian interpretation of social contract presents a dilemma: if the precontractual state is a potential war of all against all, then this state is generated either by passions, such as greed and fear, or by rationality. If it is generated by passions, then parties will still be motivated by these same passions after a contract is drawn up and will renege on the contract; if the precontractual state of war is generated by rationality, then contractors will have no more reason to comply with a contract than they did to cooperate before it was made.[62] Commentators, such as Cudd and Eftekhari, take such arguments as showing that contracting is unable to motivate morality without some preexisting "natural" inclination to morality.[63]

6 Conclusion

When we do ethics in the public square, we appeal to public reason. Given a plurality of reasonable views, public reason incorporates a requirement of equal respect for persons. A capabilities approach to

---

[59] Kyla Ebels-Duggan, "The Beginning of Community," *Philosophical Quarterly* 60(238) (2008): 50–71.

[60] *MB*, p. 624.

[61] Anita Silvers, Leslie P. Francis, "Justice Through Trust," *Ethics* 116 (2005): 40–76.

[62] Jean Hampton, *Hobbes and the Social Contract Tradition* (Cambridge: Cambridge University Press, 1986).

[63] Ann Cudd, Seena Eftekhari, "Contractarianism," in *Stanford Encyclopedia of Philosophy* ed. Edward N. Zalta, (2000), at: https://plato.stanford.edu/archives/sum2018/entries/contractarianism/.

public reason guides us to render decisions that respect human dignity through reasonable efforts to support floor-level human capabilities. While this means that political principles and policies must be justifiable to those capable of practical reason, it also entails reasonable efforts to support all capabilities at a threshold level. The question a society ought to ask is not only, "What do people think?" but also, "What can people do and be?" Appealing to capabilities carries advantages over appealing to contracting, because it includes all human beings as equals and can be used on the global stage in tandem with (or separate from) the language of international human rights. Rather than deriving individuals' worth and dignity from their capacity to reason, a capability view paints a fuller picture of shared humanity and recognizes the human dignity of those who need its protection most.

13

Public Reason and the Right to Healthcare

Michael Campbell

In the following essay I consider the prospects for deploying the concept of public reason (PR) in settling practical bioethical questions, focusing in particular on entitlements to healthcare. I begin by tracing the origins of the concept of public reason to the aspirations of the liberal political theorist to find a justification for the authority of government, which reconciles a basic belief in the autonomy of the individual with the legitimacy of the coercive institutions that create and govern the public sphere. I then consider how the concept of PR may be used in order to justify a universal entitlement to healthcare, paying particular attention to the work of Norman Daniels. Finally, I briefly consider the objection that an entitlement to healthcare grounded on such terms is insufficiently robust, because it downplays the values that healthcare encodes. I conclude by

I would like to thank Alastair Campbell, Nancy Jecker, and in particular Hon-Lam Li for helpful comments on previous versions of this chapter. I would also like to thank the participants at the Public Reason and Bioethics seminars held at the Chinese University of Hong Kong for helpful discussions of these issues.

M. Campbell (✉)
Department of Philosophy, University of Kyoto, Kyoto, Japan

H.-L. Li, M. Campbell (eds.), *Public Reason and Bioethics*,
https://doi.org/10.1007/978-3-030-61170-5_13

reflecting on whether the concept of PR can be stretched to include a more robust conception of the value of health.

The increasing sophistication of biotechnology poses acute challenges for policymaking. When innovations are controversial and hard to understand, their effects unpredictable, and their benefits and risks unevenly spread, a stable consensus on matters of ethics, law, and policy can seem all but impossible to achieve. Developments in fields such as genetic engineering and stem cell research have highlighted the way in which novel technologies may have widespread and unforeseen effects on the well-being of both present and future generations. These challenges are made even more pressing by the forces of multiculturalism and globalization. Health tourism, the transnational spread of communicable diseases, and the effects of global heating show the globalized nature of medical challenges in the twenty-first century, and so the need for international cooperation within a shared ethical framework.

Many of these issues have come to a head in the recent SARS Covid-19 global pandemic. In under a year, this disease has caused massive upheaval across the world; killing over one million people, placing healthcare systems under extreme stress, and causing the virtual shut down of economies. Although still in its infancy, this crisis has already exposed fault lines in global decision-making bodies. The World Health Organization, in particular, has come under sustained criticism for perceived defects in its initial response to the outbreak.[1] Moreover, in the race to secure sufficient supplies of personal protective equipment and to develop testing and treatment capacity, several major countries have indicated their willingness to place the needs of their own people above those of other countries; acts which threaten the possibility of a coordinated effort to tackle the spread of the disease. In these and other ways, the Covid-19 crisis risks exacerbating a general slide away from cooperation and sustainability, and toward self-interest and short-termism.

In the light of these troubling developments, the current crisis represents an opportune time to reconceptualize, reinvigorate, and reiterate a stock of common moral and political ideals concerning fundamental

[1] "Coronavirus: WHO defends coronavirus outbreak response" https://www.bbc.com/news/world-52506844 accessed on 17/11/2020.

matters of justice. It has been suggested that the political philosophy of liberalism is the ideal framework in which to undertake such a consensus-building process. Yet, evaluating this claim is made difficult by the fact that 'liberalism' is a hotly contested term.[2] Since its classical formulations in the work of thinkers such as Hobbes, Rousseau, Locke, and Hume, it has undergone multiple phases of development, growing in complexity and sophistication and spawning its own sub-genres and internecine pursuits.[3] This complexity most likely reflects liberalism's proximity to power, and its corresponding need to make accommodation for the variable needs and interests of the governing classes. The range and diversity within liberal thought means that liberalism can remain hegemonic even while conflicts play out between different governing tendencies—as, for instance, in the constant play off, within democratic discourse, between Lockean individualism and Hobbesian authoritarianism.[4]

Despite these complexities, there is some justification for the belief that liberalism provides a framework in which controversial ethical issues can be, if not solved, then at least resolved to the (at least partial) satisfaction of people with different moral outlooks. After all, the cornerstones of liberal doctrine are toleration and consensus building. And liberalism does seem to have a promising history when it comes to the mediation of conflict—the advent of liberal political theory coincided with a decrease in religious strife and intolerance within Europe, and the basic liberal ideas of the neutrality and accountability of our governing institutions

[2] In this chapter, single quotation marks do not indicate direct quotations but rather words being mentioned or being held at some slight distance. Double quotation marks indicate a direct quotation.

[3] The literature on liberalism is vast. Alongside the work of John Rawls, important modern texts include: Ronald Dworkin, *Sovereign Virtue* (Cambridge, MA: Harvard University Press, 2000); Joseph Raz, *The Morality of Freedom* (Oxford: Clarendon Press,1986); Michael Sandel, *Liberalism and the Limits of Justice* (Cambridge: Cambridge University Press, 1982); Amartya Sen, *The Idea of Justice* (Cambridge, MA: Belknap Press of Harvard University Press, 2009); and Michael Walzer, *Spheres of Justice* (NY: Basic Books, 1983).

[4] The former tendency is reflected in appeals to individual rights and liberties and the supposedly ennobling union of these in the collected 'will of the people' as expressed in legitimate government; the latter reflected in appeal to sovereign power and of a single individual as standing above the law and wielding its disciplining apparatus through displays of force. Contrast Obama's rhetoric with the infamous "torture memos" of John Yoo. See Jack Goldsmith, *The Terror Presidency: Law and Judgement Inside the Bush Administration* (New York City, New York: W. W. Norton 2007).

has shown promising results as a framework for the development of a norm-based global order.[5]

Liberalism's successes in this regard stems from its foregrounding of a conception of the individual that, like Goldilocks' third bowl of porridge, is a palatable alternative between two undesirable outcomes. The liberal conception of the rights and responsibilities of persons aims to be neither overly demanding (encoding values that a significant number of people find oppressive) nor insufficiently robust (leaving people's attitudes unchecked to the point of tolerating general conflict). For liberalism, the goal of a justificatory account of the origins of state authority is to show both that and how it is possible to have a political community where each person can live under the rule of law without compromising their capacities to formulate and to act under their own conception of the good.[6] Accordingly, an ideal political arrangement is one in which coercion has given way to constraint; any time a citizen is forced to act against their will, considerations will be available which will show the restraint to be justified in the light of our shared standards of reasonability.[7]

---

[5] See, e.g., Gillian Brock, *Global Justice: A Cosmopolitan Account* (Oxford: Oxford University Press, 2009) and Martha Nussbaum *Frontiers of Justice* (Cambridge, MA: Harvard University Press, 2006). For a critique of an attempt to found norms of global justice on a narrow Hobbesian conception of reciprocal self-interest, see Michael Campbell "Global Greed and Prudence: Reply to Alvarez", *Asian Bioethics Review* 6, no. 1 (2014): 83–95.

[6] On this picture, laws function to shape the deliberative space within which individuals operate, providing disincentives for certain courses of action (i.e., punishments for law-breaking). One can reasonably choose to break the law so long as one is willing and able to face the appropriate punishment; the law is then arranged in such a way as to make the costs of the punishments generally outweigh the expected benefits of the transgression. In this way, the law can enter into deliberation as a consideration to be weighed against others, and need not be seen as a threatening, antagonistic, or dangerous force. However, a calculation as to whether or not to obey a given law must include the fact that punishment is not merely a material cost (time, money, opportunity, etc.) but a social one as well. Since laws are an expression of social consensus, breaking the law jeopardizes one's standing with respect to society.

[7] See for instance a liberal justification of punishment, according to which punishment serves to educate and rehabilitate criminals, as well as to declare our shared commitment to the laws which govern our interactions with each other. On this view, it is unreasonable to object to a punishment handed down by the court system within a legitimate state absent special conditions. Such conditions may include (1) a mistake within the justice system (e.g., wrongful conviction), (2) an unjust or otherwise corrupted justice system (e.g., rightful conviction according to an unjust law), or (3) that one was not suitably related to the legal system under which one is now being punished (e.g., being punished for acts committed while outside of the country's legitimate sphere of influence).

Since liberalism was born out of the necessity of finding a common set of norms by which to mediate conflicts between different religious groups, the conception of the individual that it encodes is one on which the freedom of the individual to form and pursue their own conception of the good is emphasized. Thus, the basic framework of liberalism involves a sharp delineation between the authorities of faith and civil authority.[8] According to liberalism, authority concerning matters of belief may be vested by an individual in a particular institution, but the power which is so vested must be derived from a voluntary act of will on the individual's part, and so can be withdrawn. As such, the authority which spiritual or moral authorities wield must be limited, both in scope, in that the proclamations of a church can apply only to those who choose to subscribe to its tenets, and in power, in that such proclamations cannot be backed by the use of coercive means.

The limits which apply to faith or interest based associations stand in contrast to the authority of the government. Governmental institutions may issue commands that bind on individuals simply by virtue of their identification as a member of the *res publica*—a form of association whose membership, though somehow related to the free will of its subjects, is fixed without reference to an act of direct consent on their part. These commands may be enforced even if this requires suspending or curtailing the authority individuals have over matters concerning their own person.[9] Moreover, the effective operation of the state depends on a presumption that its authority persists across generations. Therefore, unlike voluntary organisations which must engage in a constant struggle to retain adherents, governmental authority possesses a degree of warranted ossification; though each generation must be allowed some freedom to think and renegotiate the social contract, its basic structure must remain relatively fixed and stable across time.

[8] For the canonical formulation of this point, see J.R. Milton & Philip Milton (Eds), *John Locke: An Essay Concerning Toleration* (Oxford: Oxford University Press, 2009).

[9] This is where liberalism differs from individualist anarchism according to which political authority is never justified, since the ultimate legitimacy of any command lies in the free choice of the individual; see Robert Paul Wolff, *In Defense of Anarchism* (Berkeley, CA: University of California Press, 1970).

Thus, liberalism, in this sense, is the conviction that the ideal form of collective association for human beings is one in which the state plays a determinate but carefully proscribed role. For liberalism, the concentration of coercive and regulative power within a government is justified in the first instance as being the only way to solve otherwise insurmountable coordination problems (which, in extremis, mean war and instability), and only insofar as the government is restricted in the scope of its ambitions to securing (a publically justified conception of) the common good.[10] The state is introduced and justified in order to avoid the twin dangers of oppression, where a dominant group uses force or the threat of force to impose a particular worldview on others, and 'anarchy', where each group must use force to defend its own worldview against others.

Between these two poles is a range of possible ways of life, different arrangements of the social world, which encode different values. Liberalism is designed to be neutral between such particular arrangements, so long as they ensure the freedom of each individual to develop and pursue his or her own conception of a valuable life. In this way, the liberal is not a skeptic about the possibility of objective value in the political sphere. Rather, the characteristic mark of a liberal is the conjunction of two beliefs; firstly, that there is a fundamental indeterminacy within our knowledge concerning the order of human goods, and secondly, that this indeterminacy is relevant for determining the appropriate form of political association for human beings. Therefore, whereas a non-liberal believes that some particular ordering (or set of orderings) of such goods justifies a particular form of political organization reflecting that overall order, a liberal believes that the lack of any secure knowledge concerning such a particular ordering justifies a particular form of political organization, one which in turn reflects our absence of certainty in any overall order.

According to the liberal, no comprehensive ranking (or set of rankings) of the different goods which human life may contain can be secure enough to justify its being backed by coercive institutions.[11] This claim

[10] For an influential game-theoretic account of these coordination problems see David Gauthier, *Morals By Agreement* (Oxford: Oxford University Press 1986).

[11] The caveat 'set of orderings' is necessary because even a theorist who believes that political structures should be instrumental for the realization of lives of a certain kind may accept that there is a

may be justified by appeal to one of a range of further views, as follows: (1) that no such set of orderings exists; or (2) that the set of appropriate orderings taken together would not yield a coherent state of affairs if all were backed equally by coercive institutions; or (3) that due to human fallibility, our beliefs concerning such an order cannot be certain enough to justify the imposition of that order on others whose own views differ from our own. Whichever position one prefers, the key idea here is that there is a certain irreducibility of the individual's perspective in determining the conditions under which their life counts as going well, that this entails a plurality of reasonable views of the good life, and that our institutions must be arranged in ways which respect this basic fact.[12]

This irreducibility of the individual's perspective is usually expressed by saying that individuals have a distinctive power, that of 'autonomy', and that this power has a distinctive kind of value. Autonomy is the ability that each individual has to determine his or her own understanding of what kind of life is worth living, and to use his or her best efforts to bring it about. The protection and promotion of this ability (and conditions which are necessary for it, such as physical security) is the prime directive of the liberal state, with any restrictions to individual self-determination to be carefully and robustly justified. According to JS Mill's classical formulation, we are each engaged in our own experiment in living, and the State's role is to provide a shared space in which such experiments can effectively be carried out.[13]

Rules and institutions, both implicit and explicit, govern this common space. Howsoever we are to determine both that set of rules, and the

---

tolerable degree of variance within a human population concerning the rankings of such goods. Even strict natural law theologians accept that certain groups (e.g., monks) may justifiably live ways of life which are at odds with certain canons of the ethical theory which underlies the political institutions. Such diverges would be unjustifiable if generally adopted, but can be justified in particular cases through special pleading borne out in the terms of the theory as a whole.

[12] On the question of whether we ought to be skeptics about the ordering of goods, or simply adopt an attitude of epistemic humility toward this question, see Thomas Nagel, "Moral Conflict and Political Legitimacy," *Philosophy and Public Affairs* 16:2 (1987); Charles Larmore, "Pluralism and Reasonable Disagreement," *Social Philosophy and Policy* 11:1 (1994); and Rawls' discussion of the "burdens of judgement" in *Political Liberalism* (henceforth "*PL*"), (New York City, NY: Columbia University Press, 1993), p. 487.

[13] See J.S. Mill, *On Liberty*, vol.18 of J. M. Robson (ed.), *Collected Works of JS Mill* (Toronto, University of Toronto Press, 1977), pp. 260–267.

character of those institutions, they must reflect both the basic principles which make common life possible, as well as the basic values which are encoded in such principles. Such values include safety, autonomy, privacy, cooperative enterprise (including commerce, industry, and collective enquiry), and (most controversially) the formation of an ongoing participative collective identity.[14] These are the foundational goods which are at least common to all different worldviews and which may be justified as necessary preconditions for the development of worldviews in the first place.

Because we live in a world of inequality and scarcity, universal and unlimited access to such foundational goods is not guaranteed. There is not even a guarantee that these goods can be made consistent with each other—trade-offs, for instance between safety and autonomy, may be inevitable.[15] As a result, even liberal states must make decisions which are guided by considerations other than simply the promotion and protection of autonomy. After all, as well as making room for competing worldviews, society must be both sustainable and stable across generations; contra Hobbes, people do not spring from the ground like mushrooms and so we must make provision for the nurturing of people into adulthood.[16] Thus, any society which makes room for a plurality of such ways of life will still have to adjudicate between claims when these conflict. For instance, if a certain way of life is resource intensive then decisions will have to be made as to how far such a way of life can be permitted and how much it must be regulated or constrained. Certain ways of life may end up dying out, either because they make unreasonable demands on the common weal or because they simply fail to adapt to the changing tastes

[14] See Hobbes' description of the goods which the social contract enables; Michael Oakeshott (ed.), *Thomas Hobbes: Leviathan* (Oxford: Blackwell, 1948 [1651]) ch. XIII; See also Judith Shklar, "The Liberalism of Fear," in *Liberalism and the Moral Life, ed. Nancy L. Rosenblum* (Harvard, MA: Harvard University Press, 1991).

[15] Plato believed that once a society takes its goals anything beyond the provision of the bare necessities of life, that strife and war become inevitable. The point is well taken, even if Plato's own list of luxuries (furniture, perfumed oils, incense, pastries, and prostitution) seems idiosyncratic. See *Republic*, 372e–373a.

[16] Richard Tuck and Michael Silverthorne (Eds), *Thomas Hobbes: On the Citizen* (Cambridge: Cambridge University Press, 1998 [1642]) p. 102.

and interests of people over time. As Rawls memorably puts it, echoing Isaiah Berlin, "there is no social world without loss."[17]

This point is of more than merely academic interest. Indeed, one of the lessons to be learned from the Covid-19 crisis, not to mention the multiple coronavirus outbreaks of the past decade, is that a prudent economic system cannot afford to chase economic growth at the expense of sustainability, even if such a pursuit seems to promise the most effective way to promote autonomous ways of life for current generations. Pestilence, like famine and the mass migration of people, are foreseeable effects of global heating, which is itself a foreseeable effect of a global system that is governed by the logic of capitalism, which dictates an inexorable and constant rise in both supply and demand. At its minimum, the idea of sustainability is encapsulated in the idea that a sustainable way of life leaves intact for future generations those common goods that are necessary for the enjoyment of human life.[18] By specifying such a range of goods we start to specify a realm of entitlements which transfer across generations and which form an unalterable background to different ways of life.[19] Such entitlements define what may be called a 'common space', a natural and social sphere to which each person has equal rights of access and enjoyment.[20]

It is in the process of negotiating the nature of this common space that the concept of PR enters. PR encapsulates the conviction that by adopting a certain discipline in discourse over the moral or political rules that regulate our common life, we will be able to justify them to all concerned parties.[21] Despite controversies over how it is to be best understood, the

---

[17] *PL*, p.197.

[18] Of course, such a heuristic is made problematic by a number of factors, including that the technological progress which comes at the expense of certain natural resources also changes the kinds of conditions under which we can live; nuclear energy would not have been possible without burning fossil fuels, but its invention has enabled us to generate energy in degrees hitherto impossible.

[19] One such attempt is encapsulated in Nussbaum's list of capabilities; see Martha Nussbaum, *Creating Capabilities* (Cambridge, MA: Harvard University Press, 2011a), and discussion in Nancy Jecker's contribution to this volume.

[20] For a discussion on issues surrounding this see Michael Otsuka, *Libertarianism Without Inequality* (Oxford: Oxford University Press, 2003).

[21] See Jonathan Quong, "On the Idea of Public Reason," in *The Blackwell Companion to Rawls*, ed. J. Mandle and D. Reidy (Oxford: Wiley-Blackwell, 2013), pp. 265–280. From this starting point debates then proliferate, with arguments over such matters as how we are to draw the boundaries

core thought in PR can be expressed as follows: when one puts forward a principle for governing the behavior of others, that principle must be expressed and justified in terms which do not depend on underlying premises upon which it would be unreasonable to expect universal agreement.[22] Our search for common intelligibility is to proceed by a process of stripping out of the public sphere considerations that are dependent on any controversial worldview. The guiding image is of the public citizen as 'thinner' than the 'thick' concept of personhood relevant to the private sphere. PR therefore encapsulates the idea that, by imposing restrictions on the form and content of our shared public deliberation we can come to a minimal outline which can serve as a common ground on which to build the structures of society.[23]

PR is valuable in part because it enables us to stand in relations of civic friendship with each other.[24] Rather than living in a society where individuals with different worldviews merely tolerate each other's presence, through making sure that our institutions are publically justified we can instead coexist in relations of mutual respect. It is, in other words, a political manifestation of the possibility of creating friendships with people who hold radically different values from our own. In the creation of public offices structured around this concept we bring into existence institu-

of our 'common life', how broadly we should understand these 'moral or political' rules, and what we are to take 'justifiability' or 'acceptability' to the governed to require. For a broader interpretation of the principle, see Gerald Gaus, *The Order of Public Reason: A Theory of Freedom and Morality in a Diverse and Bounded World* (Cambridge: Cambridge University Press, 2011). For a narrower account see Jonathan Quong, *Liberalism Without Perfection* (Oxford: Oxford University Press, 2011).

[22] Needless to say, beyond this minimal formulation, the concept remains hotly contested—and the contributions in this volume are a testament to the degree to which PR remains as controversial as the (apparently) 'first order' issues such as medically assisted suicide which it is invoked to solve. This shows, I believe, the fragility of the distinction between first order and second order issues when it comes to matters of ethics and politics. When issues of fundamental importance to a way of life are at stake, then disagreements over matters of immediate choice almost invariably spiral into debates over metaphysics, metaethics, meta-metaphysics, and so on. Different ethical convictions may even underwrite different conceptions of the nature of reason itself. For instance, someone who sees reason as a faculty designed for the articulation of insights gleaned through faith will approach practical questions differently from someone who sees it as sui generis and not beholden to any standards or supposed insights beyond itself.

[23] Importantly, these structures include regulative principles which govern the way that the society evolves over time, through standards of democratic accountability and for measuring progress.

[24] *PL*, p. 447.

tions which allow us to meet each other in a place where all are welcome and where we can lower our guard.[25] PR in this way represents an institutionalization of hospitality relations; the public sphere is a space to which each person has an equal entitlement and in which each person should feel equally at home.

However, although PR stresses tolerance, consensus building, and governance by consent, the method by which consensus is reached is not uncritical and not all possible social arrangements can pass the PR test. On this conception a just society is one which is formed through collective deliberation by equals, with an eye to securing the good of autonomy, as well as the further goods which can be derived from it. As a result, there is an obligation to ensure both that the institutions which regulate our common life are sensitive to the results of such deliberation, and that each person has the means necessary to participate in such a process.

This obligation may provide us with a starting point for the derivation of some substantive requirements on how society should be arranged. Although generally confined to more abstract matters, there is a small but growing body of work (to which this volume is a contribution) that looks to PR to shed light on matters of practical concern. In what follows I will examine in broad outline one such attempt, namely an attempt to derive an entitlement to healthcare from the requirements of PR. I will then consider some objections to such a justification, in particular, pressures which push us toward a more substantial—hence more controversial—conception of the value and importance of health, hence of the scope and force of our entitlements to healthcare.

A justification for an entitlement to healthcare may be made to turn on the connection between good health and the ability to formulate and pursue one's own conception of the good. This point has been made by Daniels, who has shown how the right to health can be derived from the requirement that society be a system which includes a reasonable balance

[25] The fact that each person is entitled to both physical security and respect while in government offices is something which we may take for granted, but it is a fragile achievement. The murder of Jamal Khashoggi in the Saudi consulate in Istanbul in 2018 provoked an international uproar in part because it struck at basic tenets of the rule of law upon which the global order is based, including the security of persons within government buildings. See "Inquiry into the Killing of Mr. Jamal Kashoggi." https://www.ohchr.org/EN/Issues/Executions/Pages/Inquiry.aspx accessed 17/11/2020.

of both fair cooperation and fair competition between people. Ensuring that each person has the ability to pursue their own conception of the good requires that people be given equal and ample opportunity to make their own decisions regarding the lifestyle they will choose to pursue, as well as the people with whom they will associate.

Such matters may be captured under the rubric of securing 'equality of opportunity'. One advantage of focusing on equality at the level of opportunity (rather than in terms of particular goods) is that it allows us to prescind, to some degree, from taking a stand on what goods people ought to value.[26] Some goods will nevertheless still be universal entitlements, by virtue of the fact that certain things (e.g., the elements on Maslow's hierarchy of needs) are necessary whatever one's particular conception of the good life.[27] In this way, the opportunity ranges open to individuals can be used as a metric for measuring their relative freedom, the chance people have to find a way of life which expresses their own particular conception of what is valuable.[28] Health can then be shown to be instrumentally valuable, in securing the connection between social justice and the preservation, promotion, and relative equalization of, opportunity ranges. As Daniels puts it:

---

[26] Another advantage is that equality of opportunity may be less demanding than equality of goods themselves, insofar as it permits inequality in outcome; we can tolerate unequal access to a given good so long as everyone had the same chance to enjoy the good in question at a relevant time. I set this aside in what follows but for discussion see G.A. Cohen, "On the currency of egalitarian justice" *Ethics* 99, no. 4 (1989): 906–944.

[27] See A.H. Maslow "A theory of human motivation" *Psychological Review* 50, 4 (1943): 370–96.

[28] An aside: Any discussion of opportunity ranges needs to take into account the fact that such ranges are both dynamic (changing over time) and nested (such that taking or failing to take a certain opportunity will change the opportunities open to one). A theory which is interested in equality of opportunity must look not only at the range of alternatives open to a person on a given occasion, but also such factors as what the ramifications of a certain decision are for the rest of the person's opportunities; how easy is it to correct a mistake, for instance. Thus, although giving people more opportunities to gamble may increase the range of options open to them on a given instance, it may actually harm their opportunity range overall. These and related cases fall within the rubric of what may be called 'coercive options'—cases where giving people more options can actually be harmful rather than beneficial. Consideration of such cases is crucial for a proper understanding of the relations between choice, freedom, and well-being. In general, people need to be given not only the opportunities to succeed, but also the support and security to be able to take chances, fail, and start again. This makes matters considerably more complex than is often presumed in appeals to equality of opportunity which do not consider what it is like to make decisions under conditions of extreme precariousness.

> The loss of function associated with disease and disability reduces the range of opportunities open to us compared to what it would be were we healthy or fully functional. By keeping people functioning normally, we protect their range of opportunities. If we have a social obligation to protect opportunity in this way, then we have a general framework for thinking about justice and health.[29]

The value of health is thus tied directly to the basic structure of society, as institutions that serve the ends of health must be regulated according to their role in determining whether social cooperation and competition are being conducted on a level playing field. This argument makes explicit appeal to Rawls' liberal framework for the equitable distribution of goods according to the principles encoded in a conception of society as a system of cooperation conducted by free and reasonable individuals. Daniels' sophisticated work refines Rawls' list of the basic goods with the important observation that health is, like wealth and income, one of the 'all-purpose means' needed in order to have a reasonable chance to determine one's own position within society. Ill-health limits the opportunities which are open to a person, sometimes drastically; it distracts attention, drains energy, and compromises agency. Thus, any society which aims to create a baseline of equality of opportunity between individuals should aim to achieve a social baseline of health. Just as an intelligent but impoverished person is a victim of injustice if they are not given the opportunity to leverage their intelligence into a career, an intelligent but diseased person is similarly wronged, if they are deprived access to the means to alleviate the burden of ill-health.

Daniels' account of the connection between healthcare and justice goes through a Rawlsian conception of justice as a matter of securing a baseline of equality in basic goods and opportunities. However, it is also possible to reformulate the argument with direct appeal to the concept of PR. In this case, a justification for healthcare will turn on the necessity of good health for being able to participate in deliberation over political matters. If, for a state to be justified, its basic structure must be justifiable to those who live under it, then individuals must enjoy the conditions

[29] Norman Daniels, *Just Health: Meeting Health Needs Fairly* (Cambridge: Cambridge University Press, 2007), p. 21, henceforth "*JH*."

which allow them to assess the state's arrangements in a balanced way. Without this precondition, the idea of 'consent of the governed' will become hollow. Chronic pain, or mental or physical impairment, shape one's deliberative field, informing calculations of reasonability and tolerability, and determining how costly (or cost-effective) it is to participate in political debate and decision-making. Just as people can be constrained by "the pressure of an inferior political or social position" to accept terms of mutual cooperation which they otherwise would not, people can similarly be constrained by pain or physical insecurity.[30] In this way, full participation in the political process—and hence a robust justification for the state's authority—requires that people are enabled to live healthy lives free from being overwhelmed by the burdens of ill-health.

This provides us with an intuitively compelling way to ground social obligations concerning the distribution of healthcare resources, as well as, more broadly, equitability in distribution of the social determinants of health. By connecting health outcomes to opportunity ranges, we have the first steps toward the creation of metrics that will enable us to determine the overall levels of justice and equality in a society with greater precision than the rough tool of measuring income and wealth inequality. Although there will always be some indeterminacy in the specification of opportunity ranges, the narrowness of the evaluative base for the value of health ensures that in the formulation of such metrics, empirical observation can largely take over from appeal to unstructured intuitions.[31] Moreover, by justifying an entitlement to healthcare without reference to controversial premises concerning the good life for human beings or our responsibilities to others, it promises a route to a set of core convictions surrounding the value and proper distribution of health which can be accepted by reasonable people irrespective of their particular moral or political outlooks.

---

[30] *PL*, 446. The reference in this formulation to 'social position' suggests also that we have to adopt a critical attitude toward positional goods, such as the "offices and positions of authority and responsibility" which Rawls mentions as basic entitlements; see John Rawls, *Justice as Fairness: A Restatement* (Harvard, MA: Harvard University Press, 2001), p. 58. but that is a matter for another day.

[31] In particular, indeterminacy arises with the weighting of opportunities; e.g. how are we to value the opportunity to own a sports car?

However, the narrowness of the base by which entitlements to healthcare are justified in this version of Daniels' system is a weakness as well as a strength. In making the base for the entitlement to health narrow, it threatens a corresponding narrowing of the scope of the entitlement to health, as well as a marginalization of the experience of ill-health in our intuitions concerning what makes it unjust.[32] In what follows I will briefly elaborate on this difficulty.

The value of health goes beyond its instrumental value in securing fair cooperation between equals. Arguably, someone who suffers from a physical ailment which could be relatively easily remedied by a third party suffers an injustice if they are not given the help of which they are in need. This is the intuition which drives Peter Singer's claim that a passer-by has a duty to rescue a drowning child, if the rescue can be effected at minimal cost and risk to the rescuer.[33] According to Singer, both the ground and the force of the obligation are derived from the cost/benefit ratio in the particular circumstance—the mere fact that a life can be saved at a minimal cost shows that there is an obligation on the passer-by to help, and the child, should he or she not receive help, can be said to be a victim of an injustice. No reference needs to be made here to some pre-existing social contract, or to the justifiability, in general terms, of rules of mutual assistance as enshrined by the law of the particular society.

Singer's case strikes a chord with us, though we should I think be skeptical of his attempt to characterize the obligation on the passer-by as being solely or even partially a matter of the cost-benefit ratio of the alternative outcomes in the situation. Rather, the appeal of the case is derived from the intuition that the relevant ground for requirements of justice is sometimes set by nothing more than our being moral beings, responsive to the needs of those around us. Our conviction concerning the moral importance of rescuing a distressed child is connected to such

---

[32] In the context of the case of Medical Assistance In Dying (MAID), this comes down to the question: does the entitlement to MAID stem from the right of the individual to have control over when and how they exit the world (and so the social contract), or does it stem from the demand that we provide individuals with access to the means necessary to address unbearable and avoidable suffering, when it is in our power to do so? If the answer is both, the question then remains as to how these two considerations are related to one another.

[33] Peter Singer, "Famine, affluence, and morality" *Philosophy and Public Affairs* 1, no. 3 (1972): 229–243.

things as our knowledge of the vulnerability of children, our understanding of what it would be like to drown, our imaginative awareness of the hurt felt by a family and a community when a young life is lost in this way, and our sense of our shared mortality.[34] This is not to say that our obligations to others, and the relevant sense of 'community' to fix questions of justice, can always be set simply by appeal to such intuitions. But it does suggest that sometimes we owe things to others based on facts which are prior to calculations about what is required by a political conception of community. In certain cases at least, the limits of community may be determined by our obligations to others, rather than vice versa.[35]

Developing this thought points toward a more robust conception of the injustice faced by those within society who languish without adequate healthcare and who suffer unnecessarily with treatable conditions. Such individuals suffer arguably a two-fold injustice; not only that of being deprived of the means to participate in society as equals, with access to a reasonable share of the 'all-purpose means' to their advancement, but also, more simply, the injustice of suffering with a preventable or treatable condition. Suffering itself is isolating, and knowledge that the suffering could be ameliorated by those around you, but is not, increases the sense of isolation and abandonment which forms part of the burden of ill-health.[36]

In this way, universal healthcare is both a precondition of justice and a manifestation of it, for medical institutions both express and enable our sense of community. Thus, the guiding ethical tenets of medical practice refer to the obligation to treat the needs of others, irrespective of extraneous considerations such as social position, financial means, and so forth. 'To each according to their needs' characterizes a core feature of the medical imperative, in that it describes the attitude with which a doctor may treat a patient, seeing their role as that of an ally to health (including mental health) and an enemy of illness and disease.

[34] See Cora Diamond, "The Importance of Being Human", *Royal Institute of Philosophy Supplement* 29 (1991): 35–62.

[35] See Raimond Gaita, *A Common Humanity: Thinking About Love and Truth and Justice* (Abingdon: Routledge, 1999).

[36] See Elaine Scarry, *The Body in Pain: The Making and Unmaking of the World* (Oxford: Oxford University Press, 1985).

Although I describe this attitude as a 'core feature' of medicine, it is a relatively modern achievement that it should be thought of as such. Such a claim can only seem truistic to those whose ethical intuitions have been shaped by an understanding of human beings as having a basic equality, such that, so to speak, there is no moral distinction between the medical needs of a prince and those of a pauper. Reflecting this conviction, Bernard Williams described the claim that medical treatment ought to be distributed on the grounds of need a "conceptual truth". He felt the point so obvious as to need no further justification.[37] His confidence in this point was no doubt a product of his having been brought up within a system of universal healthcare. That the conviction is not universally shared was made evident when Robert Nozick famously queried Williams' claim in *Anarchy State and Utopia*. In arguing that the distribution of healthcare should follow purely market forces, Nozick asked (rhetorically) whether Williams would agree that if medical services ought to be distributed according to medical need, then barbering services ought to be distributed according to barbering need as well.[38]

Nozick's objection is weak, but he does at least draw attention to the fact that this distributional principle cannot be taken as a given—for most of the history of medicine, the distribution of medical treatment has been profoundly inegalitarian, but is, for all that, still recognizably medical treatment. If it is a conceptual truth that healthcare ought to be distributed according to need, then that is so only because our concept of medicine is a morally loaded one, and so the triviality of the claim will only be apparent to those who are sufficiently attuned to the relevant moral facts. The justification of the distribution of medical treatment in terms of need must go through an understanding of multiple interrelated facts; the importance of health in a person's life, its importance for a person's ability to participate as full members of a community, and what it says of our community's values when we treat the suffering of all as of equal importance.

Medicine's connection to these values gives healthcare professionals a correlative self-image; the egalitarian principle which medicine encodes

[37] Bernard Williams, "The Idea of Equality" in his *Moral Luck: Philosophical Papers 1973–1980* (Cambridge: Cambridge University Press, 1981).

[38] Robert Nozick, *Anarchy, State, and Utopia* (NY: Basic Books, 1974), ch. 7.

enables healthcare workers to see themselves in heroic terms and their professions as noble ones, and leads to genuine expressions of empathy and gratitude toward such professionals from the public at large.[39] Although this conception of medicine is not without its pitfalls, the self-image which it fosters provides psychological benefits to both healthcare professionals and patients. On the professional side, it provides a framework which allows healthcare workers to find meaning in their work which sustains them against the unique psychological challenges which it presents (challenges which include an ever present threat of burnout, PTSD or moral injury). On the patient side, the belief that medics are guided only by concern for the protection of health enables a special kind of bond between patient and doctor, and a special kind of trust and reassurance originating from it.[40] Our vulnerability in the face of ailments is a common human trait, and when our institutions are responsive to these facts then we cement and amplify our sense of each of us as members of a community, equal in the face of our shared mortality.

However, in their reference to shared value systems and the distinctive benefits which systems enable, these reflections take us beyond a conception of the value of medicine tied to purely instrumental considerations concerning the equalization of opportunity ranges. To make sense of what makes healthcare important, it seems, we must investigate what it would mean to live in a society in which healthcare is equally distributed, as well as what role the conviction of equality in health plays in our structures of meaning-making. I have tried to suggest that this will take to us richer (and therefore more controversial) evaluative judgments than those appealed to in Daniels' framework.

Related to this is a worry that the conception of society encoded in the notion of PR is problematically idealized. When we speak of society as a system of free and equal citizens working in a system of social

---

[39] See in this context the monthly 'clap for carers' which became an impromptu British tradition during the first wave of the coronavirus pandemic. See "Coronavirus: Health workers clapped across the world for battling on the COVID-19 frontline." https://www.euronews.com/2020/03/24/coronavirus-health-workers-clapped-across-the-world-for-battling-on-the-covid-19-frontline Retrieved 17/11/2020.

[40] The importance of such trust is brought out most starkly in contexts where it has been eroded, for instance in contexts where doctors appear to be influenced by financial imperatives.

cooperation over generations, we represent social relations in terms that elide the pervasiveness of dependency, vulnerability, and inequality in human life. As Susan Okin has pointed out in the context of a critique of Rawls' *Theory of Justice*, much work in political theory tends to elide the reality that we each enter the world as entirely dependent on others for our survival and flourishing. The ideal of the generic human being which Rawls takes to underlie the ideal citizen relegates to the margins anyone whose participation in debate is necessarily shaped by their experience as marginalized or victimized.[41]

As an objection to the Rawlsian method in political philosophy in general this is not (and is not intended to be) decisive, but it does call into question the role that idealizations ought to play in guiding our thought about justice. For our purposes, the objection highlights what is lost in Rawls' simplifying assumption that the subjects of a theory of justice are all healthy, and so what it would take to correct that assumption. We may be tempted to think that the role of medicine is to keep people functioning as close to normally as we can, and, when they are ill, to return them to normal functioning as best we can.[42] Yet, this idea is misleading insofar as it invites us to think that medicine returns an unwell person to the state they were in prior to their illness. Even when medicine enables a person to recover from an illness—which is by no means guaranteed, and often is not even feasible—patients always emerge from medical treatment as changed in some respect. At the very least, he or she will have undergone certain new experiences and will have new knowledge corresponding to them.

These points are important, because an improperly idealised picture of the nature and goals of medical practice will leave it obscure both why the boundaries of medical practice extend as they do, as well as why medicine has such a distinctive place in society. Treatments such as abortion, euthanasia and palliative care are precisely moments at which medical practice serves vital needs which are at odds with a sanitised picture of medicine at aiming at 'the restoration of normal functioning'. But beyond these

---

[41] See Susan Moller Okin, *Justice, Gender and the Family* (NY: Basic Books, 1991), ch. 5.

[42] Although I speak of 'illness' as the opposite of 'health', I am speaking loosely here and do not want to commit myself one way or another to the broader question of whether we should include disabilities under the rubric of ill-health. See *JH*, pp. 38–42.

totemic issues, much uncontroversial medical practice similarly takes as its aim helping a person who is in need, without stopping to relate those needs to an idealised picture of human function. Of course, it is difficult to be any more precise than this concerning the value of medicine to an individual. The effects of medical treatment are often mixed; it is not uncommon for the experience of ill-health and recovery to leave one weaker in some respects but stronger in others, though in other cases a patient may deal with a condition where there is neither the possibility of recovery nor (from their point of view) any sense in which the experience of ill-health is a source of strength, wisdom, or insight. This makes it difficult to formulate an abstract conception of the ramifications of healthcare, which does justice to the roles which therapeutic relationships and experiences may play in a person's life. However, in general, no conception of the value of healthcare, suitable to explain the grounds of individuals entitlements to treatment, will be sufficient if it neglects or obscures these facts concerning the richness and heterogeneity of medical treatment and the complexities of its relation to human life.

Another way to get at this point is to focus on the notion of "normal functioning" which plays a central role in Daniels' account of the scope of our entitlement to healthcare. Normality is a concept with a suspect history. Granted, in certain contexts the notion of the normal is an indispensible tool for effective intervention. For instance, without the idea of the normal functioning of an organ, we would not be able to distinguish between therapies which are beneficial and those which are harmful. However, we are always in danger of allowing the concept to encroach into aspects of our lives where regulation by its standards is oppressive; as when we worry about whether our emotional reactions are normal or abnormal, or when we see some physical or psychological traits as characteristic of the normal state of humankind.

Thus, when spun out into theories of justice, the concept of 'normality' threatens to introduce structures that marginalize certain individuals and cause their voices and experiences to be treated with less weight and importance than those of the majority. This point is particularly salient when mental health is accorded its due attention. Much is rightly made of the fact of embodiment for mental health, but equally important is the reality of mindedness for physical health; psychological distress not only worsens physical suffering, but often causes it. The goal of medicine cannot be to create a population completely free from pain, disease, and disability, but must instead help us to meliorate conditions when they arise and to make bearable those conditions which cannot be fully treated. It is in this demand for help in bearing a condition that makes medicine an essentially humanistic discipline and which opens it up to the difficulties and unclarities in clarifying what is and is not 'bearable'. Hints of this difficulty can be seen in the way that Daniels discusses questions of those who we cannot bring "closer to the idealization" (and might he have said the ideal?) of normal functioning. As he says:

> We owe people health care and related social services even when they cannot be brought closer to the idealization [of normal functioning]. Terminal care and care for the seriously mentally and physically disabled are important examples. *These services raise serious issues, for example about compassion and beneficence, that go beyond questions of justice.*[43]

Needless to say, it is debatable whether questions concerning treatment of the terminally ill and the seriously disabled should be thought of as matters "beyond justice". There is a suspicion that in viewing these claims as a matter of compassion or beneficence, we underplay the extent of our obligations to the most vulnerable amongst us, or, as we might more perspicuously put it, to us at our most vulnerable.

Daniels might respond that, despite its pitfalls, the concept of normality can still have its uses, and that we should not infer from the fact that it can be misused to the idea that it has no role to play in a theory of justice. It can be important for distinguishing precisely those points at which intervention is an entitlement from those where it is a preference—for instance, in the well-worked over contrast between therapy and enhancement. Amplifying this critique we might insist that eliding this distinction would mean subsuming everything to the rubric of preference satisfaction, which in turn would be a capitulation to a form of capitalist reasoning that aims to reduce everything to the rubric of homogenized units of supply and demand. Such an approach would

---

[43] *JH*, p. 62; emphasis mine.

render invisible the significance of those preferences which are directly related to the material bases for self-respect.

This response has its point; not every way of removing the concept of the normal is of equal value, and it is true that we must retain a lively sense of the relative importance of different factors in ensuring that a person lives the best life available to them. The relative importance of different physical and psychological needs in the development of a healthy and happy person is something which can only be determined by cooperative investigation across different disciplines; and it involves, also, debate about what kinds of life we find worthy of respect. In the various disciplines through which we come to fill in our picture of the vital needs of human beings, the concept of the normal may have a role to play. However, when it comes to formulating a basic political framework, we must take care to ensure that its structure reflects a foundational belief in the equal entitlement of each person to concern, and this belief entails that each person be treated as an individual—a form of attention according to which the normality or abnormality of their condition becomes of secondary importance.

Medicine plays an important role in our individual and collective self-understanding and improvement. It is, as Alastair Campbell once put it, a 'prophetic' discipline, in that doctors often see certain of the ramifications of our way of life before others do—witness, for instance, the discovery of the negative health effects of smoking and air pollution, or the discovery of a novel coronavirus in Wuhan in 2019. Alongside this, medicine also plays a palliative and pastoral role, when the restoration of normal functioning is either no longer in question or when it is not the chief good which the medic can provide. When it comes to chronic conditions, a large part of what medicine can provide is the comfort of knowing that one has an ally in the fight against one's condition. We should, I think, view with some suspicion the tendency to see these elements of the medical profession as standing outside of the purview of questions of justice.[44]

[44] On the concept of solidarity see Ruud ter Meulen, *Solidarity and Justice in Health and Social Care* (Cambridge: Cambridge University Press, 2017). However, we must at the same time acknowledge that such a division is not entirely unmotivated—it may correspond to the distinction between justice as the ideal form of society and solidarity as its animating force.

As Raimond Gaita has argued, the language of rights and entitlements is inadequate to capture our deepest needs, or to explain our sense of injustice when those needs are violated. Nevertheless, the language of needs and entitlements must still have a role in settling questions of justice; our responsiveness to the needs of others must be made compatible with the possibility of private life—of some amount of indifference to others, special treatment, and inward concern. We are not saints, and it is at least an open question whether we ought to aspire to be.[45] For this reason, I do not think that the difficulties which I have tried to raise for Daniels' account of our entitlements to healthcare should be viewed as fatal to the aspirations to regulate our entitlements to healthcare through more neutral and less demanding conceptions of normality and the means for social advancement. We have an entitlement to being treated decently. We also appreciate treatment which goes beyond that. Just how we are to relate these intuitions to the concept of justice—how demanding the concept ought to be—is contested, and I suspect, always will be. After all, 'justice' is our name for a site of contestation over the scope and limits of our responsibilities to others, over how we imagine our society to be, and what potential we see in it and in us.

At root there is a fundamental tension between, on the one hand, our responsiveness to the moral urgency of meeting the needs of others, and on the other, the desire to make sure that we do not impose overly burdensome obligations on people. There is a tendency to set thresholds for entitlement which are such that the social nets which give protection from misfortune are sustained at minimal overall cost to others. That tendency may go along with a belief in the importance of incremental and generational redistribution to bring us closer to the ideal. Against that, there is the countervailing belief that suffering which arises even in part thanks to injustice demands wholesale and immediate rectification, even if such a course of action is costly and disruptive in the short term. Revolutionary alterations to the basic fabric of society always involve upheaval, and their consequences cannot always be clearly foreseen. Yet, such alterations are sometimes vitally necessary, for instance in the ongoing struggle against racist and patriarchal systems of oppression.

[45] See Susan Wolff, "Moral Saints", *Journal of Philosophy* 79, no. 8 (1982): 419–439.

Incremental and revolutionary approaches to practical questions both have their virtues and their vices, and some balance between them must be struck. But determining the appropriate balance will always be controversial, because the circumstances in which political matters are addressed are complex, and because slight changes in worldview can lead to very different senses of the relative importance of different matters. For this reason there will thus always be within liberalism a tendency toward democratic socialism, understood as a commitment to create a society in which everyone is able, as much as possible, to live a fulfilling life, and one free from pain, hunger, and disease.[46] At the same time, there will always be within liberalism a countervailing tendency toward individualism, and the imposition of thresholds for concern beyond which indifference is permissible.[47]

So long as PR is understood as a methodology for governing disputes, then such controversies need not be considered fatal to the aspirations of the PR theorist. Our goal is to get as close as we can to a sustainable way of life in which all people can live free from oppression and with the material bases of self-respect. It is hard enough to visualize what such a society would look like in any detail, let alone to chart a course from our current situation toward such an ideal. However, the fact that our goals are always elusive and difficult in this way need not fill us with despair. Rather, if the movement toward a more just (hence more justifiable) society is conceived of as an open-ended and ongoing task, then the plausibility of the concept of PR will depend not so much on its resilience to criticism, but rather on its adaptability to it.

[46] Thus, as G.A. Cohen has demonstrated, there is within the Rawlsian system elements which push us toward a more robust picture of the role of collective enterprise in supporting individuals in their quest to live a flourishing, happy life. See G.A. Cohen, *Rescuing Justice and Equality* (Harvard: Harvard University Press, 2008).

[47] Though, as Hon Lam Li has pointed out to me, it remains debatable how far Rawls' liberalism is individualistic, since Rawls views our natural talents as a common asset, and thinks of inequality as justified only insofar as it benefits the worst off. Despite this, however, there is still a role within liberal thought for an ineliminable value to the individual, which protects not only liberty of thought but also (more controversially) of conscience.

14

Bioethics and Public Reason: How the History of Bioethics Has Led to the Need for Some Concept of Public Reason

Alastair V. Campbell

1 Introduction

It is obvious that bioethics deals with a number of issues that are highly controversial and contentious. A recent example would be the Charlie Gard case in the UK, which led to a series of hearings in court, with the parents of this highly damaged and terminally ill child contesting the view of the hospital treating him that the only viable option was appropriate terminal care. The parents had raised a large sum of money to pay for an experimental treatment in the USA. However, the view of the hospital was that attempting this would not be in the child's best interests, as the efficacy of the intervention was totally unproven and subjecting the child to it could merely prolong his suffering. The court found in the hospital's favour, but the case provoked a major public debate, with

A. V. Campbell (✉)
Centre for Biomedical Ethics, National University of Singapore,
Singapore, Singapore

H.-L. Li, M. Campbell (eds.), *Public Reason and Bioethics*,
https://doi.org/10.1007/978-3-030-61170-5_14

advocates as diverse as the Pope and the President of the USA appearing to side with the parents.[1]

There are many other bioethics issues of this kind, including abortion, gene editing, biobanking and medically assisted suicide. These issues all provoke strong disagreement and debate, in part because they deal with fundamental questions of human vulnerability and mortality, but also because the differing views held by people reflect a wide range of divergent philosophical and theological viewpoints. Moreover, political forces can further confuse the debates. For example, in the USA a very powerful 'right to life' lobby, in part based on Christian belief, has had a massive effect on governmental research funding policy, preventing whole areas of research involving human embryos from receiving any federal funds.[2]

In this chapter, I shall give an account of the emergence of bioethics as a field of study and then describe its common features in an international context.[3] In the final section I shall suggest how some concept of public reason might be used to meet the challenges thrown up by the contentious nature of the field.

2 Origins of Bioethics

The discipline of bioethics, as we know it today, is a relative newcomer to the field. Of course, medical ethics, from which it developed, is very ancient, going back at least to the Oath of Hippocrates and possibly even earlier. But bioethics arose from the realisation that medical ethics could no longer be treated as a field internal to medicine and controlled by the medical profession. Two different post-WWII developments led to this realisation. The first included the revelations at the Nuremberg Trials of the atrocities committed by doctors in Nazi Germany (atrocities matched, as it turned out much later, by Japanese doctors in their treatment of Chinese prisoners). The second included the dramatic technological

[1] BBC News, 'Charlie Gard: The story of his parents' legal fight,' https://www.bbc.co.uk/news/health-40554462 (accessed 15/06/2021).

[2] Kathryn Claiborn, 'Federal funding for stem cell research: 15 years of indecision,' *The Journal of Clinical Investigation*, 121, no. 7 (Jul 2011): 2531.

[3] I give a fuller explanation in Alastair Campbell, *Bioethics: the Basics*, 2nd ed. (London: Routledge, 2017), Chapter One.

advances in medical treatment, of which the most challenging were transplantation of human organs and new birth technologies. These two developments led to the realisation that meeting such ethical challenges required a multidisciplinary approach, with medical specialists collaborating with philosophers, theologians, lawyers and social scientists to try to formulate answers to a whole range of new and troubling questions, beginning with the origins of human life and stretching to the definition of death. In the early 1970s the discipline found its academic home in the foundation of two multidisciplinary journals, the *Hastings Center Report* in the USA and the *Journal of Medical Ethics* in the UK. In addition, the formative text books of the period were written by theologians, not doctors—from the Roman Catholic tradition, by Bernard Haering and Richard McCormick; and from the Protestant side, by Paul Ramsey and Joseph Fletcher.

However, the character of the discipline was to change further, as governments began to look for ways to guide legislation for the myriad of medical developments and treatment innovations by setting up expert advisory committees. This involved bioethicists in new ways and resulted in several influential reports, of which typical examples are the Belmont Report[4] in the USA (dealing with research ethics) and the Warnock Report[5] in the UK (dealing with new birth technologies). These reports showed how bioethics could influence not just clinical decisions, but also public policy and legislation. In parallel with this, the field of *clinical* bioethics consultation expanded rapidly, especially in the USA, and as a result, the number of people involved professionally in bioethics escalated. Another important development was the formation of the International Association of Bioethics in the early 1990s, which encourages scholarship in the field and promotes biennial World Congresses in Bioethics, each one held in a different region of the world. To date they have been held in all continents apart from Africa and the Antarctic. The Association's core aim is to ensure the 'free and reasoned discussion' of bioethical issues.'[6]

[4] *The Belmont Report.* (US Department of Health, Education and Welfare, 1979).

[5] *Report of the Committee of Inquiry into Human Fertilisation and Embryology*, Chaired by Mary Warnock. (London: Her Majesty's Stationery Office, 1984).

[6] See iab-website.iab-secretariat.org/about-us/.

3 Plurality and Complexity

However, although bioethics is now a well-established academic field in most regions of the world, it is by no means a simple academically coherent one. This is partly because it is essentially multidisciplinary, and so there is constant debate about the best methods to use to describe and offer solutions to the complex problems it deals with. For example, is 'empirical bioethics' possible? If so, how does it avoid becoming simply a type of social study, with no authority to reach normative conclusions? Related to this problem is the plurality of philosophical and theological standpoints from which the issues are approached. This was evident from the start of the discipline. Although two of the theological pioneers were Roman Catholic, McCormick's views and Haering's (both identifiable as on the liberal wing of Catholic theology) were still at odds on a number of issues. This was even more the case with the two Protestant pioneers, Ramsay and Fletcher. The former's approach was fairly conservative, while the latter's was radically liberal. The subsequent development of the field saw similar contests between different approaches to philosophical theory. The work of Peter Singer (and others) enshrined a strongly consequentialist position,[7] regarding any other approaches as dogmatic and irrational. The rise of principlism, under the powerful influence of the numerous editions of Beauchamp and Childress's seminal work,[8] seemed to promise a more nuanced and diverse approach to ethical theory, offering a foundation of four basic principles. However, principlism failed to provide a convincing answer to how conclusions were to be reached when the principles clashed, and its whole approach was put under question by two very different accounts of ethical theory: feminist approaches to bioethics and virtue ethics. Related to these last two theories was an increasing interest in narrative ethics and care ethics. These four approaches were highly suspicious of principles, seeing them as potentially a stifling influence on moral commitment and ethical sensitivity.[9]

---

[7] Peter Singer, *Practical Ethics*, 3rd ed. (Princeton, NJ: Princeton University Press, 2011).

[8] Tom L. Beauchamp and James F. Childress, *Principles of Biomedical Ethics*, 7th ed. (New York: Oxford University Press, 2012).

[9] Campbell, *Bioethics: the Basics*, Chapters Two and Three.

In summary, bioethics is a highly complex field, with a wide array of competing theories, and this leads to the critical issue of how consensus can be achieved in a practical way that will assist both clinicians and legislators. It is only fair to point out that such complexity and clashing of theoretical assumptions are equally true of the constituent disciplines of bioethics: philosophy, theology, law and social science. All are characterised by ongoing lively debates about how practical conclusions can validly be reached and about what the starting point for theoretical debate should be. However, the debates in bioethics have been made still more heated (and apparently insoluble) by two other factors: accusations of cultural imperialism and a recognition of the pervasive and largely malign influence of economic globalisation.

The literature of bioethics has been dominated for decades by debates in Western philosophy, notably between Kantian and Utilitarian approaches. But, as the field has become more international, questions have been raised about the adequacy of such approaches for Eastern cultures, in which excessive individualism is questioned by an emphasis on duty to family and society. Moreover, the increasing secularisation of the West does not fit well in societies in which religious tradition has a strong influence on ethics. Under the aegis of the International Association of Bioethics, these concerns have led to a flowering of approaches that seek a dialogue between Eastern and Western approaches to give a richer and more culturally sensitive account of the issues encountered in all regions of the world. Of course, the challenge here is how to avoid a position that is wholly relativistic, and so gives no answer to how international solutions to these issues may be found. One obviously 'hot' area here is in different cultural approaches to the place of women in society, with the risk that abuse of basic human rights may be seen as culturally appropriate if we stress too much the current cultural beliefs and practices.

Secondly, the massive power exercised by international corporations in the field of health care has become a core issue of concern in bioethics, and this has led to calls for a more socially and politically active approach to the problems of inequity and injustice exposed by bioethical analysis. Obvious examples are the huge inequities in the distribution of health care resources in the world[10] and the distortions of health research caused by the pharmaceutical industry in its pursuit of profit.[11] If bioethics claims to be concerned with practical solutions, not just with theoretical debate, it seems that such abuses of economic power must be confronted and countered. So bioethics needs to be politically involved in some way, but how can this be done effectively and without compromising the academic integrity of the discipline?

4 A Place for Public Reason?

As is so often the case, it is much easier to describe problems than to suggest solutions! However, it seems clear that bioethics, for all its complexity and pluralism of approaches, does reveal an urgent need to define some basic rules for a 'free and rational' discussion of the crucial ethical issues it uncovers and also to point the way to achieve effective social change. The extent to which 'public reason' can be substantive as well as procedural is the subject of other papers in this symposium. However, in light of the global issues I have described, some common agreements across cultures must be found. Although we can always expect some disagreement about how values are to be cashed out in practice (both in clinical settings and in wider frameworks of legislation), we do need to agree on some core humanitarian values that we must all respect and defend. Without such a common set of moral commitments, 'public reason' will be no more than an empty phrase. But with this shared moral starting point, we can perhaps begin to define rules for cross-cultural and international debate.

[10] World Health Organisation, 'Health inequities and their causes,' https://www.who.int/features/factfiles/health_inequities/en/ (accessed 15/06/2021).

[11] Ben Goldacre, *Bad Pharma* (London: Faber and Faber, 2012).

Bibliography

馮友蘭,《中國哲學簡史》, 涂又光譯, 北京大學出版社, 1985.

———.《中國哲學史》, 下篇, 中華書局.

Abortion by the Roman Catholic Church," *Religious Tolerance* <https://www.religioustolerance.org/abo_hist_c1.htm> (n.d.).

Ackerman, Bruce, *Social Justice in the Liberal State* (New Haven: Yale University Press, 1988).

Anonymous, "Filial Piety," <https://en.wikipedia.org/wiki/Filial_piety#Adaptation> (n.d.-a).

———. "Five Relationships" (in Chinese), <https://zh.wikipedia.org/wiki/五倫> (n.d.-b).

———. "Lincoln-Douglas Debates," <https://en.wikipedia.org/wiki/Lincoln–Douglas_debates> (n.d.-c).

———. "Roman Catholicism and Abortion Access: Possible Exceptions to the Ban on".

———. "Women in Confucian Thought," <https://en.wikipedia.org/wiki/Confucianism#Women_in_Confucian_thought> (n.d.-e).

Aquinas, Thomas, *Political Writings,* ed. Robert W. Dyson (Cambridge: Cambridge University Press, 2002).

———. *Summa Theologiae*, ed. Institut d'Etudes Médiévales d'Ottawa. 5 vols., (Ottawa: Harpell's Press, 1941–1945).

H.-L. Li, M. Campbell (eds.), *Public Reason and Bioethics*,
https://doi.org/10.1007/978-3-030-61170-5

———. *Quaestiones de quolibet*, ed. René-Antoine Gauthier (Paris: Éditions du Cerf, 1996).

Aristotle, *Politica*, ed. William David Ross (Oxford: Clarendon Press, 1957).

———. *The Politics of Aristotle*, ed. Ernest Barker (London: Oxford University Press, 1958).

Arrars, J. D., "Physician-Assisted Suicide: A Tragic View," *Journal of Contemporary Health Law and Policy* 13 (1997), pp. 361–89.

Ashford, Elizabeth and Tim Mulgan, "Contractualism", *The Stanford Encyclopedia of Philosophy* (Summer 2018 Edition), ed. Edward N. Zalta, URL = <https://plato.stanford.edu/archives/sum2018/entries/contractualism/>.

Augustine, *The City of God Against the Pagans*, ed. Robert W. Dyson (Cambridge: Cambridge University Press, 1998).

Ayer, A. J., *Language, Truth, and Logic* (London: Victor Gollancz, Ltd., 1936).

Bai, Tongdong, *Against Political Equality: The Confucian Case* (Princeton: Princeton University Press, 2019).

Bao, Tong, "How Deng Xiaoping Helped Create a Corrupt China," *New York Times*, June 3, 2015 (https://www.nytimes.com/2015/06/04/opinion/bao-tong-how-deng-xiaoping-helped-create-a-corrupt-china.html).

Battin, Margaret, "Euthanasia: The Way We Do It, the Way They Do It: End-of-Life Practices in the Developed World," *Journal of Pain and Symptom Management* 6 (1991), pp. 298–305.

Beauchamp, Tom and James Childress, *Principles of Biomedical Ethics*, 7th ed. (Oxford: Oxford University Press, 2013).

Behr, Thomas C, "Luigi Taparelli D'Azeglio, SJ (1793–1862) and the Development of Scholastic Natural-Law Thought as a Science of Society and Politics," *Journal of Markets and Morality* 6 (2003), pp. 99–115.

Bell, Daniel A. and Chenyang Li, ed., *The East Asian Challenge for Democracy: Political Meritocracy in Comparative Perspective* (Cambridge: Cambridge University Press, 2013).

Berlin, Isiah, "The Pursuit of the Ideal," in Harry Hardy, ed., *The Crooked Timber of Humanity* (New York: Knopf, 1991).

Brettschneider, Corey, "The Rights of the Guilty: Punishment and Political Legitimacy." *Political Theory* 35, no. 2 (2007): 175–99.

Broadie, Sarah, *Ethics with Aristotle* (Oxford: Oxford University Press, 1999).

Brugger, E. Christian, "The First Principles of the Natural Law and Bioethics, *Christian Bioethics* 22 (2016), pp. 88–103.

Campbell, Alastair V., *Bioethics: The Basics*, 2nd ed. (London: Routledge, 2017).

Capitaine, Laura, Guido Pennings, and Sigrid Sterckx, "Why Jecker's Capabilities Approach to Rationing Is Incapable of Containing Health Care Costs," *American Journal of Bioethics* 13 (2013), pp. 22–23

Casal, Paula, "Why Sufficiency is Not Enough," *Ethics* 117 (2007), pp. 296–326.

Catholic Church, *Catechism of the Catholic Church* (2nd edn, Vatican CityWashington, DC: Libreria Editrice Vaticana—United States Catholic Conference, 1997).

Chan, Joseph, "On the Legitimacy of Confucian Constitutionalism," in *A Confucian Constitutional Order*, edited by Edmund Ryden, Daniel Bell and Ruiping Fan (Princeton: Princeton University Press, 2013).

———. *Confucian Perfectionism: A Political Philosophy for Modern Times* (Princeton: Princeton University Press, 2014).

———. "Is There A Human Right to Political Participation? A Contemporary Confucian Perspective." (unpublished paper, 2015).

Chan, Kwok-shing, "Women's Property Rights in a Chinese Lineage Village," *Modern China* 39 (2013), pp. 101–128.

Chen, M., "Debates and Analysis of Physician-assisted Suicide" (in Chinese), *Medicine and Philosophy* (*Yixue yu zhexue*) 35 (2014), pp. 15–19.

Christian Medical and Dental Society of Canada v. College of Physicians and Surgeons of Ontario, ONCA 393 (2019).

Claiborn, Kathryn, "Federal Funding for Stem Cell Research: 15 Years of Indecision," *Journal of Clinical Investigation* 121, no. 7, (2011), p. 2531.

Cohen, G. A., *Karl Marx's Theory of History: A Defence* (Princeton: Princeton University Press, 1978).

———. *Rescuing Justice and Equality* (Cambridge, MA: Harvard University Press, 2008).

Confucius, *The Analects*, translated by D. C. Lau (Hong Kong: The Chinese University of Hong Kong Press, 1992).

Congregation for the Doctrine of the Faith, "Declaration on Euthanasia Iura Et Bona (5 May 1980)," *Acta Apostolicae Sedis* 72 (1980), pp. 542–82.

———. "Declaration on Procured Abortion (18 November 1974)," *Acta Apostolicae Sedis* 66 (1974), pp. 730–47.

———. "Instruction *Dignitas Personae* on Certain Bioethical Questions (8 September 2008)," *Acta Apostolicae Sedis* 100 (2008), pp. 858–87.

———. "Instruction *Donum Vitae* on Respect for Human Life at Its Origins and for the Dignity of Procreation (22 February 1987)," *Acta Apostolicae Sedis* 80 (1988), pp. 70–102.

———. "Instruction on the Ecclesial Vocation of the Theologian *Donum Veritatis* [24 May 1990]," *Acta Apostolicae Sedis* 82, no. 1550–1570 (1990).
Crocker, David A., *Ethics of Global Development* (Cambridge: Cambridge University Press, 2008).
Crowe, Jonathan, *Natural Law and the Nature of Law* (New York: Cambridge University Press, 2019).
Crowe, Michael Bertram, *The Changing Profile of the Natural Law* (The Hague: M. Nijhoff, 1977).
Cudd, Ann and Seena Eftekhari, "Contractarianism", *The Stanford Encyclopedia of Philosophy* (Summer 2018 Edition), ed. Edward N. Zalta, URL = <https://plato.stanford.edu/archives/sum2018/entries/contractarianism/>.
Cruzan v. Missouri Department of Health, 497 U.S. 261 (1990).
Daniels, Norman, *Am I My Parents' Keeper? An Essay on Justice Between the Young and the Old* (New York: Oxford University Press, 1988).
———. *Just Health: Meeting Health Needs Fairly* (Cambridge: Cambridge University Press, 2007).
Darwall, Stephen, "Two Kinds of Respect," *Ethics* 88 (1977), pp. 36–49.
Development Programme (United Nations), *Human Development Report 1993*, United Nations, 1993.
Dillon, Robin S., "Respect", *The Stanford Encyclopedia of Philosophy (Winter 2016 Edition)*, ed. Edward N. Zalta, URL = <https://plato.stanford.edu/archives/win2016/entries/respect/>.
Division of Medical Ethics (University of South Florida College of Med.), "Physician-Assisted Suicide: The Legal Slippery Slope," *Cancer Control* 8 (2001), pp. 25–31.
Dreben, Burton, "On Rawls and Political Liberalism," in *Cambridge Companion to Rawls*, edited by Samuel Freeman (Cambridge: Cambridge University Press, 2003), pp. 316–346.
Dworkin, Gerald, *The Theory and Practice of Autonomy* (Cambridge: Cambridge University Press, 1988).
Dworkin, Ronald, "Liberalism," in *Public and Private Morality*, edited by Stuart Hampshire (Cambridge: Cambridge University Press, 1978), pp. 113–43.
———. "Neutrality, Equality and Liberalism," in*Liberalism Reconsidered*, edited by Douglas MacLean, and Claudia Mills. (Totowa: Rowman and Allanheld, 1983), pp. 1–11.
———. *Sovereign Virtue* (Cambridge, MA: Harvard University Press, 2000).
———. *Justice for Hedgehogs* (Cambridge, MA: Harvard University Press, 2011).

Dworkin, Ronald, Thomas Nagel, Robert Nozick, John Rawls, Thomas Scanlon, and Judith Jarvis Thomson, "Assisted Suicide: The Philosophers' Brief," *New York Review of Books* 44 (1997), pp. 41–7.

Ebels-Duggan, Kyla, "The Beginning of Community," *Philosophical Quarterly* 60 (2008), pp. 50–71.

Ebrey, Patricia Buckley, *Women and the Family in Chinese History* (Abingdon: Routledge, 2003).

Editors of Encyclopedia Britannica, "Popular sovereignty," <https://www.britannica.com/topic/popular-sovereignty> (1998).

Edwards, Lee, "Is China Totalitarian?" <https://www.heritage.org/asia/commentary/china-totalitarian> (2020).

Elliot, Carl, *Better Than Well: American Medicine meets the American Dream* (New York: Norton, 2004).

Emmanuel, Ezekiel J., "What is the Great Benefit of Legalizing Euthanasia," *Ethics* 109 (1999), pp. 629–42.

Engelhardt, H. Tristram Jr., *The Foundations of Bioethics*, 2nd ed (New York: Oxford University Press, 1996).

———. "Fair Equality of Opportunity Critically Reexamined: The family and the Sustainability of Health Care Systems," *Journal of Medicine and Philosophy* 37 (2012), pp. 583–602.

———. "The Family: Crucial to and Divisive in Bioethics," *International Journal of Chinese & Comparative Philosophy of Medicine* XI (2013), pp. 113–127.

English, Jane, "What Do Grown Children Owe Their Parents?," in *Ethics in Practice: An Anthology* ed. Hugh LaFollette (Oxford: Blackwell, 2002), pp. 152–155.

Fan, Ruiping, "Which Care, Whose Responsibility, and Why Family? A Confucian Account of Long Term Care for the Elderly," *Journal of Medicine and Philosophy* 32 (2007), pp. 495–517.

———. "Discourses of Confucian Medical Ethics," in *The Cambridge World History of Medical Ethics*, edited by Robert Baker, and Laurence McCullough (Cambridge: Cambridge University Press, 2009), pp. 195–201.

———. Reconstructionist Confucianism: Rethinking Morality after the West (New York: Springer, 2010).

———. "A Confucian Notion of the Common Good for Contemporary China," in *The Common Good: Chinese and American Perspectives*, edited by D. Solomon, and P.C. Lo (New York: Springer, 2014), pp. 193–218.

———, ed., *Family-Oriented Informed Consent: East Asian and American Perspectives* (Dordrecht: Springer, 2015).

———. “Nonegalitarian Social Responsibility for Health: A Confucian Perspective on Article 14 of the UNESCO *Declaration in Bioethics and Human Rights*,” *Kennedy Institute of Ethics Journal* 26 (2016), pp. 105–218.

———. “Is Confucian Harmony Foundationless? A Critical Question for Chenyang Li,” *Philosophy East & West* 67 (2017), pp. 246–256.

———. “DNA, Brain, Mind and Soul: A Confucian Perspective,” in *Interreligious Perspectives on Mind, Genes and the Self*, edited by Joseph Tham, Chris Durante, and Alberto Garcia Gomez (London: Routledge, 2019), pp. 63–74.

Fan, Ruiping. and Sun, Sihan, “To Relieve or Terminate? A Confucian Ethical Reflection on the Use of Morphine for Late-stage Cancer Patients in China,” *Developing World Bioethics* [online journal] 19 (2019), pp. 1–9, https://doi.org/10.1111/dewb.12246.

Fan, Ruiping. and Zhao, Wenqing, “Developing Confucian Virtue-based Rights,” in *Religious Perspectives on Bioethics and Human Rights*, edited by Joseph Tham, Kai Man Kwan and Alberto Garcia (Dordrecht: Springer, 2017), pp. 115–118.

Farrell, Dominic, *The Ends of the Moral Virtues and the First Principles of Practical Reason in Thomas Aquinas* (Rome: Gregorian and Biblical Press, 2012).

Feser, Edward, “Natural Law Ethics and the Revival of Aristotelian Metaphysics,” in *Cambridge Companion to Natural Law Ethics*, edited by Tom Angier (Cambridge: Cambridge University Press, 2019).

Finnis, John, *Natural Law and Natural Rights* (New York: Oxford University Press, 1980).

———. A Philosophical Case against Euthanasia,” in *Euthanasia Examined*, edited by John Keown, (Cambridge: Cambridge University Press, 1995), pp. 23–35.

———. “Public Reason, Abortion, and Cloning,” *Valparaiso University Law Review* 32 (1998), pp. 361–82.

———. “Marriage: A Basic and Exigent Good,” *The Monist* 91 (2008), pp. 388–406.

———. “Law, Morality, and ‘Sexual Orientation’,” in his *Human Rights and Common Good: Collected Essays*, Volume III (Oxford: Oxford University Press, 2011a).

———. *Natural Law and Natural Rights* (2nd edn, Oxford: Oxford University Press, 2011b).

Fisher, Anthony, *Catholic Bioethics for a New Millennium* (Cambridge: Cambridge University Press, 2012).

Foot, Philippa, *Natural Goodness* (Oxford: Clarendon, 2001).

Francis, J. Beckwith, "Natural Law, Catholicism, and the Protestant Critique: Why We Are Really Not That Far Apart," *Christian bioethics: Non-Ecumenical Studies in Medical Morality* 25 (2019), pp. 154–168, <https://doi.org/10.1093/cb/cbz001>.

Frankfurt, Harry, *The Importance of What We Care About,* (Cambridge: Cambridge University Press, 1988).

Freeman, Samuel, *Rawls* (New York: Routledge, 2007).

Froelich, Gregory, "The Equivocal Status of *Bonum Commune*," *The New Scholasticism* 63 (1989), pp. 38–57.

Fu, Shihlin Ema, "The Economic Morality of Leadership: The Confucian Ethics that Affected Emperor Kangxi of the Qing Dynasty" (MA thesis); https://citeseerx.ist.psu.edu/viewdoc/download?doi=10.1.1.940.509&rep=rep1&type=pdf.

Fullam, Lisa, "Sex in 3-D: A Telos for a Virtue Ethics of Sexuality," in *Journal of the Society of Christian Ethics* 27 (2007), pp. 151–170.

Fung, Yu-lan, *History of Chinese Philosophy* (Princeton: Princeton University Press, 1953).

———. *A Short History of Chinese Philosophy* (New York: Macmillan, 1960).

Gaita, Raimond, *A Common Humanity: Thinking About Love and Truth and Justice* (New York: Routledge, 1999).

———. *Good and Evil: An Absolute Conception*, 2nd ed. (Macmillan, 2004).

Ganzini, Linda et al., "Physicians' Experiences with the Oregon Death with Dignity Act," *New England Journal of Medicine* 342 (2000), pp. 557–63.0

Ganzini, Linda, "Oregon Physicians' Attitudes about and Experiences with End-of-Life Care since Passage of the Oregon Death with Dignity Act," *The Journal of the American Medical Association*, 285 (2001), pp. 2363–9.

Gaus, Gerald, "Liberal Neutrality: A Compelling and Radical Principle," in *Perfectionism and Neutrality,* edited by Steven Wall and George Klosko (Lanham: Rowman and Littlefield, 2003), pp. 137–65.

———. "The Moral Foundations of Liberal Neutrality", in *Contemporary Debates in Political Philosophy,* edited by Thomas Christiano and John Christman (Aufl. Hoboken: Blackwell Publishing, 2009).

———. *The Order of Public Reason: A Theory of Freedom and Morality in a Diverse and Bounded World* (Cambridge: Cambridge University Press, 2011).

———. "On Being Inside Social Morality and Seeing it," *Criminal Law and Philosophy* 9 (2016), pp. 141–153.

Gauthier, David, *Morals by Agreement* (Oxford: Oxford University Press, 1986).

Gawande, Atul, Being Mortal*: Medicine and What Matters in the End* (New York: Metropolitan Books, 2014).

Geach, Peter Thomas, "Good and Evil," *Analysis* 17 (1956), pp. 23–42. George, Robert P. *Natural Law and Public Reason* (Washington, DC: Georgetown University Press, 2000).

Geach, Robert P. "Public Morality, Public Reason," *First Things* 167 (2006), pp. 21–26.

George, Robert P. and Christopher Wolfe, "Natural Law and Liberal Public Reason," *The American Journal of Jurisprudence* 42 (1997), pp. 31–49.

Geuss, Raymond, *The Idea of A Critical Theory: Habermas and the Frankfurt School* (Cambridge: Cambridge University Press, 1981).

Giubilini, Alberto, "Objection to Conscience: An Argument against Conscience Exemptions in Healthcare," *Bioethics* 31 (2017), pp. 400–08.

Goldacre, B., *Bad Pharma* (London: Faber and Faber, 2012).

Gómez-Lobo, Alfonso and John Keown, *Bioethics and the Human Goods: An Introduction to Natural Law Bioethics* (Washington, DC: Georgetown University Press, 2015).

Gormally, Luke, *Euthanasia, Clinical Practice, and the Law* (London: Linacre Centre, 1994).

Grisez, Germain, Joseph Boyle, and John Finnis, "Practical Principles, Moral Truth, and Ultimate Ends," *American Journal of Jurisprudence* 32 (1987), pp. 99–151.

Habermas, Jürgen "Reconciliation Through the Public Use of Reason," *Journal of Philosophy* 92 (1995), pp. 109–131.

Haldane, John, "Public Reason, Truth, and Human Fellowship: Going Beyond Rawls," *Journal of Law, Philosophy and Culture* 1 (2007), pp. 175–90.

Hampton, Jean, *Hobbes and the Social Contract Tradition* (Cambridge: Cambridge University Press, 1986).

Harman, Gilbert, "Moral Relativism Defended," *Philosophical Review* 84 (1975), pp. 3–22.

Hart, H. L. A., *The Concept of Law*, (Oxford: Clarendon Press, 1961).

———. *Law, Liberty and Morality* (Stanford: Stanford University Press, 1963).

———. *Punishment and Responsibility* (Oxford: Oxford University Press, 1968).

Her Majesty's Stationery Office (HMSO), *Report of the Committee of Inquiry into Human Fertilisation and Embryology,* Chaired by Mary Warnock, HMSO, London, 1984.

Hester, Marcus, "Aristotle on the Function of Man in Relation to Eudaimonia," *History of Philosophy Quarterly* 8, no. 1 (January 1991), pp. 3–14.

Hibbs, Thomas S., "MacIntyre, Aquinas, and Politics," *The Review of Politics* 66 (2004), pp. 357–83.

Hittinger, Russell, "The Situation of Natural Law in Catholic Theology," *Nova et Vetera* 9 (2011), pp. 657–70.

———. "Two Modernisms, Two Thomisms: Reflections on the Centenary of Pius X's Letter against the Modernists," *Nova et Vetera* 5 (2007), pp. 843–80.

Hobbes, Thomas, *Leviathan*, edited by Michael Oakeshott (Oxford: Blackwell, 1948 (1651)).

Hospital Authority, *HA Guidelines on Life-sustaining Treatment in the Terminally Ill* (Hong Kong: Hospital Authority, 2002), digitized by the Hospital Authority 2002, <https://www.ha.org.hk/haho/ho/adm/124655e.pdf>.

Huang, Yukon, "The Truth about Chinese Corruption," *Diplomat, Carnegie Endowment for International Peace*, May 29, 2015 (https://carnegieendowment.org/2015/05/29/truth-about-chinese-corruption-pub-60265).

Hume, David, *Treatise of Human Nature: A Critical Edition*, edited by David Fate Norton and Mary J. Norton (Oxford: Clarendon Press, 2007).

Hussain, Waheed, "The Common Good", *The Stanford Encyclopedia of Philosophy* (Spring 2018 Edition), ed. Edward N. Zalta, URL = <https://plato.stanford.edu/archives/spr2018/entries/common-good/>.

Irwin, Terence, *The Development of Ethics: A Historical and Critical Study.* 3 vols (Oxford: Oxford University Press, 2007–2009).

Jecker, Nancy S., *Ending Midlife Bias: New Values for Old Age* (Oxford: Oxford University Press, 2020).

———. "Age-Related Inequalities in Health and Healthcare," *Developing World Bioethics* 18 (2017), pp. 144–155.

Kaczor, Christopher Robert, *Proportionalism: For and Against.* Milwaukee, (WI: Marquette University Press, 2000).

———. *Proportionalism and the Natural Law Tradition* (Washington, DC: Catholic University of America Press, 2002.)

Kamm, F. M., *Bioethical Prescriptions* (New York: Oxford University Press, 2013).

———. "Four-Step Arguments for Physician-Assisted Suicide and Euthanasia," in *Bioethical Prescriptions*, pp. 54–76.

Kant, Immanuel, *Foundations of Metaphysics of Morals*, edited by Lewis W. Beck (New York: Macmillan, 1959).

———. *Critique of Pure Reason*, trans. Paul Guyer and Allen Wood (Cambridge: Cambridge University Press, 1997 (1781, 1787)).

———. *An Answer to the Question: What is Enlightenment?*, in *Kant's Practical Philosophy*, edited by Mary Gregor (Cambridge: Cambridge University Press, 1996 (1784)).

———. Kant's Practical Ethics, edited by Mary Gregor (Cambridge: Cambridge University Press, 1996a (1784)).

———. *Groundwork for the Metaphysics of Morals*, ed. Allen Wood (New Haven: Yale University Press, 2002 (1785)) (with an updated edition in 2018).

———. *What Does It Mean to Orient Oneself in Thinking?*, in *Religion and Rational Theology*, edited & tran. by Allen wood (Cambridge: Cambridge University Press, 1996b (1786)).

———, *Critique of Practical Reason*, in *Practical Philosophy*, edited and tran. by Mary Gregor (Cambridge: Cambridge University Press, 1996c (1788)).

———. *Critique of the Power of Judgment*, edited by Paul Guyer (Cambridge: Cambridge University Press, 2000 (1790)).

———. *On the Common Saying: That May be correct in Theory, but it is of No Use in Practice*, in *Kant's Practical Philosophy*, edited & tran. by Mary Gregor (Cambridge: Cambridge University Press, 1996d (1793))

———. *Toward perpetual peace*, in *Kant's Practical Philosophy*, edited & tran. by Mary Gregor (Cambridge: Cambridge University Press, 1996e (1795)).

———. *The Metaphysics of Morals*, in *Practical Philosophy*, edited & tran. by Mary Gregor (Cambridge: Cambridge University Press, 1996f (1797)).

———. *Anthropology from a Pragmatic Point of View*, in *Anthropology, History, and Education*, edited by Robert Louden (Cambridge: Cambridge University Press, 2007 (1798)).

———. *The Conflict of the Faculties*, in *Religion and Rational Theology*, edited by Mary Gregor and Robert Anchor (Cambridge: Cambridge University Press, 1996g (1798)).

Kaufman, Alexander, "Hegel and the Ontological Critique of Liberalism," *American Political Science Review* 91 (4) (1997), pp. 807–17.

Kenny, Anthony, *The Aristotelian Ethics*, (Oxford: Oxford University Press, 1978).

Kim, Sungmoon, *Public Reason Confucianism* (Cambridge University Press, 2016).

Kittay, Eva Feder, "At the Margins of Moral Personhood," *Ethics* 116 (2005), pp. 100–131.

———. *Love's Labor* 2nd ed., (New York: Routledge, 2019).

Kripke, Saul, *Naming and Necessity* (Cambridge, MA: Harvard University Press, 1980).

Kuhn, T. S., *The Structure of Scientific Revolutions* 3rd ed., (Chicago: University of Chicago Press, 1996).

Kumar, Rahul, "Reasonable Reasons in Contractualist Moral Argument," *Ethics* 114 (2003a), pp. 6–37.

———. "Who Can Be Wronged?," *Philosophy and Public Affairs* 31 (2003b), pp. 99–118.

Kymlicka, Will, "Liberal Individualism and Liberal Neutrality," *Ethics* 99 (1989a), pp. 883–905.

———. *Liberalism, Community and Culture* (Oxford: Oxford University Press, 1989b).

Lam, R. C. S. and Wai-Tong Chien, "Attitudes of Acutely Ill Patients towards Euthanasia in Hong Kong," *The Open Nursing Journal* 1 (2007), pp. 1–5.

Larmore, Charles, *Patterns of Moral Complexity* (Cambridge: Cambridge University Press, 1987).

———. "Pluralism and Reasonable Disagreement," *Social Philosophy and Policy* (1994), pp. 61–79.

———. "The Moral Basis of Political Liberalism," *Journal of Philosophy*: 96 (1999), pp. 99–625.

Lee, Patrick, and Robert P. George, *Body-Self Dualism in Contemporary Ethics and Politics* (Cambridge: Cambridge University Press, 2008).

Lewis, V. Bradley, "The Common Good against the Modern State? On Macintyre's Political Philosophy," *Josephinum Journal of Theology* 16 (2009), pp. 357–78.

Li, Hon-Lam, "Abortion and Degrees of Personhood: Understanding Why the Abortion Problem (and the Animal Rights Problem) are Irresolvable," *Public Affairs Quarterly* 11 (1997), pp. 1–19.

———. "Animal Research, Non-vegetarianism, and the Moral Status of Animals—Understanding the Impasse of the Animal Rights Problem," *Journal of Medicine and Philosophy*, vol. 27, no. 5 (2002), pp. 589–615.

———. "Towards Quasi-Vegetarianism," in Hon-Lam Li and Anthony Yeung, eds., *New Essays in Applied Ethics* (Basingstoke, UK: Palgrave Macmillan, 2007), pp. 64–90.

———. "Engelhardt on the Family," *International Journal of Chinese and Comparative Philosophy of Medicine* (2013), pp. 153–60.

———. "Contractualism and Punishment." *Criminal Justice Ethics* 34, no. 2 (2015), pp. 177–209, DOI: https://doi.org/10.1080/0731129X.2015.1067959.

———. "What We Owe to Terminally Ill Patients: The Option of Physician-Assisted Suicide," *Asian Bioethics Review* 8, no. 3 (2016), pp. 224–243. https://doi.org/10.1353/asb.2016.0019.

———. "Contractualism and the Death Penalty," *Criminal Justice Ethics*, 36, no. 2 (2017), pp. 52–182, DOI: https://doi.org/10.1080/0731129X.2017.1358912.

———. "Public Reason as The Way for Dialogue," *The American Journal of Bioethics*, vol. 20, no. 12 (2020a), 29–31. DOI: https://doi.org/10.1080/15265161.2020.1832618.

Li, Hon-Lam, Nancy S. Jecker, and Roger Yat-Nork Chung, "Reopening Economies during the COVID-19 Pandemic: Reasoning about Value Tradeoffs," *The American Journal of Bioethics*, vol. 20, no. 7 (2020b), pp. 136–138, DOI: https://doi.org/10.1080/15265161.2020.1779406.

Li, Jiange, et al., "Social Security Reform in China: Issues and Options," January 27, 2005. (https://economics.mit.edu/files/691).

Li, Zhuoyao, "The public conception of morality in John Rawls' political liberalism," *Ethics and Global Politics* 9 (2016), DOI: https://doi.org/10.3402/egp.v9.28679

Liang, Shuming, *The Essentials of Chinese Culture* (*zhongguo wenhua yaoyi*) (Shanghai: Shanghai Renmin Press, 1949).

Lincoln, A., "Debate at Ottawa, Illinois," <https://www.nps.gov/liho/learn/historyculture/debate1.htm> (1858).

Lippert-Rasmussen, Kasper, *Luck Egalitarianism* (London: Bloomsbury Academic, 2015).

Lisska, Anthony J., *Aquinas's Theory of Natural Law: An Analytic Reconstruction* (New York: Oxford University Press, 1996).

Liu, X., and Yang, J., "Right to Die: Myth of Autonomy and Limitation of Rights" (in Chinese) *Medicine and Philosophy* (*Yixue yu zhexue*) 36 (2015), pp. 22–24.

Lo, Ping-cheung, *Starry Heaven and Morality: Ethics and Its Applications* (in Chinese) (Hong Kong: Joint Publishing Co. Ltd., 1993).

———. "Confucian Ethic Death with Dignity and Its Contemporary Relevance," *The Journal of the Society of Christian Ethics* 19 (1999a), pp. 313–333.

———. "Confucian Views on Suicide and Their Implications for Euthanasia," in *Confucian Bioethics*, edited by Ruiping Fan (Dordrecht: Kluwer Academic Publishers, 1999b), pp. 69–102.

Marquis, Don, "Why Abortion is Immoral," *Journal of Philosophy* 86 (1989), pp. 183–202.

McCarthy, Michael, Mary Homan and Michael Rozier, "There's No Harm in Talking: Re-establishing the Relationship between Theological and Secular Bioethics," *The American Journal of Bioethics*, (forthcoming).

MacIntyre, Alasdair, "Epistemological Crises, Dramatic Narrative and the Philosophy of Science," *The Monist* 60 (1977), pp. 453–72.

———. *After Virtue* (University of Notre Dame Press, 1981).

———. *Whose Justice? Which Rationality?* (Notre Dame: University of Notre Dame Press, 1988).

———. *Three Rival Versions of Moral Enquiry: Encyclopedia, Genealogy, and Tradition: Being Gifford Lectures Delivered in the University of Edinburgh in 1988* (Notre Dame: University of Notre Dame Press, 1990).

———. "Politics, Philosophy and the Common Good," in *The Macintyre Reader*, edited by Kelvin Knight (Notre Dame: University of Notre Dame Press, 1998), pp. 235–52.

———. *Dependent Rational Animals: Why Human Beings Need the Virtues* (Chicago: Open Court, 1999).

———. "Intractable Moral Disagreements," *Intractable Disputes About the Natural Law: Alasdair Macintyre and Critics*, edited by in Lawrence Cunningham (Notre Dame: University of Notre Dame Press, 2009), pp. 1–52.

———. *Ethics in the Conflicts of Modernity: An Essay on Desire, Practical Reasoning, and Narrative* (New York: Cambridge University Press, 2016).

Marx, Karl, *Capital* vol. 1 (New York: International Publishers, 1967).

———. *The German Ideology*, edited by C. J. Arthur (New York: International Publishers, 1970).

———. "Critique of the Gotha Program," in *Marx Selections*, edited by Allen W. Wood, (New York: Macmillan, 1988).

———. *The Poverty of Philosophy* (Amherst, NY : Prometheus Books, 1995).

Marx, Karl, and Friedrich Engels, *Manifesto of the Communist Party* (London: Lawrence & Wishart, 1943 (1848)).

Mattison, William C III. and John Berkman, *Searching for a Universal Ethic: Multidisciplinary, Ecumenical, and Interfaith Responses to the Catholic Natural Law Tradition* (Grand Rapids, MI: William B. Eerdmans Publishing Company, 2014).

May, William E., *Catholic Bioethics and the Gift of Human Life* (Huntington, IN: Our Sunday Visitor, 2000).

McMahan, Jeff, "Self-Defense and the Problem of the Innocent Attacker," *Ethics* 104 (1994), pp. 252–90.

Meulen, Ruud, *Solidarity and Justice in Health and Social Care* (Cambridge: Cambridge University Press, 2017).

Mill, John Stuart, *On Liberty* (Indianapolis: Bobbs-Merrill, 1956).

———. "On Liberty," in *Essays on Politics and Society*, edited by John M. Robson (Toronto: University of Toronto Press—Routledge & Kegan Paul, 1977), pp. 213–310.

Miller, Richard W., *Analyzing Marx* (Princeton: Princeton University Press, 1984).

Mills, Charles W., *The Racial Contract* (Ithaca: Cornell University Press, 1997).

Mulhall, Stephen, and Adam Swift, *Liberals and Communitarians*, 2nd ed. (Oxford: Blackwell, 1996).

Mulhall, Stephen. "Rawls and Communitarianism," in *The Cambridge Companion to Rawls*, edited by Samuel Freeman (Cambridge: Cambridge University Press, 2002), pp. 460–87.

Murphy, Mark, "The Natural Law Tradition in Ethics", *The Stanford Encyclopedia of Philosophy* (Summer 2019 Edition), ed. Edward N. Zalta, URL = <https://plato.stanford.edu/archives/sum2019/entries/natural-law-ethics/>.

Murry, Charles, *Coming Apart: The State of White America 1960–2010* (New York: Crown Forum, 2012).

Nagel, Thomas. *Mortal Questions*. (New York: Cambridge University Press, 1979).

———. *The View from Nowhere* (Oxford: Oxford University Press, 1986).

———. "Moral Conflict and Political Legitimacy," *Philosophy and Public Affairs*, 16 (1987), pp. 215–240.

———. *Equality and Partiality* (New York: Oxford University Press, 1995).

———. *The Last Word* (New York: Oxford University Press, 1997).

———. "Rawls and Liberalism," in Samuel Freeman, ed., *Cambridge Companion to Rawls* (Cambridge: Cambridge University Press, 2003), pp. 62–85.

Nozick, Robert, *Anarchy, State, and Utopia* (New York: Basic Books, 1974).

Nussbaum, Martha C., *Aristotle's De Motu Animalium* (Princeton: Princeton University Press, 1978).

———. "Capabilities and Human Rights," *Fordham Law Review* 66 (1997), pp. 273–300.

———. "Capabilities and Disabilities," *Philosophical Topics* 30 (2002), pp. 133–165.

———. "Beyond the Social Contract," *Oxford Development Studies* 32 (2004), pp. 3–18.

———. *Frontiers of Justice* (Cambridge: Cambridge University Press, 2006).

———. *Creating Capabilities* (Cambridge, MA: Harvard University Press, 2011a).

———. "Perfectionist Liberalism and Political Liberalism", *Philosophy and Public Affairs* 39 (2011b), pp. 3–45.

———. "Political Liberalism and Global Justice," *Journal of Global Ethics* 11 (2015), pp. 68–79.

Obama, Barack. "Obama's Speech to UK Parliament, in full, with analysis," <https://www.bbc.com/news/uk-politics-13549927> (2011).

Oderberg, David, "The Metaphysical Foundations of Natural Law," in, *Natural Moral Law in Contemporary Society* edited by Holger Zaborowski (Washington, DC: Catholic University of America Press, 2010), pp. 44–75.

Okin, Susan, "Justice, Gender, and the Family," *Hypatia* 8 (1993), pp. 209–214.

O'Neill, Onora, "Autonomy, Coherence and Independence," *Liberalism, Citizenship and Autonomy*, edited by David Milligan & William Watts Miller (Aldershot: Avebury, 1992), pp. 209–29.

———. *Autonomy and Trust in Bioethics* (Cambridge: Cambridge University Press, 2002).

———. *Constructing Authorities: Reason, Politics and Interpretation in Kant's Philosophy* (Cambridge: Cambridge University Press, 2015).

The Organisation for Economic Co-operation and Development, *Health Care Systems Efficiency and Policy Settings*, <https://doi.org/10.1787/9789264094901-en> (2010).

Osborne, Thomas, "Macintyre, Thomism and the Contemporary Common Good," *Analyse & Kritik* 30 (2008), pp. 75–90.

Otsuka, Michael, "Killing the Innocent in Self-Defense," *Philosophy & Public Affairs* 23, no. 1 (Winter 1994), pp. 74–94.

Pacholczyk, Tadeusz, "When Pregnancy Goes Awry: Ectopic Pregnancies," *Catholic Education Resource Center*, <https://www.catholiceducation.org/en/science/ethical-issues/when-pregnancy-goes-awry-ectopic-pregnancies.html> (2009).

Parfit, Derek, *Reasons and Persons* (Oxford: Oxford University Press, 1984).

Paterson, Craig, *Assisted Suicide and Euthanasia: A Natural Law Ethics Approach* (Abingdon: Routledge, 2017).

Pateman, Carole, *The Sexual Contract* (Stanford: Stanford University Press, 1989).

Patten, Alan, *Equal Recognition: The Moral Foundations of Minority Rights* (Princeton: Princeton University Press, 2014).

John Paul II, *Evangelium vitae. Acta Apostolicae Sedis* 87 (1995), pp. 401–522.

Piccinin, Antonella, "Rawls and Catholicism: Towards Reconciliation?," *Cultural and Religious Studies* 7 (2019), pp. 50–56.

Pogge, Thomas, *World Poverty and Human Rights* (Cambridge: Polity Press, 2002).

Piketty, Thomas, *Capital in the 21st Century* (Cambridge, MA: Belknap Press, 2013).

Quong, Jonathan, "The Scope of Public Reason," *Political Studies* 52 (2004), pp. 233–250.

———. *Liberalism without Perfection* (Oxford: Oxford University Press, 2011).

———. "The Idea of Public Reason," in *A Companion to Rawls*, edited by Jon Mandle and David A. Reidy (John Wiley & Sons, 2014), pp. 265–280.

———. "Public Reason", *The Stanford Encyclopedia of Philosophy* (Spring 2018 Edition), ed. Edward N. Zalta, URL = <https://plato.stanford.edu/archives/spr2018/entries/public-reason/>.

Rauscher, Frederick, "Kant's Social and Political Philosophy", *The Stanford Encyclopedia of Philosophy* (Spring 2017 Edition), ed. Edward N. Zalta, URL = <https://plato.stanford.edu/archives/spr2017/entries/kant-social-political/>.

Rasmussen, Douglas B. and Douglas J. Den Uyl, *Norms of Liberty: A Perfectionist Basis for Non-Perfectionist Politics* (University Park, PA: Pennsylvania State University Press, 2005).

Railton, Peter, "The Affective Dog and Its Rational Tale: Intuition and Attunement," *Ethics* 124 (2014), pp. 813–859.

Rawls, John, *A Theory of Justice* (Cambridge, MA: Harvard University Press, 1971).

———. "Fairness to Goodness," *Philosophical Review* 84 (1975), pp. 536–54.

———. "Justice as Fairness: Political not Metaphysical," *Philosophy and Public Affairs* 14 (1985), pp. 223–251.

———. *Political Liberalism* (New York: Columbia University Press, 1993).

———. "The Idea of Public Reason Revisited," *The University of Chicago Law Review* 64 (1997), pp. 765–807.

———. *The Law of Peoples* (Cambridge, MA: Harvard University Press, 1999a).

———. "The Idea of Public Reason Revisited," in his *The Law of Peoples* (Cambridge, MA: Harvard University Press, 1999b), pp. 131–180

———. *Justice as Fairness: A Restatement* (Cambridge, MA: Belknap Press, 2001).

Raz, Joseph, "Facing Diversity: The Case of Epistemic Abstinence," *Philosophy & Public Affairs* (1990), pp. 3–46.

———. *The Morality of Freedom* (Oxford: Oxford University Press, 1986).

———. "Facing Up: A Reply," *Southern California Law Review* 62 (1988), pp. 1153–235.

Rhonheimer, Martin, "The Political Ethos of Constitutional Democracy and the Place of Natural Law in Public Reason: Rawls's '*Political Liberalism*' Revisited," in *The Common Good of Constitutional Democracy: Essays in Political Philosophy and on Catholic Social Teaching*, edited by Martin Rhonheimer (Washington, D.C.: Catholic University of America Press, 2013a), pp. 191–264.

———. "Rawlsian Public Reason, Natural Law, and the Foundation of Justice: A Response to David Crawford," in *The Common Good of Constitutional Democracy: Essays in Political Philosophy and on Catholic Social Teaching*, edited by Martin Rhonheimer (Washington, D.C.: Catholic University of America Press, 2013b), pp. 265–91.

Rorty, Richard, "The Priority of Democracy to Philosophy," in *The Virginia Statute for Religious Freedom: Its Evolution and Consequences in American History*, edited by Merrill D. Peterson, and Robert C. Vaughn (New York: Cambridge University Press, 1991), pp. 257–82.

Sandel, Michael J., "Review of Political Liberalism," *Harvard Law Review* 107 (1994), pp. 1765–1794.

———. *Liberalism and the Limits of Justice* 2nd ed. (Cambridge: Cambridge University Press, 1998).

Savulescu, Julian and Udo Schuklenk, "Doctors Have No Right to Refuse Medical Assistance in Dying, Abortion or Contraception," *Bioethics* 31 (2017), pp. 162–70.

Scarry, Elaine, *The Body in Pain: The Making and Unmaking of the World* (Oxford: Oxford University Press, 1985).

Schuklenk, Udo and Ricardo Smalling, "Why Medical Professionals Have No Moral Claim to Conscientious Objection Accommodation in Liberal Democracies," *Journal of Medical Ethics* 43 (2017), pp. 234–40.

Sen, Amartya, *The Idea of Justice* (Cambridge, MA: Belknap Press of Harvard University Press, 2011).

———. "Capabilities, Lists, and Public Reason," *Feminist Economics* 10 (2004), pp. 77–80.

Sgreccia, Elio, *Personalist Bioethics* (Philadelphia: The National Catholic Bioethics Center, 2012).

Strauss, Leo, "What Is Liberal Education?," in *An Introduction to Political Philosophy: Ten Essays*, edited by Hilail Gildin (Detroit: Wayne State University Press, 1989), pp. 4–9.

Scanlon, T. M., *What We Owe to Each Other* (Cambridge, MA: Belknap Press of Harvard University, 1998).

———. "Contractualism and Utilitarianism," in *Moral Discourse and Practice*, edited by Stephen Darwall, Allan Gibbard, and Peter Railton (Oxford: Oxford University Press, 1996), pp. 267–286.

———. "Rawls on Justification," in *The Cambridge Companion to Rawls*, edited by Samuel Freeman (Cambridge: Cambridge University Press, 2002), pp. 139–167.

———. *Moral Dimensions* (Cambridge, MA: Harvard University Press, 2009).

———. *Being Realistic about Reasons* (Oxford: Oxford University Press, 2014).

———. "Some Main Points in Rawls' Theory of Justice," *The Journal of Ethical Reflections*, Summer 2020, vol. 1, no. 2, pp. 31–45.

Shklar, Judith, "The Liberalism of Fear," in *Liberalism and the Moral Life,* edited by Nancy L. Rosenblum (Cambridge, MA: Harvard University Press, 1989).

Sher, George, *Beyond Neutrality: Perfectionism and Politics* (Cambridge: Cambridge University Press, 1997).

Shields, Liam, *Just Enough: Sufficiency as a Demand of Justice* (Edinburgh: Edinburgh University Press, 2016).

Shuai, J., "Autonomous Choice of Right to Life," (in Chinese) *Jishou University Journal* (*Jishou daxue xuebao*) 39 (2018), pp. 45–47.

Shum, Desmond, *Red Roulette* (New York: Scribner, 2021).

Silvers, Anita and Lesile Pickering Francis, "Justice Through Trust: Disability and the 'Outlier Problem' in Social Contract Theory," *Ethics* 116, no. 1 (2005): 40–76.

Singer, Peter, *Famine, Affluence, and Morality* (New York: Oxford University Press, 1972).

———. *Practical Ethics*, 3rd ed. (Princeton: Princeton University Press, 2011).

Smith, Stephen W., "Evidence for the Practical Slippery Slope in the Debate of Physician-Assisted Suicide and Euthanasia," *Medical Law Review* 13 (2005), pp. 17–44.

Steinbock, Bonnie, "The Case for Physician Assisted Suicide: Not (Yet) Proven," *Journal of Medical Ethics* 31 (2005), pp. 235–41.

Stevenson, Charles, *Facts and Values* (New Haven: Yale University Press, 1963).

Sun, Y., "Terminally Ill Patients' Medical Autonomy: Focusing on Latest Legislation beyond Mainland China," (in Chinese) *Journal of Southwest University of Political Science & Law* (*Xinan zhengfa daxue xuebao*) 19 (2017), pp. 65–73.

Taylor, Charles, "Atomism," in his *Philosophy and the Human Sciences* (Cambridge: Cambridge University Press, 1985), pp. 187–210.

Thomson, Judith Jarvis, "A Defense of Abortion," *Philosophy & Public Affairs*, 1 (1971), pp. 47–66.

———. "Self-defense," *Philosophy & Public Affairs* 20, no. 4 (1991), pp. 283–310.

———. "Abortion," *Boston Review*. <https://bostonreview.mit.edU/BR20.3/thomson.html> (1995).

———. "Physician-Assisted Suicide: Two Arguments," *Ethics* 109 (1999), pp. 497–518.

———. *Normativity* (Chicago: Open Court, 2008).

Thum, Ping Tjin, "A Brief History of Elections in Singapore," <https://newnaratif.com/research/a-brief-history-of-elections-in-singapore/> (2020).

Topping, Alexandra, "Assisted Dying," *The Guardian*, <https://www.theguardian.com/society/2014/oct/19/right-to-die-campaigner-starved-herself-jean-davies> (19 October 2014).

Tuschling, Burkhard, "*Rationis Societas*: Remarks on Kant and Hegel", in *Kant's Philosophy of Religion Reconsidered*, edited by Philip J. Rossi and Michael Wreen (Bloomington: Indiana University Press, 1991).

Velleman, J. David, "A Right of Self-Termination?" *Ethics* 109 (1999), pp. 606–28.

de Vitoria, Francisco, *Political Writings*, ed. Anthony R. Pagden and Jeremy Lawrance (Cambridge: Cambridge University Press, 1991).

Waldron, Jeremy, "Autonomy and Perfectionism in Raz's *Morality of Freedom*," *Southern California Law Review* 62 (1989), pp. 1097–152.

———. "Legislation and Moral Neutrality," in his *Liberal Rights* (Cambridge: Cambridge University Press, 1993), pp. 143–67.

———. "Is Dignity the Foundation of Human Rights," in *Philosophical Foundations of Human Rights*, edited by Rowan Cruft, S. Matthew Liao, and Massimo Renzo, (Oxford: Oxford University Press, 2015).

Wall, Steven, *Liberalism, Perfectionism and Restraint* (Cambridge: Cambridge University Press, 1998).

———. "Is Public Justification Self-defeating?", *American Philosophical Quarterly* 39, no. 4 (Oct 2002), pp. 385–394.

Walzer, Michael, *Spheres of Justice* (New York: Basic Books, 1983).

———. *Interpretation and Social Criticism* (Cambridge, MA: Harvard University Press, 1987).

Wang, Qingjie, "The Confucian Filial Obligation and Care for Aged Parents," in *Confucian Bioethics*, edited by Ruiping Fan, (Dordrecht: Kluwer Academic Publishers, 1999), pp. 235–256.

Wang, Z., "Ethical Controversy on Abandoning Treatment and Euthanasia," (in Chinese). *Chinese Medical Ethics* 18 (2005), pp. 61–63.

Warren, Mary Ann, "On the Moral and Legal Status of Abortion," *The Monist* 57 (1973), pp. 43–61.

Watson, Gary, "Free Agency," *Journal of Philosophy* 72 (1975), pp. 205–20.

Weithman, Paul, *Why Political Liberalism? On John Rawls's Political Turn* (Oxford: Oxford University Press, 2010),

Wendt, Fabian (2019) "Rescuing Public Justification from Public Reason Liberalism," in *Oxford Studies in Political Philosophy*, vol. 5, edited by Steven Wall, Peter Vallentyne, and David Sobel (Oxford: Oxford University Press), pp. 39–64.

West, Cornel, *Race Matters* (Boston: Beacon Press, 1993).

Wildes, Kevin Wm, *Moral Acquaintances: Methodology in Bioethics* (Notre Dame: University of Notre Dame Press, 2000).

Williams, Alan, "Intergenerational Equity: An Exploration of the 'Fair Innings' Argument," *Health Economics* 6 (1997), pp. 117–32.

Wolff, Robert Paul, *In Defense of Anarchism* (University of California Press, 1970).

Wood, Allen, "Hegel's Ethics," *The Cambridge Companion to Hegel*, edited by Frederick Beiser (Cambridge: Cambridge University Press, 1993), pp. 211–233.

———. *Kant's Ethical Thought* (Cambridge: Cambridge University Press, 1999).

———. *Kantian Ethics* (New York: Cambridge University Press, 2008).

———. *Formulas of the Moral Law* (Cambridge: Cambridge University Press, 2017).

Wu, H., "Challenge and criticism: Euthanasia should not be legalized" (in Chinese). *Medicine and Law* (*Yixue yu faxue*) 8 (2016), pp. 7–11.

Xunzi, "An Exhortation to Learning," in *Xunzi: The Complete Text*, edited by Eric L. Hutton, (Princeton: Princeton University Press, 2014), pp. 1–8.

Yu, E., "Respect for the Elderly and Family Responsibility: Confucian Response to the Old Age Allowance Policy in Hong Kong," in *The Family, Medical Decision-Making and Biotechnology: Critical Reflections on Asian Moral Perspectives*, edited by Lee Shui Chuen (Dordrecht: Springer, 2007), pp. 197–206.

Young, Iris Marion, "Rawls's Political Liberalism," *Journal of Political Philosophy 3* (1995), pp. 181–190.

Zhang, Weiwei, *The China Wave: Rise of a Civilizational State* (Singapore: World Scientific Press, 2012).

Index[1]

A

B

[1] Note: Page numbers followed by 'n' refer to notes.

H.-L. Li, M. Campbell (eds.), *Public Reason and Bioethics*,
https://doi.org/10.1007/978-3-030-61170-5

The manufacturer's authorised representative in the EU is Springer Nature Customer Service Centre GmbH, Europaplatz 3, 69115 Heidelberg, Germany. If you have any concerns regarding our products, please contact ProductSafety@springernature.com

Printed and bound by CPI Group (UK) Ltd, Croydon, CR0 4YY
30/09/2026
02225457-0019